VARIABLE SPEED AC DRIVES WITH INVERTER OUTPUT FILTERS

VARIABLE SPEED AC DRIVES WITH INVERTER OUTPUT FILTERS

Jaroslaw Guzinski
Gdansk University of Technology, Poland

Haitham Abu-Rub
Texas A&M University at Qatar, Qatar

Patryk Strankowski
Gdansk University of Technology, Poland

This edition first published 2015
© 2015 John Wiley & Sons, Ltd

Registered Office
John Wiley & Sons, Ltd, The Atrium, Southern Gate, Chichester, West Sussex, PO19 8SQ, United Kingdom

For details of our global editorial offices, for customer services and for information about how to apply for permission to reuse the copyright material in this book please see our website at www.wiley.com.

The right of the author to be identified as the author of this work has been asserted in accordance with the Copyright, Designs and Patents Act 1988.

All rights reserved. No part of this publication may be reproduced, stored in a retrieval system, or transmitted, in any form or by any means, electronic, mechanical, photocopying, recording or otherwise, except as permitted by the UK Copyright, Designs and Patents Act 1988, without the prior permission of the publisher.

Wiley also publishes its books in a variety of electronic formats. Some content that appears in print may not be available in electronic books.

Designations used by companies to distinguish their products are often claimed as trademarks. All brand names and product names used in this book are trade names, service marks, trademarks or registered trademarks of their respective owners. The publisher is not associated with any product or vendor mentioned in this book.

Limit of Liability/Disclaimer of Warranty: While the publisher and author have used their best efforts in preparing this book, they make no representations or warranties with respect to the accuracy or completeness of the contents of this book and specifically disclaim any implied warranties of merchantability or fitness for a particular purpose. It is sold on the understanding that the publisher is not engaged in rendering professional services and neither the publisher nor the author shall be liable for damages arising herefrom. If professional advice or other expert assistance is required, the services of a competent professional should be sought.

Library of Congress Cataloging-in-Publication data applied for.

A catalogue record for this book is available from the British Library.

ISBN: 9781118782897

Set in 10/12pt Times by SPi Global, Pondicherry, India
Printed and bound in Singapore by Markono Print Media Pte Ltd

1 2015

Dedicated to my parents, my wife Anna and my son Jurek
—Jaroslaw Guzinski

Dedicated to my parents, my wife Beata, and my children Fatima, Iman, Omar, and Muhammad
—Haitham Abu-Rub

Dedicated to my parents Renata and Władysław, and my girlfriend Magdalena
—Patryk Strankowski

Contents

Foreword		xi
Acknowledgments		xiii
About the Authors		xiv
Nomenclature		xvi

1 Introduction to Electric Drives with LC Filters — 1
  1.1 Preliminary Remarks — 1
  1.2 General Overview of AC Drives with Inverter Output Filters — 2
  1.3 Book Overview — 4
  1.4 Remarks on Simulation Examples — 5
  References — 6

2 Problems with AC Drives and Voltage Source Inverter Supply Effects — 9
  2.1 Effects Related to Common Mode Voltage — 9
    2.1.1 Capacitive Bearing Current — 15
    2.1.2 Electrical Discharge Machining Current — 15
    2.1.3 Circulating Bearing Current — 15
    2.1.4 Rotor Grounding Current — 17
    2.1.5 Dominant Bearing Current — 17
  2.2 Determination of the Induction Motor CM Parameters — 18
  2.3 Prevention of Common Mode Current: Passive Methods — 20
    2.3.1 Decreasing the Inverter Switching Frequency — 20
    2.3.2 Common Mode Choke — 21
    2.3.3 Integrated Common Mode and Differential Mode Choke — 23
    2.3.4 Common Mode Transformer — 25
    2.3.5 Machine Construction and Bearing Protection Rings — 26
  2.4 Active Systems for Reducing the CM Current — 27

	2.5	Common Mode Current Reduction by PWM Algorithm Modifications	28
		2.5.1 Three Nonparity Active Vectors	30
		2.5.2 Three Active Vector Modulation	32
		2.5.3 Active Zero Voltage Control	32
		2.5.4 Space Vector Modulation with One Zero Vector	36
	2.6	Simulation Examples	39
		2.6.1 Model of Induction Motor Drive with PWM Inverter and CMV	39
		2.6.2 PWM Algorithms for Reduction of CMV	44
	2.7	Summary	46
	References		46
3	**Model of AC Induction Machine**		**49**
	3.1	Introduction	49
		3.1.1 T-Model of Induction Machine	50
	3.2	Inverse-Γ Model of Induction Machine	53
	3.3	Per-Unit System	54
	3.4	Machine Parameters	56
	3.5	Simulation Examples	59
	References		63
4	**Inverter Output Filters**		**65**
	4.1	Structures and Fundamentals of Operations	65
	4.2	Output Filter Model	71
	4.3	Design of Inverter Output Filters	74
		4.3.1 Sinusoidal Filter	74
		4.3.2 Common Mode Filter	80
	4.4	dV/dt Filter	83
	4.5	Motor Choke	85
	4.6	Simulation Examples	86
		4.6.1 Inverter with LC Filter	86
		4.6.2 Inverter with Common Mode and Differential Mode Filter	90
	4.7	Summary	95
	References		96
5	**Estimation of the State Variables in the Drive with LC Filter**		**97**
	5.1	Introduction	97
	5.2	The State Observer with LC Filter Simulator	99
	5.3	Speed Observer with Simplified Model of Disturbances	103
	5.4	Speed Observer with Extended Model of Disturbances	106
	5.5	Speed Observer with Complete Model of Disturbances	107
	5.6	Speed Observer Operating for Rotating Coordinates	109
	5.7	Speed Observer Based on Voltage Model of Induction Motor	114
	5.8	Speed Observer with Dual Model of Stator Circuit	122
	5.9	Adaptive Speed Observer	125
	5.10	Luenberger Flux Observer	129

		5.11	Simulation Examples	130
			5.11.1 Model of the State Observer with LC Filter Simulator	130
			5.11.2 Model of Speed Observer with Simplified Model of Disturbances	133
			5.11.3 Model of Rotor Flux Luenberger Observer	136
		5.12	Summary	138
		References		138
6	**Control of Induction Motor Drives with LC Filters**			**141**
	6.1	Introduction		141
	6.2	A Sinusoidal Filter as the Control Object		141
	6.3	Field Oriented Control		143
	6.4	Nonlinear Field Oriented Control		148
	6.5	Multiscalar Control		156
		6.5.1 Main Control System of the Motor State Variables		157
		6.5.2 Subordinated Control System of the Sinusoidal Filter State Variables		160
	6.6	Electric Drive with Load-Angle Control		166
	6.7	Direct Torque Control with Space Vector Pulse Width Modulation		178
	6.8	Simulation Examples		186
		6.8.1 Induction Motor Multiscalar Control with Multiloop Control of LC Filter		186
		6.8.2 Inverter with LC Filter and LR Load with Closed-Loop Control		194
	6.9	Summary		198
	References			198
7	**Current Control of the Induction Motor**			**201**
	7.1	Introduction		201
	7.2	Current Controller		203
		7.2.1 Predictive Object Model		207
		7.2.2 Costs Function		208
		7.2.3 Predictive Controller		208
	7.3	Investigations		208
	7.4	Simulation Examples of Induction Motor with Motor Choke and Predictive Control		210
	7.5	Summary and Conclusions		216
	References			217
8	**Diagnostics of the Motor and Mechanical Side Faults**			**218**
	8.1	Introduction		218
	8.2	Drive Diagnosis Using Motor Torque Analysis		218
	8.3	Diagnosis of Rotor Faults in Closed-Loop Control		233
	8.4	Simulation Examples of Induction Motor with Inverter Output Filter and Load Torque Estimation		235
	8.5	Conclusions		239
	References			239

9	**Multiphase Drive with Induction Motor and an LC Filter**	**241**
	9.1 Introduction	241
	9.2 Model of a Five-Phase Machine	243
	9.3 Model of a Five-Phase LC Filter	246
	9.4 Five-Phase Voltage Source Inverter	247
	9.5 Control of Five-Phase Induction Motor with an LC Filter	253
	9.6 Speed and Flux Observer	255
	9.7 Induction Motor and an LC Filter for Five-Phase Drive	257
	9.8 Investigations of Five-Phase Sensorless Drive with an LC Filter	257
	9.9 FOC Structure in the Case of Combination of Fundamental and Third Harmonic Currents	262
	9.10 Simulation Examples of Five-Phase Induction Motor with a PWM Inverter	266
	References	269
10	**General Summary, Remarks, and Conclusion**	**271**

Appendix A	**Synchronous Sampling of Inverter Output Current**	**273**
	References	276
Appendix B	**Examples of LC Filter Design**	**277**
	B.1 Introduction	277
Appendix C	**Equations of Transformation**	**282**
	References	285
Appendix D	**Data of the Motors Used in Simulations and Experiments**	**286**
Appendix E	**Adaptive Backstepping Observer**	**289**

Marcin Morawiec

	E.1 Introduction	289
	E.2 LC Filter and Extended Induction Machine Mathematical Models	290
	E.3 Backstepping Speed Observer	292
	E.4 Stability Analysis of the Backstepping Speed Observer	298
	E.5 Investigations	304
	E.6 Conclusions	305
	References	307
Appendix F	**Significant Variables and Functions in Simulation Files**	**308**
Index		**311**

Foreword

The converter-fed electric drive technologies have grown fast and matured notably over the last few years through the advancement of technology. Therefore, it is my great pleasure that this book, *Variable Speed AC Drives with Inverter Output Filters*, will perfectly fill the gap in the market related to design and modern nonlinear control of the drives fed from the inverters equipped with output filters. Such filters are installed mainly for reducing high dv/dt of inverter pulsed voltage and achieving sinusoidal voltage and currents waveforms (sinusoidal filter) on motor terminals. As a result, noises and vibrations are reduced and the motor efficiency is increased. These advantages, however, are offset by the complication of drive control because with inverter output filter there is a higher order control plant.

The book is structured into ten chapters and five appendices. The first chapter is an introduction, and general problems of AC motors supplied from voltage source inverter (VSI) are discussed in Chapter 2. The idealized complex space-vector models based on T and Γ equivalent circuits and its presentation in state space equations form for the AC induction machine are derived in Chapter 3. Also, in this chapter, definitions of per-unit system used in the book are given. The detailed overview, modeling, and design of family of filters used in inverter-fed drives: sinusoidal filter, common mode filter, and dV/dt filter are presented in Chapter 4. Next, in Chapter 5, several types of state observers of induction machine drive with output filter are presented in detail. These observers are necessary for in-depth studies of different sensorless high-performance control schemes presented in Chapter 6, which include: field oriented control (FOC), nonlinear field oriented control (NFOC), multiscalar control (MC), direct load angle control (LAC), direct torque control with space vector pulse width modulation (DTC-SVM). Chapter 7, in turn, is devoted to current control and basically considers the model predictive stator current control (MPC) of the induction motor drive with inductive output filter implemented and investigated by authors. A difficult, but important issue of fault diagnosis in the induction motor drives (broken rotor bars, rotor misalignment, and eccentricity) are studied in Chapter 8, which presents methods based on frequency analysis and artificial intelligence (NN) and adaptive neuro-fuzzy inference system (ANFIS). In Chapter 9, the results of analyzing, controlling, and investigating the classical three-phase drives with inverter output filter are generalized for five-phase inductive machines, which are characterized by several important advantages such as higher torque density, high fault tolerance, lower torque pulsation and noise, lower current losses, and reduction of the rated current of power converter devices. Chapter 10 gives a short summary and final conclusions that underline the main topics and achievements of the book. Some special aspects are presented in appendices (A to F): synchronous sampling of inverter current (A), examples of LC filter design (B), transformations of equations (C), motor data used in the book (D), adaptive back stepping observer (E), and significant variables and functions used in simulation files (F).

This book has strong monograph attributes and discusses several aspects of the authors' current research in an innovative and original way. Rigorous mathematical description, good illustrations, and a series of well-illustrated MATLAB®-Simulink models (S Functions written in C language included). Simulation results in every chapter are strong advantages which makes the book attractive for a wide spectrum of researches, engineering professionals, and undergraduate/graduate students of electrical engineering and mechatronics faculties.

Finally, I would like to congratulate the authors of the book because it clearly contributes to better understanding and further applications of converter-fed drive systems.

MARIAN P. KAZMIERKOWSKI, IEEE Fellow
Institute of Control and Industrial Electronics
Faculty of Electrical Engineering
Warsaw University of Technology, Poland

Acknowledgments

We would like to take this opportunity to express our sincere appreciation to all the people who were directly or indirectly helpful in making this book a reality. Our thanks go to our colleagues and students at Gdansk University of Technology and Texas A&M University at Qatar. Our special thanks go to Professor Zbigniew Krzeminski who has given us a lot of interesting and helpful ideas.

We are indebted to our family members for their continuous support, patience, and encouragement without which this project would not have been completed. We would also like to express our appreciation and sincere gratitude to the Wiley staff for their help and cooperation.

We are also grateful to the National Science Centre (NSC) for the part of the work that was financed by them as part of the funds allocated based on the agreement No. DEC-2013/09/B/ST7/01642. Special thanks also go to Texas A&M University, Qatar, for funding the language revision, editing, and other related work.

Above all, we are grateful to the almighty, the most beneficent and merciful who provides us confidence and determination in accomplishing this work.

Jaroslaw Guzinski, Haitham Abu-Rub, and Patryk Strankowski

About the Authors

Jaroslaw Guzinski received M.Sc., Ph.D., and D.Sc. degrees from the Electrical Engineering Department at Technical University of Gdansk, Poland in 1994, 2000, and 2011, respectively. From 2006 to 2009 he was involved in European Commission Project PREMAID Marie Curie, "Predictive Maintenance and Diagnostics of Railway Power Trains," coordinated by Alstom Transport, France. Since 2010, he has been a consultant in the project of integration of renewable energy sources and smart grid for building unique laboratory LINTE^2. In 2012 he was awarded by the Polish Academy of Sciences—Division IV: Engineering Sciences for his monograph *Electric drives with induction motors and inverters output filters—selected problems*. He obtained scholarships in the Socrates/Erasmus program and was granted with three scientific projects supported by the Polish government in the area of sensorless control and diagnostic for drives with LC filters.

He has authored and coauthored more than 120 journal and conference papers. He is an inventor of some solutions for sensorless speed drives with LC filters (three patents). His interests include sensorless control of electrical machines, multiphase drives (five-phase), inverter output filters, renewable energy, and electrical vehicles. Dr Guzinski is a Senior Member of IEEE.

Dr Haitham Abu-Rub holds two PhDs, one in electrical engineering and another in humanities. Since 2006, Abu-Rub has been associated with Texas A&M University–Qatar, where he was promoted to professor. Currently he is the chair of Electrical and Computer Engineering Program there and the managing director of the Smart Grid Center—Extension in Qatar. His main research interests are energy conversion systems, including electric drives, power electronic converters, renewable energy, and smart grid.

Abu-Rub is the recipient of many international awards, such as the American Fulbright Scholarship, the German Alexander von Humboldt Fellowship, the German DAAD Scholarship, and the British Royal Society Scholarship. Abu-Rub has published more than 200 journal and conference papers and has earned and supervised many research projects. Currently he is leading many potential projects on photovoltaic and hybrid renewable power generation systems with different types of converters and on electric drives. He has authored and coauthored several books and book chapters. Abu-Rub is an active IEEE senior member and serves as an editor in many IEEE journals.

Patryk Strankowski received the BSc degree in electrical engineering and the MSc degree in automation systems from the Beuth University of Applied Science, Berlin, Germany in 2012 and 2013, respectively. During his bachelor studies he was involved in the Siemens scholarship program, where he worked for customer solutions at the Department of Automation and Drives.

He is currently working toward his PhD degree at Gdansk University of Technology in Poland. His main research interests include monitoring and diagnosis of electrical drives as well as sensorless control systems and multiphase drives.

Nomenclature

Vectors are denoted with bold letters, for example, $\mathbf{u}_s$.

Latin letters

$a_1, a_2, \ldots a_6$	Coefficients of motor model equations
ABC	Three phase reference frame
C_{s0}	Common mode motor capacitance
d, q	Orthogonal coordinates of rotating reference system with angular speed of rotor flux vector
$\mathbf{e}$	Motor electromotive force
f	Frequency
f_2	Slip frequency
f_{imp}	Inverter modulation frequency
f_n	Nominal frequency
f_{rez}	Resonance frequency
f_r	Rotor rotation frequency
f_s	Stator voltage and current frequency
i	Current
i_1	Inverter output current
i_c	Filter capacitor current
I_n	Nominal current
$\mathbf{i}_s$	Stator current
J	Inertia
K, L	Orthogonal coordinates of rotating reference system with angular speed of stator voltage vector
$k_1, \ldots k_6, k_A, k_B, k_{1L}, k_{2L}$	Observer gain variables
L_σ	Total leakage inductance of motor
$l_{\sigma s}, l_{\sigma r}$	Leakage inductance of stator winding and rotor
L_m	Mutual inductance of stator and rotor

L_{s0}	Motor inductance for common mode
M	Mutual inductance
m_0	Load torque
m_1, m_2	Multiscalar control system variables
m_e	Electromagnetic torque
n	Speed of motor shaft
p	Number of motor pole pairs
Q	Quality factor of resonance circuit
R_0	Circuit resistance of common mode
R_r	Rotor circuit resistance
R_s	Stator circuit resistance
R_{s0}	Motor resistance of common mode
S	Speed direction sign
S_b, S_x	Observer stabilizing magnitude
$t_0, \ldots, t_6$	Sequence switching time of inverter voltage vectors
T_{imp}	Inverter impulse period
t_r	Voltage rise time on motor terminals
T_r	Rotor circuit time constant
T_{Sb}, T_{KT}, T_{Sx}	Inertial filters time constants
U	Voltage
u_α, u_β	Voltage vector components in α, β frame
u_0	Common mode voltage
u_1, u_2	Auxiliary variables of multiscalar control system
U_d	Inverter supply voltage
u_f	Inverter output voltage
u_L	Voltage drop on filter choke
U_n	Nominal voltage
$\mathbf{u}_s$	Stator voltage
u_U, u_V, u_W	Inverter or motor phase voltages U, V, W
UVW	Inverter output phase notation
$\mathbf{U}_{w0}, \mathbf{U}_{w1}, \ldots \mathbf{U}_{w7}$	Output voltage vector of inverter
w_σ	Coefficient in motor model equations
w_t	Coefficient depending on the pulse width modulator
$\mathbf{x}$	Vector variable of nonlinear object state
x, y	Orthogonal coordinates of rotating reference frame with arbitrary chosen angular speed ω_a
$x_{11}, x_{12}, x_{21}, x_{22}$	Multiscalar variables
XYZ	Filter output phase description
Z	Impedance
Z_0	Characteristic filter impedance

Greek letters

α, β	Orthogonal coordinates of fixed reference frame
δ	Load angle
δ^*	Reference load angle
ξ_d	Damping coefficient

ξ	Disturbance
ρ_{us}	Stator voltage vector position angle in αβ system
$\rho_{\psi r}$	Rotor linked flux vector position angle
σ	Total coefficient of motor leakage
σ_s, σ_r	Leakage coefficient of stator windings and rotor
τ	Relative time (time in pu)
τ_s'	Time constant of stator circuit
ψ_0	Magnetic flux in the common mode choke core
ψ_r	Rotor flux
ψ_s	Stator flux
ω_2	Slip pulsation
ω_a	Angular speed of arbitrary chosen reference frame
ω_i	Stator current pulsation
ω_r	Angular speed of motor shaft
ω_u	Stator voltage pulsation
$\omega_{\psi r}$	Rotor flux pulsation

Abbreviations

CM	Common mode
DSPC	Direct speed control
DTC	Direct torque control
EMF	Electromotive force
FFT	Fast Fourier transformation
FOC	Field oriented control
IGBT	Insulated gate bipolar transistor
IM	Induction motor
PE	Earth potential
PI	Proportional–plus–integral controller
PWM	Pulse width modulation
SVM	Space vector modulation
THD	Total harmonic distortion

1

Introduction to Electric Drives with LC Filters

1.1 Preliminary Remarks

The basic function of electric drives is to convert electrical energy to mechanical form (in motor mode operation) or from mechanical form to electrical energy (in generation mode). The electric drive is a multidisciplinary problem because of the complexity of the contained systems (Figure 1.1).

It is important to convert the energy in a controllable way and with high efficiency and robustness. If we look at the structure of global consumption of electrical energy the significance is plain. In industrialized countries, approximately two thirds of total industrial power demand is consumed by electrical drives [1, 2].

The high performance and high efficiency of electric drives can be obtained only in the case of using controllable variable speed drives with sophisticated control algorithms [3, 4].

In the industry, the widely used adjustable speed electrical drives are systems with an induction motor and voltage inverter (Figure 1.2). Their popularity results mainly from good control properties, good robustness, high efficiency, simple construction, and low cost of the machines [5].

Simple control algorithms for induction motors are based on the V/f principle. Because the reference frequency changes, the motor supply voltage has to be changed proportionally. In more sophisticated algorithms, systems such as field-oriented, direct torque, or multiscalar control have to be applied [6, 7]. Simultaneously, because of the estimation possibilities of selected controlled variables, for example, mechanical speed, it is possible to realize a sensorless control principle [7–10]. The sensorless speed drives are beneficial to maintain good robustness. Unfortunately, for sophisticated control methods, knowledge of motor parameters as well as high robustness of the drives against changes in motor parameters is required.

Variable Speed AC Drives with Inverter Output Filters, First Edition. Jaroslaw Guzinski, Haitham Abu-Rub and Patryk Strankowski.
© 2015 John Wiley & Sons, Ltd. Published 2015 by John Wiley & Sons, Ltd.

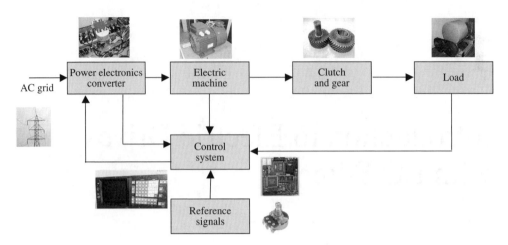

Figure 1.1 General structure of an electrical drive

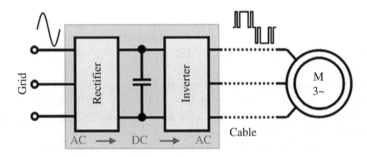

Figure 1.2 Electrical drive with voltage inverter and AC motor

1.2 General Overview of AC Drives with Inverter Output Filters

The inverter output voltage has a rectangular shape and is far from the sinusoidal one. Also, the use of semiconductor switches with short switching times causes high rates of rises of dV/dt voltages that initiate high levels of current and voltage disturbance [4, 11]. For this reason, it is necessary to apply filters between the inverter and the motor (Figure 1.3).

The introduction of a filter at the inverter output disables the proper operation of advanced drive control systems because doing this introduces more passive elements (inductances, capacitances, and resistances), which are not considered in the control algorithm [4, 12, 13]. This irregularity is caused by amplitude changes and phase shifts between the first current component and the motor supply voltage, compared to the currents and voltages on the inverter output. This causes the appearance in the motor control algorithm of inaccurate measured values of current and voltage at the standard measuring points of the inverter circuit. A possible solution to this issue is the implementation of current and voltage sensors at the filter output. However, this solution is not applied in

Introduction to Electric Drives with LC Filters

Figure 1.3 AC motor with voltage inverter and inverter output filter

industry drive systems because the filter is an element connected to the output of the inverter. The implementation of external sensors brings an additional cable network and that increases the susceptibility of the system to disturbances, reduces the system reliability, and increases the total cost of the drive.

A better solution is to consider the structure and parameters of the filter in the control and estimation algorithms. This makes it possible to use the measurement sensors that are already installed in the classical voltage inverter systems.

The addition of the filter at the voltage inverter output is beneficial because of the limitation of disturbances at the inverter output by obtaining sinusoidal voltage and current waveforms. Noises and vibrations are reduced and motor efficiency is increased. Furthermore, output filters reduce overvoltages on the motor terminals, which are generated through wave reflections in long lines and can result in accelerated aging of insulation. Several filter solutions are also used for limiting motor leakage currents, ensuring a longer failure-free operation time of the motor bearings.

The application of an inverter output filter and its consideration in the control algorithm is especially beneficial for various drive systems such as cranes and elevators. In that application, a long connection between motor and inverter is common.

The limitation of disturbances in inverter output circuits is an important issue that is discussed in numerous publications [14–18]. To limit such current and voltage disturbances, passive or active filters are used [4, 15]. The main reasons for preferring passive filters are especially the economic aspects and the possibility of limiting current and voltage disturbances in drive systems with high dV/dt voltage.

The control methods presented so far in the literature (e.g., [8–10, 19–27]) for an advanced sensorless control squirrel cage motor are designed for drives with the motor directly connected to the inverter. Not using filters in many drives is the result of control problems because of the difference between the instantaneous current and voltage values at the filter output and the current and voltage values at the filter input. Knowledge of this values is needed in the drive system control [28, 29]. A sensorless speed control in a drive system with an induction motor is most often based on the knowledge of the first component of the current and voltage. The filter can be designed in such a way that it will not significantly influence the fundamental components and will only limit the higher harmonics. However, most output filter systems introduce a voltage drop and a current and voltage phase shift for the first harmonic [4, 30]. This problem is important especially for sinusoidal filters, which ensure sinusoidal output voltage and current waveforms.

Another problem that has received attention in the literature [16, 30–37] is the common mode current that occurs in drive systems with a voltage inverter. The common mode current flow reduces the motor durability because of the accelerated wear of bearings. This current might also have an effect on the wrong operation of other drives included in the same electrical grid and can cause rising installation costs, which could lead to the need for an increase in the diameter of earth wire. Such problems come from both the system topology and the applied pulse width modulation in the inverter, which are independent of the main control algorithms. Modifying the modulation method can cause a limitation of the common mode current [4, 30, 38].

This book presents the problems related to voltage-inverter-fed drive systems with a simultaneous output filter application. The authors have presented problems and searched for new solutions, which up to now, have not been presented in the literature. Therefore, this book introduces, among other topics, new state observer structures and control systems with LC filters.

The problem of drive systems with output filters, justifying the need for their application, is also explained. Moreover, the aim of this book is to present a way to control a squirrel cage induction motor and estimation of variables by considering the presence of the output filter, especially for drive systems without speed measurement.

Other discussed topics are several motor control structures that consider the motor filter as the control object. Such solutions are introduced for nonlinear-control drive systems and field-orientated control with load-angle control. Predictive current control with the presence of a motor choke is also analyzed. Solutions for systems with the estimation of state variables are presented, and the fault detection scheme for the mechanical part of the load torque transmission system is shown. Thus, for diagnostic purposes, state observer solutions were applied for drive systems with a motor filter.

The main points to be discussed are:

- A motor filter is an essential element in modern inverter drive systems.
- The introduction of a motor filter between the inverter and motor terminals changes the drive system structure in such a way that the drive system might operate incorrectly.
- The correct control of the induction motor, especially for sensorless drives, requires consideration of the filter in the control and state variable estimation process.

Some of the presented problems in the book also refer to drive systems without filters. Those problems are predictive current control using the state observer, fault diagnostics using a state observer in rotating frame systems, and decoupled field-orientated control with load-angle control.

1.3 Book Overview

Chapter 2 presents the problems of voltage and current common mode. The common mode is the result of voltage inverter operation with pulse width modulation in addition to the motor parasitic capacitances. The equivalent circuit of the common mode current flow is presented and explained extensively. Furthermore, attention is paid to the bearing current, whose types are characterized by a fundamental method. The main ways to limit the common mode current are mentioned, taking into consideration the application of common mode chokes. Additionally, a way of determining the motor parameters for common mode is shown. A considerable part of the chapter is dedicated to the active method of limiting the common mode through the modification of the pulse width modulation scheme. Some comments on synchronous sampling of inverter output current are also included in Appendix A.

Chapter 3 presents the motor model of a squirrel cage motor used for simulation research. The induction motor model dependencies are also used for analysis and presentation of the state estimators and control algorithms. The equations of transformations are given in Appendix C. The examples of data of the motors used in simulations and experiments are in Appendix D.

Chapter 4 contains selected output filter structures of the voltage inverter. The equivalent circuit of the output filter in the orthogonal frame is presented. The analysis of the obtained model makes it possible to conclude that only the sinusoidal filter has an influence on the motor control and variable estimation. Furthermore, this chapter contains a description of how to choose the filter elements for the complex filter structure, sinusoidal filter, common mode filter, and motor chokes. The examples of LC filter design are presented in Appendix B.

Chapter 5 demonstrates the problem of state variables estimation for drive systems with a sinusoidal filter. Several observers are presented, considering the installation of a sinusoidal filter. These include a state observer with a filter simulator, a speed observer in a less complicated version, an extended and full disturbance model, a speed observer in a rotating orthogonal frame, a speed observer based on a voltage model of the induction motor, and an adaptive speed observer. The presented systems make it possible to calculate the rotational motor speed, rotor and stator flux, and other required state variables of the control process. A supplement to chapter 5 is Appendix E in which the adaptive type backstepping observer [39] is presented.

Chapter 6 contains the control of an induction motor considering a sinusoidal filter. The problem is presented for the influence of the filter on an electric drive control operating in a closed loop without a speed sensor. Among the controls discussed, the following methods are included: classical field-orientated control, decoupled nonlinear field-orientated control, multiscalar nonlinear control, and nonlinear decoupled operation with load-angle control. Structures and dependencies are presented for further control methods, comparing the system operation for both situations, with and without consideration of the presence of the filter.

Chapter 7 presents a description of predictive motor current control for a drive system with a motor choke. To control the motor current, a controller was used in which the electromotive force of the motor was determined directly in the state observer dependencies.

Chapter 8 contains the diagnostic task of the chosen fault appearances in the drive system with an induction motor, voltage inverter, and motor choke. The fault diagnosis mainly concentrates on detection of failures in the mechanical torque transmission system and rotor bar faults. The diagnostic method in this chapter is based on the analysis of the calculated electromagnetic and load torques of the motor. Moreover, the chapter presents the fault diagnostic problem of a motor operating in a closed loop control structure, which is based on the analysis of chosen internal signals of the control system.

Chapter 9 presents a five-phase induction motor drive with an LC filter. The solutions presented in previous chapters for a three-phase system are adapted to a multiphase drive.

The last chapter, Chapter 10, contains a summary of the book.

1.4 Remarks on Simulation Examples

Generally, simulations of electric drives and power electronics converters could be done in universal simulation software with some standard models (e.g., MATLAB/Simulink, PSIM, TCAD, CASPOC, etc.) or in dedicated software written by researchers (e.g., in C or C++ language). Both solutions have advantages and disadvantages [40, 41]. The concept of simulation in C language is attractive if one wants to keep good transfer of the models between different simulation software applications where all of them have the possibility of creating user-oriented blocks.

The simulation examples are an integral part of the book. The examples are prepared for a MATLAB/Simulink base without a requirement for any additional toolboxes. The principal part of each simulation is Simulink S-Functions for particular models written in C language. The complete C files of the simulation model are included in the book. With that approach, the examples used are not limited to MATLAB/Simulink. Because of the simple structure of the files and ANSI C standard, C files could be used in other simulation software that has the ability to define user blocks. The compilation of C files in MATLAB/Simulink requires the C compiler corresponding to the reader's particular version of MATLAB/Simulink. The book companion simulation examples were prepared in MATLAB/Simulink version 2014b and exported to previous version 2013a (files with the extension .slx) and simultaneously exported to version R2007b (files with the extension .mdl). The examples are described at the end of each chapter. The main structure and basic results are presented. The use of simulation examples requires basic knowledge of C language. With the knowledge given in the examples, it is possible to convert models to other simulation software. For each example, the particular results are presented.

The list of most used variables and functions is given in Appendix F.

References

[1] Mirchevski S. Energy efficiency in electric drives. *Electronics Journal*. 2012; **16** (1): 46–49.
[2] Siemens AG. Energy-efficient drives: Answers for industry. 2009. Germany. Available from: https://w3.siemens.com/mcms/water-industry/en/Documents/Energy-Efficient_Drives.pdf.
[3] Bose BK. *Power electronics and motor drives—Advances and trends*. San Diego: Elsevier/Academic Press; 2006.
[4] Abu-Rub H, Iqbal A, Guzinski J. *High performance control of AC drives with MATLAB/Simulink models*. Chichester, UK: John Wiley & Sons, Ltd; 2012.
[5] Bose BK. Power electronics and motor drives recent progress and perspective. *IEEE Transactions on Industrial Electronics*. 2009; **56** (2): 581–588.
[6] Kaźmierkowski MP, Krishnan R, Blaabjerg F. *Control in power electronics*. San Diego: Academic Press; 2002.
[7] Krzeminski Z, Lewicki A, Wlas M. Properties of sensorless control systems based on multiscalar models of the induction motor. *COMPEL: The International Journal for Computation and Mathematics in Electrical and Electronic Engineering*. 2006; **25** (1): 195–206.
[8] Rajashekara K, Kawamura A, Matsuse K. *Sensorless control of AC motor drives*. New York: IEEE Industrial Electronics Society, IEEE Press; 1996.
[9] Orłowska-Kowalska T, Wojsznis P, Kowalski Cz: Comparative study of different flux estimators for sensorless induction motor drive. *Archives of Electrical Engineering*. 2000; **49** (1): 49–63.
[10] Orłowska-Kowalska T, Wojsznis P, Kowalski Cz. Dynamical performances of sensorless induction motor drive with different flux and speed observers. Ninth European Conference on Power Electronics and Applications (EPE'2001). Graz, Austria. August 27–29, 2001.
[11] Enlayson P. Output filters for PWM drives with induction motors. *IEEE Industry Applications Magazine*. 1998 (Jan/Feb); **4** (1): 46–52.
[12] Guziński J. Closed loop control of AC drive with LC filter. Thirteenth International Power Electronics and Motion Conference EPE-PEMC, 2008, Poznań, Poland. September 1–3, 2008.
[13] Salomaki J. Hikkanen M. Luomi J. Sensorless control of induction motor drives equipped with inverter output filter. IEEE International Conference on Electric Machines and Drives, IEMDC'05, San Antonio, TX. May 15–18, 2005.
[14] Hanigovszki N, Poulsen J, Blaabjerg F. Performance comparison of different output filter topologies for ASD. European Conference on Power Electronics and Applications, EPE'03, Toulouse, France. September 2–4, 2003.

[15] Moreira A, Lipo T. Modeling and evolution of dv/dt filters for AC drives with high switching speed. Ninth European Conference on Power Electronics and Applications, EPE'2001, Graz, Austria. August 27–29, 2001.
[16] Muetze A, Binder A. High frequency stator ground currents of inverter-fed squirrel-cage induction motors up to 500 kW. Tenth European Conference on Power Electronics and Applications EPE'03, Toulouse, France. September 2–4, 2003.
[17] Popescu M, Bitoleanu A. The influence of certain PWM methods on the quality of input energy of the asynchronous motor and frequency converters driving system. Tenth International Power Electronics and Motion Control Conference, EPE–PEMC 2002, Dubrovnik, Croatia. September 9–11, 2002.
[18] Akagi H. New trends in active filters for power conditioning. *IEEE Transactions on Industry Applications*. 1996; **32** (6): 1312–1322.
[19] Abu-Rub H, Guzinski J, Krzeminski Z, Toliyat HA. Speed observer system for advanced sensorless control of induction motor. *IEEE Transactions on Energy Conversion*. 2003; **18** (2): 219–224.
[20] Abu-Rub H, Guzinski J, Toliyat HA. An advanced low-cost sensorless induction motor drive. *IEEE Transactions on Industry Applications*. 2003; **39** (6): 1757–1764.
[21] Abu-Rub H, Oikonomou N. Sensorless observer system for induction motor control. Thirty-ninth IEEE Power Electronics Specialists Conference, PESC0'08, Rodos, Greece. June 15–19, 2008.
[22] Adamowicz M, Guzinski J. Control of sensorless electric drive with inverter output filter. Fourth International Symposium on Automatic Control AUTSYM 2005, Wismar, Germany. September 22–23, 2005.
[23] Holtz J. Sensorless vector control of induction motors at very low speed using a nonlinear inverter model and parameter identification. *IEEE Transactions on Industry Applications*. 2002; **38** (4):1087–1095 4.
[24] Holtz J. Sensorless control of induction motor drive. *Proceedings of the IEEE*. 2002; **90** (8): 1358–1394.
[25] Holtz J. Sensorless control of induction machines—with or without signal injection? *IEEE Transactions on Industrial Electronics*. 2006; **53** (1): 7–30.
[26] Kubota H, Sato I, Tamura Y, Matsuse K, Ohta H, Hori Y. Regenerating-mode low-speed operation of sensorless induction motor drive with adaptive observer. *IEEE Transactions on Industry Applications*. 2002; **38** (4): 1081–1086.
[27] Tsuji M, Chen S, Izumi K, Ohta T, Yamada E. A speed sensorless vector-controlled method for induction motor using q-axis flux. Second International Power Electronics and Motion Control Conference, IPEMC'97, Hangzhou, China. November 3–6, 1997.
[28] Seliga R, Koczara W. Multiloop feedback control strategy in sine-wave voltage inverter for an adjustable speed cage induction motor drive system. Tenth European Conference on Power Electronics and Applications, EPE'2001, Graz, Austria. August 27–29, 2001.
[29] Seliga R, Koczara W. High quality sinusoidal voltage inverter for variable speed ac drive systems. Tenth International Power Electronics and Motion Control Conference, EPE–PEMC 2002, Dubrovnik, Croatia. September 9–11, 2002.
[30] Guzinski J. *Selected problems of induction motor driver with inverter output filters. Monograph No. 115.* Gdansk, Poland: Gdansk University of Technology; 2011.
[31] Akagi H. Prospects and expectations of power electronics in the 21st century. Power Conversion Conference, PCC'02, Osaka, Japan. April 2–5, 2002.
[32] Akagi H, Hasegawa H, Doumoto T. Design and performance of a passive EMI filter for use with a voltage-source PWM inverter having sinusoidal output voltage and zero common-mode voltage. *IEEE Transactions on Power Electronics*. 2004; **19** (4): 1069–1076.
[33] Busse DF, Erdman J, Kerkman R, Schlegel D, Skibinski G. The effects of PWM voltage source inverters on the mechanical performance of rolling bearings. Eleventh Annual Applied Power Electronics Conference, APEC'96, San Jose, CA. March 3–7, 1996.
[34] Busse DF, Erdman JM, Kerkman RJ, Schlegel DW, Skibinski .L: An evaluation of the electrostatic shielded induction motor: a solution for rotor shaft voltage buildup and bearing current. *IEEE Transactions on Industry Applications* 1997; **33** (6): 1563–1570.
[35] Cacciato M, Consoli A, Scarcella G, Testa A. Reduction of common-mode currents in PWM inverter motor drives. *IEEE Transactions on Industry Applications*. 1999; **35** (2): 469–476.
[36] Kikuchi M, Kubota H. A novel approach to eliminating common-mode voltages of PWM inverter with a small capacity auxiliary inverter. Thirteenth European Conference on Power Electronics and Applications EPE'09. Barcelona, Spain. September 8–10, 2009.
[37] Muetze A, Binder A. Practical rules for assessment of inverter-induced bearing currents in inverter-fed ac motors up to 500 kW. *IEEE Transactions on Industrial Electronics*. 2007; **54** (3): 1614–1622.

[38] Choi HS, Cho BH Power factor pre-regulator (PFP) with an improved zero-current-switching (ZCS) PWM switch cell. PCC'02, Osaka, Japan. April 2–5, 2002.

[39] Morawiec M, Guzinski J. Sensorless control system of an induction machine with the Z-type backstepping observer. IEEE Twenty-third International Symposium on Industrial Electronics, ISIE 2014, Istanbul, Turkey. June 1–4, 2014.

[40] Yusivar F, Wakao S. Minimum requirements of motor vector control modeling and simulation utilizing C MEX S-function in MATLAB/Simulink. Fourth IEEE International Conference on Power Electronics and Drive Systems, PEDS 2001, Indonesia. October 22–25, 2001.

[41] Ji Y-H, Kim JG, Park SH, Kim J-H, Won C-Y. C-language based PV array simulation technique considering effects of partial shading. IEEE International Conference on Industrial Technology, ICIT 2009, Gippsland, Australia. February 10–13, 2009

2

Problems with AC Drives and Voltage Source Inverter Supply Effects

2.1 Effects Related to Common Mode Voltage

Accelerated degradation of the bearings of an induction motor (IM) operating with voltage inverters is an effect of parasitic current flow and is defined as a *bearing current*. The first reports of bearing currents were published nearly 100 years ago [1]. The observed phenomena were reported only for high-power machines as a result of magnetic asymmetry [2–5]. Its value is negligible compared with the bearing currents that occur in machines with inverter type supplies [6]. Bearing failure is now the most common failure of AC machines operating in adjustable speed drive (ASD). Because of the high number of ASDs, this type of failure requires special attention. The bearing current in modern ASDs is closely connected with the appearance of the common mode (CM) voltage resulting from the operation of a voltage inverter with pulse width modulation (PWM). For that operation, long-term current flow through bearings with a current density greater than that allowed for a bearing's rolling elements can completely destroy the bearings. It is reported that a current density of $J_b \geq 0.1$ A/mm^2 has no noticeable impact on the life of the bearing, but a current density of $J_b \geq 0.7$ A/mm^2 can significantly shorten its lifetime [6].

Bearing currents has long been an issue, having been observed many years ago in case of high-power low-poles machines with a sinusoidal supply [7]. The reason is that classical bearing current is asymmetric of the machine's magnetic path. If compared with the bearing current in a drive with a power electronic converter, the classical bearing current can be neglected [2].

In electrical drives with a converter-type supply, a main cause of the bearing current is the common mode voltage (CMV). That CMV results from the use of voltage inverters operating with PWM. In case of motor supply, the CMV is defined as the voltage between star connected motor neutral point potential and protective earth (PE) potential. In Figure 2.1, an equivalent circuit of the typical system of three-phase frequency converter, supply line, and motor is shown [8].

Variable Speed AC Drives with Inverter Output Filters, First Edition. Jaroslaw Guzinski, Haitham Abu-Rub and Patryk Strankowski.
© 2015 John Wiley & Sons, Ltd. Published 2015 by John Wiley & Sons, Ltd.

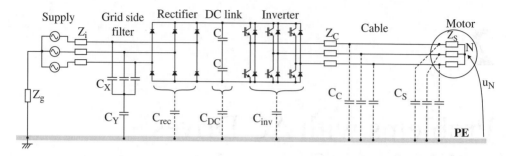

Figure 2.1 An equivalent circuit of the typical system of three phase frequency converter, supply line, and motor [8]

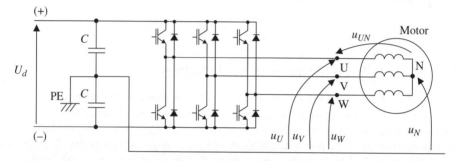

Figure 2.2 Notation of inverter output voltages related to DC link capacitors midpoint

In Figure 2.1, the CMV is indicated as u_N. Dotted lines indicate parasitic capacitances between elements of the electrical circuit and PE. The detailed analysis of the circuit is difficult because of the lack of full information on some parameters of circuit components.

For CM analysis of u_N, it is assumed that reference potential is the midpoint of the DC input circuit (Figure 2.2). It is accepted because of the low impedance of the CM series circuit. It is assumed that the reference point is connected with ground PE.

According to Figure 2.2, the expressions for the voltages u_U, u_V, and u_W are as follows:

$$u_U = u_{UN} + u_N \tag{2.1}$$

$$u_V = u_{VN} + u_N \tag{2.2}$$

$$u_W = u_{WN} + u_N \tag{2.3}$$

which leads to:

$$u_U + u_V + u_W = u_{UN} + u_{VN} + u_{WN} + 3u_N \tag{2.4}$$

because in a three-phase system

$$u_{UN} + u_{VN} + u_{WN} = 0 \tag{2.5}$$

Problems with AC Drives and Voltage Source Inverter Supply Effects

so the CMV is:

$$u_N = \frac{u_U + u_V + u_W}{3} \quad (2.6)$$

In the voltage inverters operating with PWM used now, the most popular modulation algorithm is vector modulation. It is widely known as *space vector modulation (SVM)*. For a three-phase inverter during SVM operations, the inverter output voltage is adequate for the inverter switches state. For six switches there are $2^6 = 64$ combinations, but only eight have a technical sense. Between them, six states are noted as active and two as passive. During the active state, the output voltage is non-zero; for the passive state, the output voltage is zero; and the load terminals are shorted. The inverter output voltages corresponding with each of eight vectors are listed in Table 2.1.

Unfortunately if the inverter DC midpoint is assumed as reference potential, it is not easy to see the waveform on oscilloscope. It is because in most of industrial inverters, the DC midpoint is not easy accessible. So for practical reasons, the positive or negative DC potential is assumed as reference potential as presented in Figure 2.3.

Table 2.1 The components of the inverter output voltage vectors referenced to the negative potential of the inverter input circuit (Figure 2.2)

Value	Binary notation of the inverter switches states and corresponding voltage values[a]							
	100	110	010	011	001	101	000	111
u_U	$\frac{U_d}{2}$	$\frac{U_d}{2}$	$-\frac{U_d}{2}$	$-\frac{U_d}{2}$	$-\frac{U_d}{2}$	$\frac{U_d}{2}$	$-\frac{U_d}{2}$	$\frac{U_d}{2}$
u_V	$-\frac{U_d}{2}$	$\frac{U_d}{2}$	$\frac{U_d}{2}$	$\frac{U_d}{2}$	$-\frac{U_d}{2}$	$-\frac{U_d}{2}$	$-\frac{U_d}{2}$	$\frac{U_d}{2}$
u_W	$-\frac{U_d}{2}$	$-\frac{U_d}{2}$	$-\frac{U_d}{2}$	$\frac{U_d}{2}$	$\frac{U_d}{2}$	$\frac{U_d}{2}$	$-\frac{U_d}{2}$	$\frac{U_d}{2}$
u_N	$-\frac{U_d}{6}$	$\frac{U_d}{6}$	$-\frac{U_d}{6}$	$\frac{U_d}{6}$	$-\frac{U_d}{6}$	$\frac{U_d}{6}$	$-\frac{U_d}{2}$	$\frac{U_d}{2}$

[a] The binary value means that for adequate state the upper switch is 1, on and 0, off. The bottom switch is the opposite state.

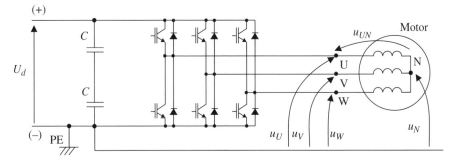

Figure 2.3 Notation of inverter output voltages related to inverter DC link negative potential

Table 2.2 Inverter output vectors components related to PE potential (Figure 2.3)

Notation	Inverter switching state							
	100	110	010	011	001	101	000	111
u_U	U_d	U_d	0	0	0	U_d	0	U_d
u_V	0	U_d	U_d	U_d	0	0	0	U_d
u_W	0	0	0	U_d	U_d	U_d	0	U_d

Table 2.3 Inverter CMV related to PE potential (Figure 2.3) for each switching state

Notation	Inverter switching state							
	100	110	010	011	001	101	000	111
u_N	$\dfrac{U_d}{3}$	$\dfrac{2U_d}{3}$	$\dfrac{U_d}{3}$	$\dfrac{2U_d}{3}$	$\dfrac{U_d}{3}$	$\dfrac{2U_d}{3}$	0	U_d

According to Figure 2.3, the Table 2.1 changes to the form presented in Table 2.2. In Table 2.3, the inverter CMV is given for each switching state.

If analysis of CMV is done in relation to the DC link voltage, positive or negative potential no difference is done to shape of value of u_N instantaneous values.

The CMV u_N is used in analysis of natural three-phase abc system. In case of orthogonal transformation and use of $\alpha\beta 0$, a corresponding value to u_N is u_0. That difference is because $\alpha\beta 0$ coordinates are used and will be discussed for PWM, motor model, and estimation and control algorithms.

For $\alpha\beta 0$ coordinates and power-invariant transformation, the inverter output voltage components are as given in Table 2.4. Based on analysis of Table 2.4, it is clear that inverter output voltage has CM component. The CMV waveform has significant high dV/dt changes. The peak-to-peak value of CMV in $\alpha\beta 0$ coordinates is equal to $\sqrt{3}U_d$.

In a real system, the zero-voltage component can be observed by measuring the voltage at the neutral point of the motor stator windings (Figure 2.4).

The occurrence of CM output voltage of the inverter and the presence of parasitic capacitance in the motor (Figure 2.5) and other ASD components causes the flow of zero sequence current (Figure 2.6).

Voltage drop across the serial impedance of the cable for CM is small compared with the value of u_0. Therefore, it is possible to provide an equivalent circuit for a CM current as shown in Figure 2.7; voltage u_0 is given directly on the cable input (i.e., the inverter is represented as the source of u_0). Figure 2.7 shows the equivalent circuit for the CM current for the motor and feeder cable.

The elements noted in the structure shown in Figure 2.7 are as follows:

- **Cable parameters:**
  R_c, cable resistance
  L_c, cable inductance
  C_c, cable capacitance

Problems with AC Drives and Voltage Source Inverter Supply Effects

- **Motor parameters:**
    C_{wf}, motor winding to frame capacitance
    C_{wr}, motor winding to rotor capacitance
    C_{rf}, motor rotor to frame capacitance
    C_b, motor bearings equivalent capacitance
    R_b, motor bearings equivalent resistance
    S_w, switch modeling breakdown of the bearing lubrication film

The electrical cable impedance has a voltage drop that is small if compared with the u_0 value. As a result of that, the CM equivalent circuit takes the form presented in Figure 2.7. In Figure 2.7, the voltage u_0 is applied to the cable input, which means that inverter is source of CMV. However in the case of motor choke or CM transformer, the voltage will differ from that in Table 2.4. If the difference is high enough, the CM current will be strongly depreciated.

Table 2.4 Inverter output voltage component in abc and αβ0 references[a]

Notation	Inverter switching states							
	100	110	010	011	001	101	000	111
u_U	U_d	U_d	0	0	0	U_d	0	U_d
u_V	0	U_d	U_d	U_d	0	0	0	U_d
u_W	0	0	0	U_d	U_d	U_d	0	U_d
u_N	$\dfrac{U_d}{3}$	$\dfrac{2U_d}{3}$	$\dfrac{U_d}{3}$	$\dfrac{2U_d}{3}$	$\dfrac{U_d}{3}$	$\dfrac{2U_d}{3}$	0	U_d
u_0 (αβ0 coordinates)	$\dfrac{U_d}{\sqrt{3}}$	$\dfrac{2U_d}{\sqrt{3}}$	$\dfrac{U_d}{\sqrt{3}}$	$\dfrac{2U_d}{\sqrt{3}}$	$\dfrac{U_d}{\sqrt{3}}$	$\dfrac{2U_d}{\sqrt{3}}$	0	$\sqrt{3}U_d$
u_α	$\dfrac{\sqrt{2}U_d}{\sqrt{3}}$	$\dfrac{U_d}{\sqrt{6}}$	$-\dfrac{U_d}{\sqrt{6}}$	$-\dfrac{\sqrt{2}U_d}{\sqrt{3}}$	$-\dfrac{U_d}{\sqrt{6}}$	$\dfrac{U_d}{\sqrt{6}}$	0	0
u_β	0	$\dfrac{U_d}{\sqrt{2}}$	$\dfrac{U_d}{\sqrt{2}}$	0	$-\dfrac{U_d}{\sqrt{2}}$	$-\dfrac{U_d}{\sqrt{2}}$	0	0

[a] Voltages related to dc link voltage negative potential (Figure 2.1)

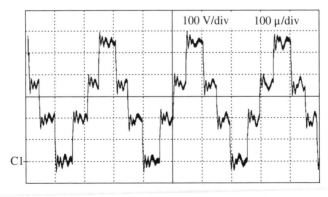

Figure 2.4 The voltage u_N waveform in the case of a voltage inverter type supply

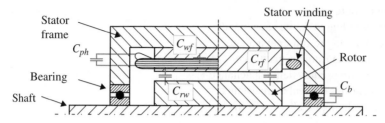

Figure 2.5 Parasitic capacitances in induction motor

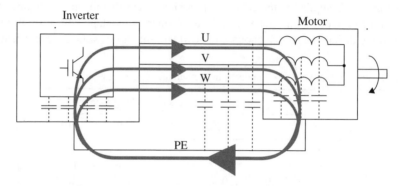

Figure 2.6 The circuit for common mode current of converter model presented in Figure 2.3

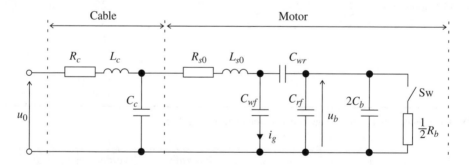

Figure 2.7 Common-mode equivalent circuit for induction motor and motor cable [9]

CMV with high values of dV/dt causes a current bearing and shaft voltage in the motor. As shown in the inverter supply drives [9], there are several types of bearing currents, such as:

- capacitive bearing current
- electric discharge bearing current
- circulating bearing current or shaft current, related to the shaft voltage effect
- rotor ground current

The primary reason for each of these types of currents is high dV/dt values at the motor terminals. But each type of current is related to a different motor physical phenomenon.

2.1.1 Capacitive Bearing Current

Capacitive bearing current, i_{bcap}, is flowing in electrical circuits where bearings equivalent capacitance, C_b, exists. The maximal value of this current is 5–10 mA in the case of a bearing temperature of $T_b \approx 25\ °C$ and a motor mechanical speed of $n \geq 100$ rpm [9]. An increase in either temperature, T_b, or motor speed or both will result in an increase in i_{bcap} as well. However the value of i_{bcap} is relatively small compared with other components of the CM current, so usually i_{bcap} is assumed to be harmless in terms of the motor bearing life.

2.1.2 Electrical Discharge Machining Current

Electrical discharge machining (EDM) current, i_{bEDM}, is a result of the breakdown of bearing oil film. The oil film is a thin insulating layer of the lubricant with a dielectric strength of the order of 15 kV/mm. The oil film thickness depends on the bearing type and size [10] and for a typical motor, bearing is close to 0.5 μm, which corresponds to a voltage breakdown of approximately 7.5 V [9]. The bearing voltage, u_b, corresponds to the CM voltage, u_0, according to the voltage distribution in a capacitive voltage divider with C_{wr} and C_{rf}. If the voltage, u_b, exceeds the oil film breakdown stress then an impulse of machine discharging current appears. According to the circuit structure shown in Figure 2.5 it corresponds to S_W switch on state. In accordance with the data given [6,11], maximal values of i_{bEDM} take values in the range 0.5–3 A. The bearing voltage u_b is independent of the motor size [12], which causes the i_{bEDM} current to be more dangerous especially for small power motors. This is because of the lower elastic contact surface between bearing balls and races, which increases the current density.

2.1.3 Circulating Bearing Current

Circulating bearing current and shaft voltage are related with current flow through the motor stator winding and frame capacitance. It is a high-frequency grounding current, i_g. The i_g flow excites the magnetic flux, ψ_{circ}, which circulates through the motor shaft. The flux ψ_{cir} induces the shaft voltage, u_{sh}. If u_{sh} is large enough, then the oil film in the bearings breaks down and the circulating bearing current, i_{bcir}, appears. The circuit for i_{bcir} flow contains the motor frame, shaft, and both bearings (Figure 2.8).

According to the data given [6], the maximal value of i_{bcir} is in the range of 0.5–20 A, depending on motor size; the largest values are observed for high-power motors. The measurement of i_{bcir} requires the use of a Rogowski coil on the motor shaft. The coil must be placed as far as possible from the stator coil out-hang (Figure 2.9a) [13].

The measurement of i_{bcir} in the way presented in Figure 2.9a is precise. However it is rather impossible to provide it in the standard motors at the workplace. So a practical solution is that the simplified measurement of i_{bcir} is only possible with a good galvanic connection between motor shaft and motor frame (Figure 2.9b, case A). In the same figure, it is also presented how measure the shaft grounding current (Figure 2.9b, case B) [14].

Measurement of i_{bcir} is complex, requiring special equipment and access inside the motor, so for practical reasons i_{bcir} could also be estimated on the basis of knowledge of the grounding current, i_g, measurement and type of bearings used in the motor. In the case of a standard motor with a pair of conventional bearings, the circulating bearing current is [6]:

$$i_{bcir(max)} \leq 0.4 \cdot i_{g(max)} \qquad (2.7)$$

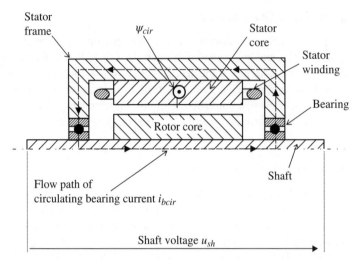

Figure 2.8 The flow path of the circulating bearing current i_{bcir} [19]

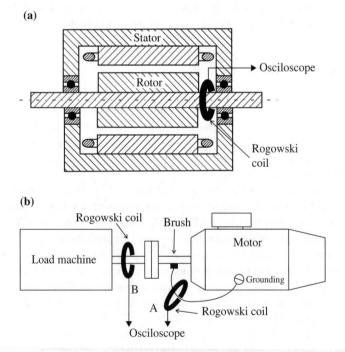

Figure 2.9 The methods of Rogowski coil location: (a) for circulating bearing current i_{bcir} precise measurement [13] and (b) for i_{bcir} simplified measurement (A) and shaft grounding current measurement (B) [14]

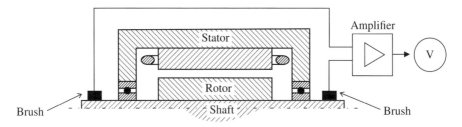

Figure 2.10 Shaft voltage measurement

The estimation of shaft voltage in a simple way is also possible [6]. It was observed empirically, based on a series of tests, that the shaft voltage, u_{sh}, is proportional to the grounding current and length of the stator core, l_{Fe}. Obviously, the length, l_{Fe}, is proportional to the motor frame size, H. The grounding current, i_g, is proportional to stator winding-to-frame parasitic capacitance, C_{wf}, which is proportional to the square of the motor frame size, which finally leads to the empirical relation [6]:

$$u_{sh} = i_g l_{Fe} \sim H^2 \cdot H = H^3 \tag{2.8}$$

The measurement of shaft voltage is complex. It requires the use of high-quality brushes on both ends of the motor shaft to assure a good contact area. Because of the small value of u_{sh}, it has to be amplified (Figure 2.10).

The shaft voltage, measured in laboratory condition, is in range from millivolts to several volts [7,13]. If u_{sh} is high, it is recommended to protect the motor using conducting brushes for a low-impedance connection between the shaft and grounding potential. Some commercially solutions are proposed [15].

2.1.4 Rotor Grounding Current

The appearance of rotor grounding current, i_{rg}, is possible only when the motor rotor has a galvanic connection with the earth potential through the driven load. If the impedance of the stator-rotor electric circuit is significantly lower than stator-frame impedance, then part of the total grounding current i_g flows as total grounding current, i_{rg}. The amplitude of i_{rg} can reach large values and quickly destroy the motor bearings [6].

2.1.5 Dominant Bearing Current

The dominant bearing current is dependent on the motor mechanical size, H. The proper relation has been formulated [6] and the dominant components are:

- machine discharging current i_{bEDM} if $H < 100$ mm
- both machine discharging current, i_{bEDM}, and circulating bearing current, i_{bEDM}, if 100 mm < $H < 280$ mm
- circulating bearing current, i_{bcir}, if $H > 280$ mm

When the motor size increases, the motor circulating bearing current increases simultaneously.

2.2 Determination of the Induction Motor CM Parameters

Knowledge of the motor's CM parameters is indispensable for modeling of the inverter type drive. As outlined in Section 2.1, the detailed CM circuit of the motor is complex (see Figure 2.5). An additional difficulty is the variability of some parameters of the circuit. In the literature [6, 9, 16–18], the analytical dependencies used for calculating the parameters of the circuit shown in Figure 2.5 are presented. However, the calculations are complex and require a lot of motor data, which are difficult to access from datasheets and simple measurements, although some of the results are presented in the literature [1].

For practical reasons, in most cases, the simpler motor CM circuit is considered, as presented in Figure 2.11 [12,18,19].

Inductance, L_0, and resistance, R_0, are the leakage inductance and substitute resistance, respectively, of the motor stator windings. These parameters are easy to measure in the configuration presented in Figure 2.13.

The measurement of parasitic capacitance C_0 needs deeper analysis. As shown in Figure 2.5, the equivalent capacitance, C_0, which is used in the simplified model of CM circuit is combination of a few parasitic elements: C_{wr}, C_{wf}, C_{rf}, and C_b. The topology of the circuit can also change in the case of oil film breakdown. Additionally, the bearing capacitance is non-linear, depending mainly on the motor speed. These phenomena can lead to measurement errors when typical measuring equipment such as an electronic RLC bridge meter is used. This equipment operates at low-voltage power and frequency, which may differ significantly from the value and frequency of CM voltage. Therefore, the C_0 capacitance measurement should be performed with a frequency and voltage similar to those observed during normal operation of the inverter.

The solution to the measurement of C_0 could be the creation of a serial resonance circuit (Figure 2.12b). The supply source of the circuit could be an inverter, which supplies the motor during normal operation. In this way, it is possible to measure C_0 at the same voltage and frequency as those which appear during normal operation of the drive system. This approach requires access to a PWM algorithm to generate a square wave voltage waveform with a constant modulation factor of 0.5 and the possibility of changing the modulation frequency.

The inverter is connected to a resonant circuit L_{res} and C_{res} including dumping resistance, R_{res}. The measurement is done at a frequency, f_{res}, corresponding to the frequency component of the common voltage, that is, the inverter switching frequency, f_{imp}.

The value of C_{res} must be significantly larger than the expected motor CM capacitance, C_0. If this condition is fulfilled, the impact of C_0 on the resonant frequency is negligible. Because the value of C_0 is in the range of nanofarads, the capacitor, C_{res}, should have a value of microfarads. The resistance, R_{res}, should be chosen to limit the inverter output current to a safe level and to provide a good quality factor of the resonant circuit.

The sample waveforms of current and voltages collected by the measurement test bench shown in Figure 2.12b are presented in Figure 2.13.

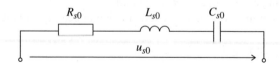

Figure 2.11 Simplified structure of the motor CM circuit

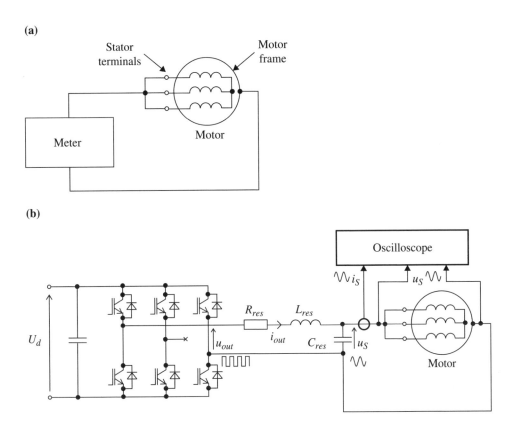

Figure 2.12 Electrical circuits for measurement the CM parameters of the motor with LCR meter (a) and in resonance circuit (b) with inverter

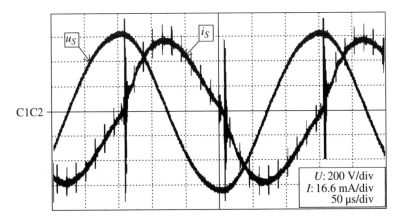

Figure 2.13 Examples of the u_s and i_s waveforms measured by the test bench presented in Figure 2.10 (motor: 1.5 kW, $U_{S(RMS)} = 430$ V, $I_{0(MAX)} = 46$ mA, $f_{res} \approx f_{imp} = 3.3$ kHz, $C_0 = 2.7$ nF)

Table 2.5 Induction motor capacitances, C_{s0}, measured using the test bench and the serial resonance method

H (mm)	P_n (kW)	p (–)	U_n Y (V)	n_n (rpm)	C_0 (nF)
90	1.5	2	300	1420	2.9
90	1.5	2	400	1410	3.53
112	1.5	8	400	720	4.57
132	10	2	173	1455	6.4

In Table 2.5, the experimentally measured values of C_0 for some typical industrial induction motors are given. The motor capacitances C_0 given in Table 2.5 were measured for a rotor at standstill. In the exact model of the CM circuit the bearing capacitance is a function of motor speed [16,17]. So for measurement purposes the tested motor should be driven at different speeds using an external machine. With that test the characteristic $C_0 = f(\omega_r)$ should be created. However, C_b is small in comparison to other parasitic capacitances, and its influence on the total CM capacitance of the motor can be negligible. A typical value of the bearing capacity is close to 190 pF and is much smaller than the dominant capacitances between the stator windings and the motor frame for the motor sizes from 80 to 315. For that motor that value is in the range of a few nanofarads up to tens of nanofarads [1,17].

2.3 Prevention of Common Mode Current: Passive Methods

The first and most important thing that must be implemented to reduce the influence of CMV and current on the system is a proper cabling and earthing system. Manufacturers of converters recommend the use of symmetrical multicore motor cables, which prevents the CM at fundamental frequency. Also a short, low-impedance path for the return of CM current to the inverter must be provided. The best way to do that is to use shielded cables where shield connections have to be made with 360-degree termination on both sides. Also, a high-frequency bonding connection must be made between the motor and load machine frame and the earth. It is recommended that flat braided strips of copper wire should to be used and the strip should be at least 50 mm wide [14].

If these conditions are fulfilled, the elimination or reduction of the high frequency CM current can be done by increasing the impedance or by using a specially designed motor. Some solutions are shown in Table 2.6.

2.3.1 Decreasing the Inverter Switching Frequency

Decreasing the inverter switching frequency is the simplest way of reducing the CM current. Most industrial inverters offer the possibility of changing that value within a wide range. With a decrease in f_{imp} the dV/dt is not changed and the CM current peaks will not decrease. However their frequency is reduced and the value of total RMS CM current is decreased. The disadvantage of decreasing f_{imp} is that the total harmonic distortion (THD) of the motor supply current is decreased as well.

Table 2.6 Methods for CM current reduction and elimination [6]

Reduction of electric discharge current, i_{bEDM}
Ceramic bearings
Common mode passive filters
Systems for active compensation of CMV
Decreasing the inverter switching frequency
Motor shaft grounding by using brushes
Conductive grease in the bearings

Reduction of circulating bearing current i_{bcir}
Common mode choke
Systems for active compensation of CMV
Decreasing inverter switching frequency
Use of one or two insulated bearings
Use of one or two ceramic bearings
dV/dt filter

Reduction of rotor grounding current, i_{gr}
Common mode choke
Systems for active compensation of CMV
Decreasing inverter switching frequency
Use of one or two insulated bearings
Use of one or two ceramic bearings
Shielded cable for motor supply

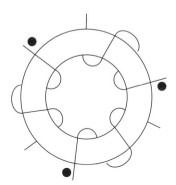

Figure 2.14 Common mode choke for a three-phase system

2.3.2 Common Mode Choke

The most commonly used component for limiting the CM current is CM choke (Figure 2.14). CM choke is constructed with three symmetrical windings on a toroidal core. The mutual inductance, M, between the windings is identical. The choke inductance is negligible for differential modem (DM) current because the total flux in the core is zero for three-phase symmetrical currents. However, the inductance of CM choke, L_{CM}, is significant for a CM current circuit. The equivalent electrical circuit for a system with CM choke is presented in Figure 2.15.

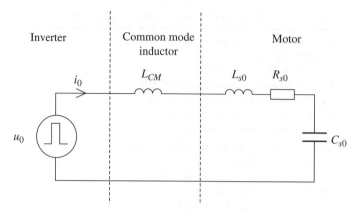

Figure 2.15 The equivalent electrical circuit with CM choke use

The design process for CM choke selection requires previous measurement of the CM voltage, u_0, which allows the choke core flux to be specified [20]:

$$\psi_0 = \frac{1}{N_{CM}} \int u_0 dt \qquad (2.9)$$

where N_{CM} is the number of turns of one CM winding.

The magnetic flux density of the CM choke is:

$$B_0 = \frac{\psi_0}{A_{CM}} = \frac{1}{A_{CM} N_{CM}} \int u_0 dt \qquad (2.10)$$

where A_{CM} is the CM choke core cross-sectional area.

The value of B_{CM} must be less than the saturation value B_{sat} of the core material. Nowadays the manufacturers offer materials dedicated to CM cores with $B_{sat} = 1\text{--}1.2$ T.

Inductance of a CM choke with known core dimensions is determined by the relationship:

$$L_{CM} = \frac{\mu S_{CM} N_{CM}^2}{l_{Fe0}} \qquad (2.11)$$

where: l_{Fe0} is the average path of the flux in the CM choke core.

The maximum value of the zero sequence current, I_{0max}, is proportional to the ratio of l_{Fe0}/N_{CM} [20]. A decrease in the core size reduces the flux path of l_{Fe0}, increases the number of turns N_{CM}, and finally decreases I_{0max}. At the same, the maximal windings dimensions should be taken into account.

The search for the optimal CM choke is a complex iterative process, requiring proper equipment for measurement of u_0 and I_{0max}. Therefore, magnetic materials manufacturers suggest specific solutions and complete CM reactor cores fitted for specific motor power. Particularly good properties are characterized by a toroidal core made of nanocrystalline materials.

For high-power motors the use of several simple CM cores is a good solution. In that case, the windings of CM choke is simply the motor cable ($N_{CM} = 1$). This construction is easy to use in industry

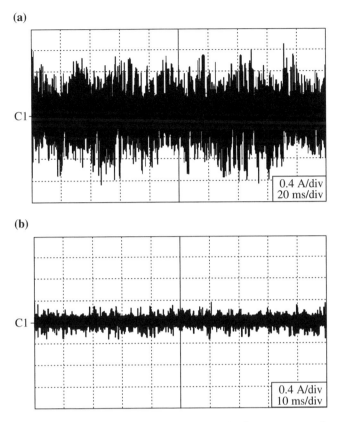

Figure 2.16 CM current measured in PE wire for a 1.5-kW induction motor operating with an inverter *without* a CM choke (**a**) and *with* a CM choke (**b**) for switching frequency 3.3 kHz

because CM cores are installed directly on motor cable and no any terminals have to be used. It is worth to see that only the shields have to be cut off in the place where the cores are installed.

Figure 2.16 presents waveforms of CM current that show the effects of CM choke use. In Figure 2.16a an electric drive with a 1.5-kW induction motor is operating without a CM choke. One can see current pulses of up to 1.4 A, which have been strongly limited in the case of CM choke use with inductance, $L_{CM} = 14$ mH (see Figure 2.16b).

2.3.3 Integrated Common Mode and Differential Mode Choke

In some inverter applications, it is possible to use an integrated CM and DM choke [21]. The aim of the integration of the two chokes is to minimize the geometrical dimensions of the reactor and reduce the cost of materials by reducing the amount of copper used. In Figures 2.17 and 2.18 the choke structure, equivalent circuit, and implementation in the converter are presented.

The integrated choke is made in the form of two coils wound on two toroidal cores. The dimensions of the toroidal cores are such that the outer diameter of core 2 is slightly smaller than the inner diameter of core 1. Core 1 is made of a magnetic material with high permeability, whereas core 2 has low magnetic permeability. A suitable difference in permeability is

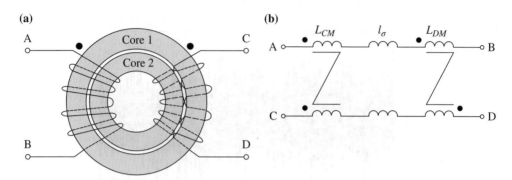

Figure 2.17 The integrated CM and DM choke: **(a)** structure [18] and **(b)** equivalent electrical circuit

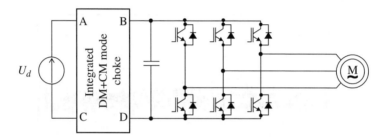

Figure 2.18 Toroidal shape integrated common mode and differential mode choke, the way of installation in the electrical drive circuit

achieved of the ferrite for core 1 and iron powders for core 2. Winding AB is wound on both cores simultaneously, and winding CD is wound on both cores in the form of an interlaced figure-eight shape. The number of turns of windings AB and CD is the same; that is, $N_{AB} = N_{CD} = N$. If leakage inductance is negligible then the ratio of L_{CM} and L_{DM} is dependent on the magnetic reluctance $\mathfrak{R}_1$, $\mathfrak{R}_2$ of the cores:

$$\frac{L_{CM}}{L_{DM}} = \frac{\mathfrak{R}_2}{4\mathfrak{R}_1} \tag{2.12}$$

The disadvantage of an integrated choke is the higher cost of manufacture of the windings because of the complex interleaving of winding CD. The limitation of the choke use is that it can only be applied in the DC circuit of the inverter. So the CM part of the choke can only limit the CM voltage forced by external source, for example, by grid-side inverter operation.

The more practical solution of the integrated CM and DM choke has been presented elsewhere [22]. The structure of the choke and connection diagram is presented in Figure 2.19.

The choke presented in Figure 2.18a has four windings. Windings A1-A2 and B1-B2, placed on outer columns of the core, create conventional choke and operate for differential mode only. The winding C1-C2 and D1-D2, placed in the central column core, are for CM. To keep CM mode inductance as high as possible, there is no air gap in central column. Based on ideas presented [22], the integrated choke for three phase system has been proposed elsewhere [23].

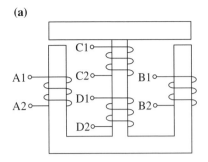

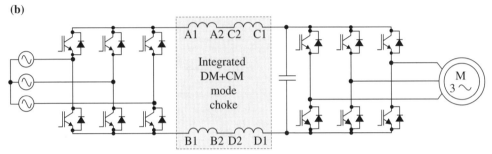

Figure 2.19 Integrated CM and DM choke with EI shape core: (a) structure and (b) connection diagram [22]

2.3.4 Common Mode Transformer

The advantages of using a CM choke could be improved through the use of a CM transformer [18]. The construction of a CM transformer is nearly the same as that of a CM choke but a fourth additional winding is added (Figure 2.20).

In Figure 2.20a, inductances, $l_{\sigma t}$, are leakage inductances of the primary and secondary windings of the CM transformer. Magnetizing inductance is identified as L_t. In the equivalent circuit of the CM transformer, the primary winding comprises three windings connected in parallel, U1-U2, V1-V2, and W1-W2, as indicated in Figure 2.20a.

The structure of the equivalent circuit of the inverter, CM transformer, and motor is presented in Figure 2.21.

The resistance, R_t, operates as damping resistance in the resonance circuit. If leakage inductances, $l\sigma_t$, are omitted, the CM current relation is as follows [18]:

$$I_0(s) = \frac{sL_t C_{s0} + R_t U_d}{s^3 L_t L_{s0} C_{s0} + s^2 (L_t + L_{s0}) C_{s0} R_t + sL_t + R_t} \tag{2.13}$$

Analysis of the relationship (Equation 2.13) allows the selection of R_t, which can reduce both the peak and RMS value of the CM current. Accordingly, resistance R_t satisfying this condition should be within the acceptable range [18]:

$$2Z_{00} \leq R_t \leq \frac{1}{2} Z_{0\infty} \tag{2.14}$$

where Z_{00} and $Z_{0\infty}$ are characteristic impedances of the electrical circuit presented in Figure 2.21, for $R_t = 0$ and $R_t = \infty$, respectively.

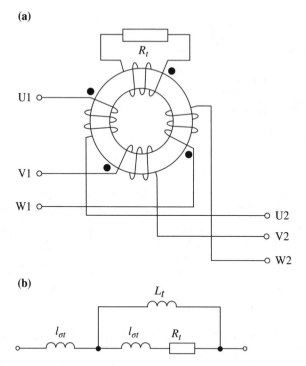

Figure 2.20 Common mode transformer: (a) topology and (b) equivalent circuit. $l_{\sigma t}$, transformer leakage inductance

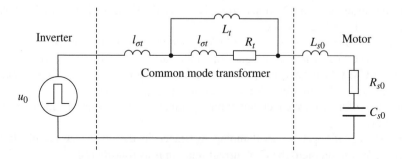

Figure 2.21 The structure of the equivalent circuit of the inverter, CM transformer, and motor

2.3.5 Machine Construction and Bearing Protection Rings

Suitable design of electrical machines could also reduce or eliminate the bearing currents. Such construction solutions could be as follows [6]:

- shielding the rotor by placing a metal shield between the stator and the rotor and connecting the shield to earth potential; this solution is known as an *electrostatically shielded induction motor* [16].

- using insulated bearings in which the outer surface of the outer race is covered with a non-conductive oxide layer.
- using hybrid bearings in which ceramic rolling elements are used.

Another market solution for bearing protection is the installation of bearing protection rings [15]. Round-shaped rings with brushes inside are connected to PE and installed on one or both sides of the motor shaft. This installation creates a low-impedance path between the motor shaft and earth. So the shaft current flows to earth directly without bearing reactance.

Among the solutions, only the insulated bearings are commercially available. Others solutions are still under development.

2.4 Active Systems for Reducing the CM Current

Besides the passive methods for the reduction of CM current, active systems are also used. These are active serial filters [21, 24] (Figure 2.22) or low-power auxiliary inverters [7, 26] (Figure 2.23).

The practical solution presented in Figure 2.22 requires a pair of complementary bipolar transistors. Control of the transistors is done without the use of an inverter digital control system. The transistors are directly triggered by a CMV signal. Unfortunately because of the lack of suitable high-speed high-voltage bipolar transistors, the solution is limited to low-power electrical systems.

This disadvantage is eliminated in the system presented in Figure 2.23. The auxiliary inverter uses the same kind of transistors as the main inverter. Only the power range is smaller because of the value of the CM current. To eliminate the CM current, the auxiliary inverter generates an opposite CM voltage. Such auxiliary inverter operation almost completely eliminates the CM current. However, a small CM voltage may still be present in the system as a result of differences in the transistor switching parameters of both inverters and irregularities in the dead time compensation algorithm. Correction by the dead time compensation algorithm allows complete elimination of the CM voltage [7].

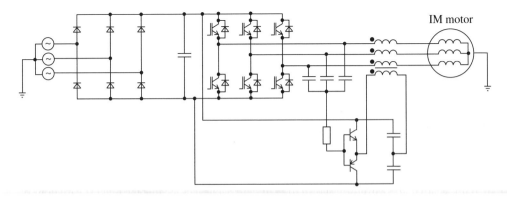

Figure 2.22 Active filter for CM current reduction [24, 25]

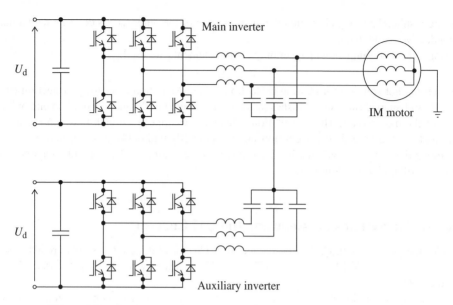

Figure 2.23 Application of additional voltage inverter to eliminate the CMV

2.5 Common Mode Current Reduction by PWM Algorithm Modifications

Both passive and converter-based classical methods to reduce CM current are relatively expensive and require the installation of additional elements or modifications of the converter topology.

However, the reduction of CM current in the drive system is possible in a simpler way that does not require complex changes in the inverter topology. It is possible to reduce CM current by changing the PWM algorithms. The PWM has to be changed to eliminate or reduce the CM voltage generated by the inverter. Various proposals for such modification of the PWM are presented in the literature [27, 28].

Nowadays the space vector PWM method is widely used in voltage inverters. With SVM, the CMV could be reduced by:

- removal of zero or passive vectors
- use of a nonzero or active vectors which have the same value of the CMV.

The CMV reduction possibilities can be explained using data given in Table 2.7.

Active even vectors (P) are $\mathbf{U}_{w6}$, $\mathbf{U}_{w3}$, and $\mathbf{U}_{w5}$, whereas odd (NP) are $\mathbf{U}_{w4}$, $\mathbf{U}_{w2}$, and $\mathbf{U}_{w1}$. The definition *odd* or *even* results from notation presented in Figure 2.24 where graphical vectors representation is showed. For each odd even vectors, the CMV u_0 is equal to $2/\sqrt{3}U_d$, whereas for odd vectors it is smaller, that is, $1/\sqrt{3}U_d$ (half of even vectors value).

In the case of an inverter operating with a classical SVM, the waveform voltage at the star point of the load related to negative potential of the DC link is as presented in Figure 2.25.

With regard to the waveform shown in Figure 2.25, one can see that the highest variations of u_N are between zero vectors, being from 0 up to 540 V. Therefore, the elimination of these vectors can significantly reduce CMV and accordingly the CM current. Simultaneously, it can be

Table 2.7 CMV for voltage inverter output vectors

Vectors	Inverter switching states							
	Active						Passive	
Notation	U_{w4}	U_{w6}	U_{w2}	U_{w3}	U_{w1}	U_{w5}	U_{w0}	U_{w7}
Binary representation	100	110	010	011	001	101	000	111
Decimal representation	4	6	2	3	1	5	0	7
Following vectors number	1	2	3	4	5	6	0	7
u_N (abc coordinates)	$\frac{1}{3}U_d$	$\frac{2}{3}U_d$	$\frac{1}{3}U_d$	$\frac{2}{3}U_d$	$\frac{1}{3}U_d$	$\frac{2}{3}U_d$	0	U_d
u_0 ($\alpha\beta 0$ coordinates)	$\frac{1}{\sqrt{3}}U_d$	$\frac{2}{\sqrt{3}}U_d$	$\frac{1}{\sqrt{3}}U_d$	$\frac{2}{\sqrt{3}}U_d$	$\frac{1}{\sqrt{3}}U_d$	$\frac{2}{\sqrt{3}}U_d$	0	$\sqrt{3}U_d$
Vector name[a]	NP	P	NP	P	NP	P	Z	Z

[a] Vector name: Nonparity (NP), active odd; Parity (P), active even; Z, passive.

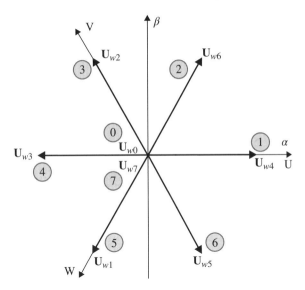

Figure 2.24 Graphic representation of the inverter output vectors

observed that if the consecutive active vectors have the same value of the u_0, the CMV will be direct current voltage and the current will not flow through the motor parasitic capacitances.

On the basis of the analysis of the relationship between voltage vectors and the corresponding values of u_0, it is possible to propose modifications of PWM that can lead to the reduction of bearing currents. This solution was proposed, where modification of the SVM algorithm was done by elimination of zero vectors and use of parity or nonparity active vectors only [29,30].

The problem that occurs when using only parity active vectors or nonparity active vectors is that the inverter maximal output voltage is limited. The problem is underlined when the inverter operation is limited to the modulation region only. This is because in the classical SVM algorithm, the output voltage vector u^*_{out} is generated for each modulation period as a combination of the two active and two passive vectors (Figure 2.26).

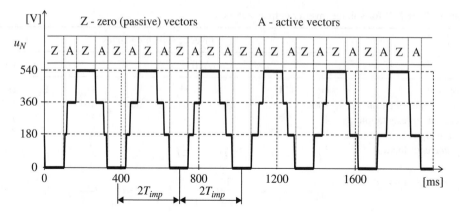

Figure 2.25 The CMV for the inverter operating with classical SVM; DC link voltage $U_d = 540$ V

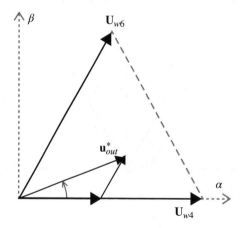

Figure 2.26 Generation of the inverter output voltage by the classical SVM algorithm

The use of active vectors only in the SVM modulator makes it possible to generate the maximum amplitude of the inverter output voltage as shown in Figure 2.27.

2.5.1 Three Nonparity Active Vectors

If the zero vectors are eliminated, the output inverter voltage can be generated using, for example, only nonparity active vectors (3NPAV), as presented in Figure 2.28.

In Figure 2.28, the inverter output voltage vector, u^*_{out}, is generated using vectors U_{w4}, U_{w2}, and U_{w1}, which are indicated in Table 2.4 as nonparity vectors. The switching times of particular vectors denoted as t_4, t_1, and t_2 are defined as:

$$\mathbf{u}^*_{out} \cdot T_{imp} = \mathbf{U}_{w4} t_4 + \mathbf{U}_{w4} t_2 + \mathbf{U}_{w4} t_1 \qquad (2.15)$$

$$T_{imp} = t_4 + t_2 + t_1 \qquad (2.16)$$

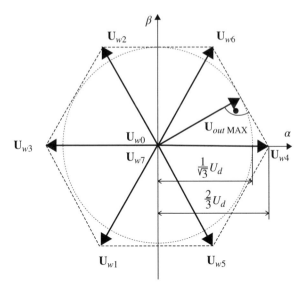

Figure 2.27 The range of inverter output voltage with classical SVM

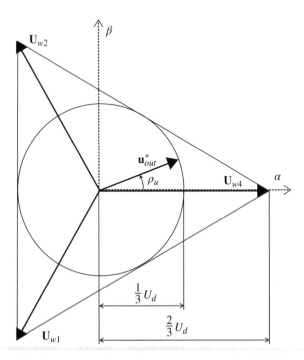

Figure 2.28 The creation of inverter output voltage in the 3NPAV algorithm

The vector switching times can be determined from the relations [31]:

$$t_4 = \frac{1}{3}\left(1 + \frac{2U_{out}^*}{U_d}\left(\cos\left(\frac{\pi}{3} + \rho_u\right) + \sin\left(\frac{\pi}{6} + \rho_u\right)\right)\right) \quad (2.17)$$

$$t_2 = \frac{1}{3}\left(1 - \frac{2U_{out}^*}{U_d}\cos\left(\frac{\pi}{3} + \rho_u\right)\right) \quad (2.18)$$

$$t_1 = \frac{1}{3}\left(1 - \frac{2U_{out}^*}{U_d}\sin\left(\frac{\pi}{6} + \rho_u\right)\right) \quad (2.19)$$

where U_{out}^* and ρ_u are the magnitude and angle position of the reference inverter output voltage vector $\mathbf{U}_{out}^*$.

The use of nonparity active vectors leads to a constant value of the CMV:

$$u_0 = \frac{U_d}{\sqrt{3}} \quad (2.20)$$

and obviously does not allow CM current flow.

The main drawback of the modified PWM is that the maximum value of the inverter output phase voltage is limited to the value $U_d/3$, which leads to deformation of the motor current at higher ranges of motor speed.

2.5.2 Three Active Vector Modulation

The inverter output voltage can be increased simultaneously with CMV reduction using a method similar to the standard SVM. The modification of SVM uses all active vectors; no zero vectors are used [29, 30]. The modified modulation method is called three active vector modulation (3AVM) [31]. In 3AVM, a voltage vector plane is divided into six sectors (Figure 2.29).

If both parity and nonparity active vectors are used, the CMV value has a value $U_d/3$ when the output vector goes from one sector to the adjacent one. Therefore, the CMV frequency is equal to six times the frequency of the first harmonic of the inverter output voltage. This is a much lower frequency of u_0 than appears in classical SVM. Therefore, despite variation in u_0, the CM current is significantly reduced. Simultaneously the amplitude of the inverter output voltage is 15.5% higher in comparison with the 3NPAV algorithm.

2.5.3 Active Zero Voltage Control

Another PWM method for decreasing CM current is active zero voltage control (AZVC). The method is characterized by the replacement of zero vectors by two reverse active vectors (AZVC-2) or one active vector (AZVC-1) [31, 32].

The principle of modulation by the AZVC-2 method is presented in Figure 2.30. For the case shown in Figure 2.30, the output voltage vector is the result of the particular active vectors $\mathbf{U}_{w4}$ and $\mathbf{U}_{w6}$. The angular position of the output voltage vector is determined by

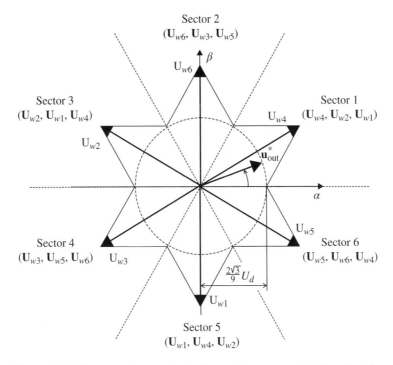

Figure 2.29 Vectors and sectors in the three active vector modulation algorithm

times t_4 and t_6 respectively. Both vectors correspond to active vectors with standard SVM. In comparison with SVM, instead of a zero vector the reverse active vectors U_{w5} and U_{w2} are generated. Because the values of t_5 and t_2 are the same, the reverse vectors do not change the angle position of the output voltage vector, only the output voltage vector length is reduced. Thus, in method AZVC-2 the switching time calculations are as follows:

$$\mathbf{u}^*_{out} \cdot T_{imp} = \mathbf{U}_{w4} t_4 + \mathbf{U}_{w6} t_6 + \mathbf{U}_{w2} t_2 + \mathbf{U}_{w5} t_5 \qquad (2.21)$$

$$t_4 = T_{imp} \cdot \frac{u^*_{out\,\alpha} \cdot U_{w6\beta} - u^*_{out\beta} \cdot U_{w6\alpha}}{U_d \cdot w_t} \qquad (2.22)$$

$$t_6 = T_{imp} \cdot \frac{-u^*_{out\,\alpha} \cdot U_{w4\beta} + u^*_{out\beta} \cdot U_{w4\alpha}}{U_d \cdot w_t} \qquad (2.23)$$

$$t_2 = t_5 = \frac{1}{2}\left(T_{imp} - t_4 - t_6\right) \qquad (2.24)$$

where:

$$w_t = U_{w4\alpha} \cdot U_{w6\beta} - U_{w4\beta} \cdot U_{w6\alpha} \qquad (2.25)$$

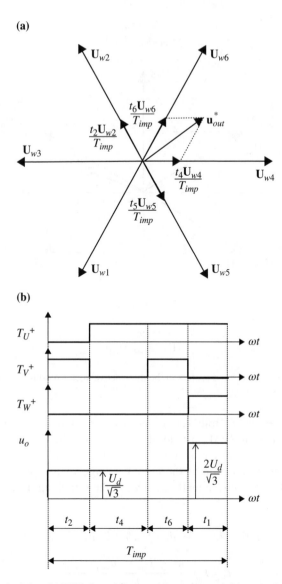

Figure 2.30 The principle of AZVC-2 modulation: (a) vectors and (b) control signals (T_U^+, T_V^+, T_W^+) and CMV waveform u_o

The principle of modulation by AZVC-1 is illustrated in Figure 2.31 [28]. For the case shown in Figure 2.31, the vectors $\mathbf{U}_{w4}$ and $\mathbf{U}_{w3}$ have been properly chosen to assure the decreasing of the magnitude of the output vector $\mathbf{U}_{out}^*$. These vectors are respectively attached to the times t_4 and t_3^*, where $t_4^* = t_3$. The sum of times t_4^* and t_3 is equivalent to t_0 in classical SVM. In AZVC-1 the reverse vectors are chosen such that one of them must be the same as one of the active vectors determining the angle position of $\mathbf{u}_{out}^*$; that is, it is U_{w4} for the combination presented in Figure 2.31. In the method AZVC-1 the relation between vectors and switching times is:

$$\mathbf{u}_{out}^* \cdot T_{imp} = \mathbf{U}_{w4} t_4 + \mathbf{U}_{w6} t_6 + \mathbf{U}_{w3} t_3 + \mathbf{U}_{w4} t_4^* \tag{2.26}$$

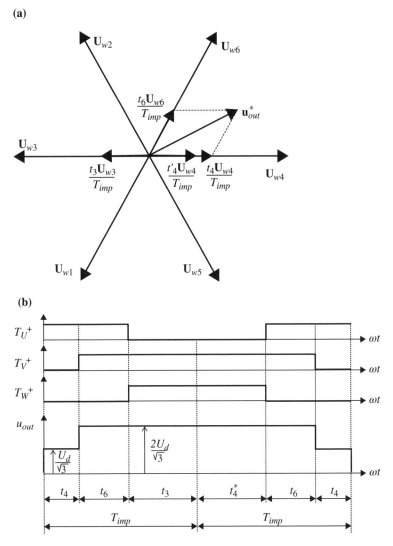

Figure 2.31 The principle of AZVC-1 modulation: (a) vectors and (b) control signals (T_U^+, T_V^+, and T_W^+) and CMV waveform u_o

The switching times t_4 and t_6 are determined by Equations 21.22 and 21.23, whereas t_4^* and t_3 take the values:

$$t_3 = t_4^* = \frac{1}{2}(T_{imp} - t_4 - t_6) \tag{2.27}$$

Both AZVC methods make it possible to obtain the same maximum output voltage of the inverter as the SVM method. In comparison with the classical SVM, both the amplitude and frequency of the CMV are reduced and finally the CM current is also limited.

2.5.4 Space Vector Modulation with One Zero Vector

The previously presented methods operate without zero vectors. This is a disadvantage, because it leads to problems with inverter output current measurement. In the voltage inverter operating with classical SVM the output current has to be sampled synchronously with SVM. Simultaneously the sampling instant must be in the middle of the zero vectors period. If the samples are taken in that instant, the instantaneous value of the current is nearly equal to the fundamental harmonic for the inverter output current, $i_{out\,1har}$, as presented in Figure 2.32 (see Appendix A for a more detailed discussion) [33–35].

The reduction of CM current while leaving the synchronous measurement unchanged is possible if only one of the zero vectors is eliminated, U_{w0} or U_{w7}. Thus, the modified SVM algorithm is called the <Space Vector Modulation with One Zero vector (SVM1Z) modulator. An example of the control signals and output voltage waveforms for SVM1Z is presented in Figure 2.33.

The example shown in Figure 2.33 consists of a sequence including one zero vector, U_{w0}, and two active vectors, U_{w4} and U_{w6}. When U_{w6} is switched off, the same zero vector, U_{w0}, is applied. The number of transistors switching for one PWM cycle increases but will not influence synchronous sampling of the inverter output current.

If one zero vector is eliminated, the amplitude of the CMV is reduced by 33%. Unfortunately dV/dt is increased for the instant of transition from active vector to zero vector, that is, from U_{w6} to U_{w0}. The comparison can be observed in Figure 2.34.

The quantitative effect of the SVM1Z algorithm on CM current reduction is difficult to evaluate in an analytical manner. This is because a model of the electrical machine CM is non-linear and not perfectly defined. For these reasons, the usefulness of the SVM1Z method needs to be tested in practical applications.

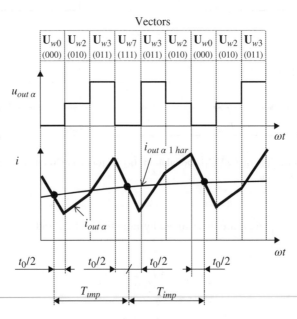

Figure 2.32 The relationship between current sampling instant and inverter output vectors in classical SVM

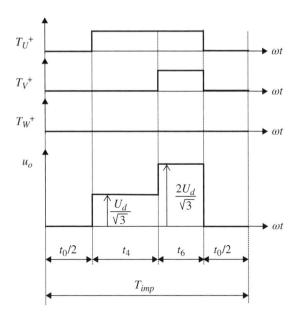

Figure 2.33 An example of control signals and inverter CMV waveforms in the SVM1Z method

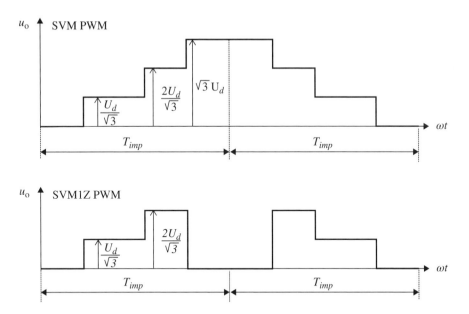

Figure 2.34 The comparison of CMV waveform in case of SVM PWM and SVM1Z PWM

The use of SVM1Z methods can bring significant benefits in the case of CM filter use. With increasing frequency of u_o, the core of the CM choke will have a smaller diameter, or if the diameter is kept constant the CM current will decrease.

A comparison of the SVM, AZVC-2, and SVM1Z methods is presented in Figures 2.35 to 2.37.

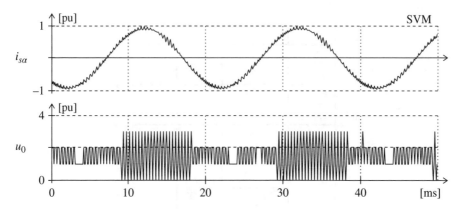

Figure 2.35 The motor current and CMV waveforms in the case of classical SVM control

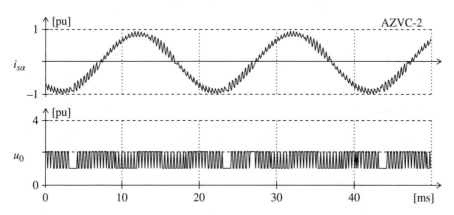

Figure 2.36 The motor current and CMV waveforms in the case of AZVC-2 control

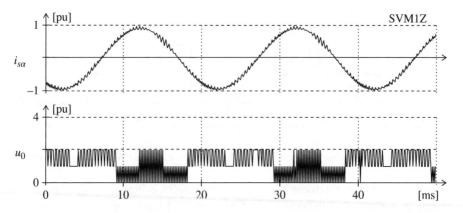

Figure 2.37 The motor current and CMV waveforms in the case of SVM1Z control

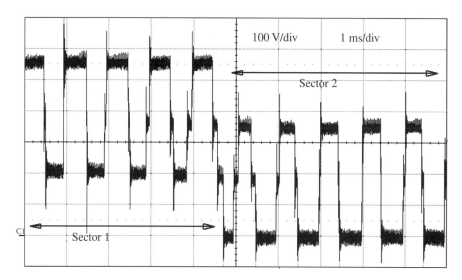

Figure 2.38 The motor current and CMV waveforms in the case of SVM1Z control with different zero vectors for output voltage position sectors

As can be seen in the waveforms presented in Figures 2.35 to 2.37, the shape of the motor current is the same with SVM and SVM1Z. In the case of SVM1Z the amplitude of the CMV decreases. The smallest CMV occurs in the case of AZVC-2 control, but the current shape is the worst.

The practical implementation of SVM1Z is simpler than that of NPAV, AVM, and AZVC. At the same time, the inverter output voltage is not reduced. So SVM1Z is implemented in some industrial inverters nowadays (Figure 2.38).

In the case of SVM1Z in each of the position sectors of the inverter output voltage vector, different zero vectors could be used. This minimizes the switching of transistors when the output vector crosses the border between the sectors.

2.6 Simulation Examples

2.6.1 Model of Induction Motor Drive with PWM Inverter and CMV

The scheme of the MATLAB®/Simulink model of induction motor drive (*Chapter_2\Simulation_2_6_1\Simulator.slx*) is presented in Figure 2.39.

The *Simulator.slx* model consists of two S-functions, references, and scopes. The S-functions are for a three-phase inverter with a space vector PWM algorithm (*PWM_SFUN*) and induction motor model (*MOTOR_SFUN*). Both blocks are Simulink S-functions and implemented using C language. The C files are noted as follows:

- *declarations.h*, list of predefined values
- *motor.h*, declaration of global variables for induction motor model
- *motor.c*, source file of induction motor model

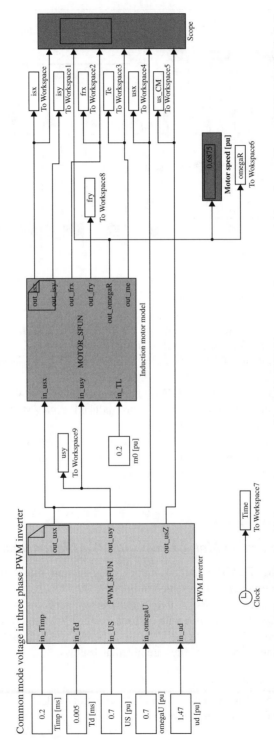

Figure 2.39 MATLAB®/Simulink model of induction motor drive with PWM inverter and common mode voltage, *Chapter_2\Simulation_2_6_1\ Simulator.slx*

Problems with AC Drives and Voltage Source Inverter Supply Effects

Table 2.8 Base values for per unit system used in MATLAB®/Simulink model

Definition	Description
$U_b = \sqrt{3} U_n$	Base voltage
$I_b = \sqrt{3} I_n$	Base current
$Z_b = U_b / I_b$	Base impedance
$T_b = (U_b I_b p) / \omega_0$	Base torque
$\Psi_b = U_b / \omega_0$	Base flux
$\omega_b = \omega_0 / p$	Base mechanical speed
$L_b = \Psi_b / I_b$	Base inductance
$J_b = T_b / (\omega_b \omega_0)$	Base inertia
$\tau = \omega_0 t$	Relative time

Where $\omega_0 = 2\pi f_n$ is nominal grid pulsation.

- *pwm.h*, declaration of global variables for SVPWM algorithm of inverter model
- *pwm.c*, source file of SVPWM inverter model.

The reference values are:

- *US*, module of the commanded inverter output voltage
- *omegaU*, inverter output voltage pulsation.

The inverter function parameters are:

- *ud*, DC link voltage
- *Timp*, switching period of transistors (in milliseconds)
- *Td*, dead time (in milliseconds).

The disturbance is:

- *m0*, motor load torque.

All variables and parameters are in per unit system, where the fundamental base values are presented in Table 2.8.

The induction motor drive is working according to the V/f = const principle. Both voltage (*US*) and frequency (*omegaU*) can be set independently. No ramp is added for stator voltage pulsation, so the motor direct start-up is done. The machine model it is induction motor (P_n = 1.5 kW, U_n = 380 V, four poles). The motor parameters are written in the *pwm.h* file.

The examples of simulation results are presented in Figures 2.40 and 2.41.

The motor starts with reference frequency $\omega_u = 0.7$ pu (i.e., 55 Hz), $|U_s| = 0.7$ pu (i.e., 266 V), and DC link voltage $U_d = 1.47$ pu (i.e., 560 V). Figure 2.40 presents the motor direct start-up. In Figure 2.41, the inverter output voltage of α and β components, as well as the CMV, are presented with zoom of 0.2 to 0.24 s. The CMV is related to the negative terminal of the inverter DC link.

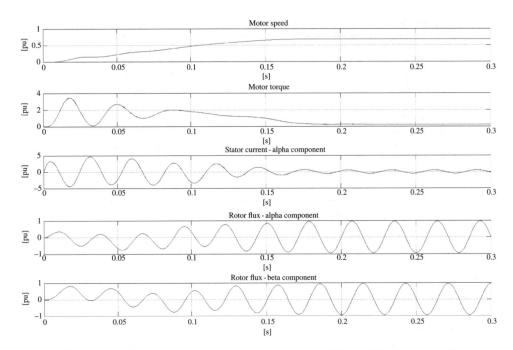

Figure 2.40 Simulation results for motor direct start up (MATLAB® script *graph.m*)

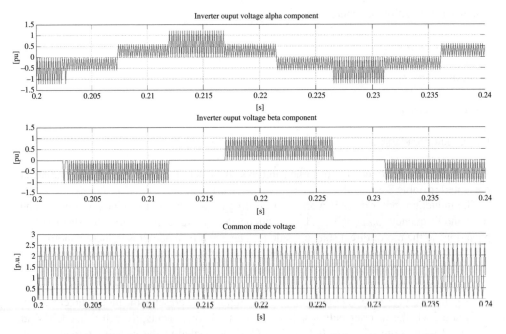

Figure 2.41 CMV waveforms in PWM inverter (MATLAB® script *graph_CM.m*)

Problems with AC Drives and Voltage Source Inverter Supply Effects

The PWM function, defined in *pwm.c*, is:

```
void PWM(double Timp,double Td,double US,double roU,double
  ud,double hPWM)
```

The input values of *PWM()* are as described previously, *hPWM* is the simulation step. The main part of the function is:

```
if (impuls>Timp)
    {
    if (roU>2*PI) roU-=2*PI;
    if (roU<0) roU+=2*PI;
    NroU=floor(roU/(PI/3));
    switch(NroU)
        {
        case 0: idu1=1; idu2=2; sector=1; z1=0; n1=1; n2=2;
        z2=0; break;
        case 1: idu1=2; idu2=3; sector=2; z1=0; n1=2; n2=3;
        z2=0; break;
        case 2: idu1=3; idu2=4; sector=3; z1=0; n1=3; n2=4;
        z2=0; break;
        case 3: idu1=4; idu2=5; sector=4; z1=0; n1=4; n2=5;
        z2=0; break;
        case 4: idu1=5; idu2=6; sector=5; z1=0; n1=5; n2=6;
        z2=0; break;
        case 5: idu1=6; idu2=1; sector=6; z1=0; n1=6; n2=1;
        z2=0; break;
        }
    USX=US*cos(roU); USY=US*sin(roU);
    wt=ux[idu1]*uy[idu2]-uy[idu1]*ux[idu2];
    t1=Timp*(USX*uy[idu2]-USY*ux[idu2])/(ud*wt);
    t2=Timp*(-USX*uy[idu1]+USY*ux[idu1])/(ud*wt);

    t0=(Timp-t1-t2)/2;
    cycle=cycle*(-1);

    // signum of the current for dead time compensation
    if(iap>0) sgn_ia_comp=1; else sgn_ia_comp=-1;
    if(ibp>0) sgn_ib_comp=1; else sgn_ib_comp=-1;
    if(icp>0) sgn_ic_comp=1; else sgn_ic_comp=-1;

    if (cycle>0)
        {
        if (sgn_ia_comp>0) comp_ia=-1; else comp_ia=0;
        if (sgn_ib_comp>0) comp_ib=-1; else comp_ib=0;
        if (sgn_ic_comp>0) comp_ic=-1; else comp_ic=0;
```

```
switch(NroU)
    {
    case 0:
        time_a=t0           +comp_ia*Td;
        time_b=t0+t1            +comp_ib*Td;
        time_c=t0+t1+t2 +comp_ic*Td;
    break;
```

... the rest part of the function is in pwm.c file

It can be seen that the position of the reference voltage vector is defined by *roU* angle. Based on actual *roU*, the sector (*sector* = *1–6*) of the reference voltage position is calculated. For each sector two passive ($z1$, $z2$) and two active ($n1$, $n2$) vectors are found. In the next step, the components of the reference vector (*USX*, *USY*) are calculated using trigonometric functions, the denominator coefficient *wt* is found and finally the active vectors switching times ($t1$, $t2$) are determined. The passive vector switching time, $t0$, complements the switching period *Timp*, and is evenly divided by the first and second passive vectors. The switching period *Timp* is for vectors sequence: $z1$-$n1$-$n2$-$z2$, in next period the sequence is reversed into $z2$-$n2$-$n1$-$z1$ what can be recognized using variable *cycle* (*1* or −*1*). The PWM function has an embedded procedure for dead time compensation based on motor phase current direction sign, for example, *sgn_ia_comp* for phase A current.

The final values from PWM function are variables *time _a*, *time _b*, and *time_c*, which determine time of the switching ON/OFF transistors in appropriate inverter phase. Depending on current direction, the times are unchanged (positive current direction) or reduced with dead time, *Td*.

The use of the simulation requires that before simulation start, each of the S-functions have to be built. The final waveforms can be seen using the m-files: *graph.m* (Figure 2.40) and *graph_CM.m* (Figure 2.41).

2.6.2 PWM Algorithms for Reduction of CMV

MATLAB®/Simulink model of PWM algorithms for reduction of CMV is presented. Three different PWM simulation are:

- SVM, classical space vector PWM (*Simulator_SVM.slx*)
- SVM1Z, space vector PWM with one passive vector use (*Simulator_SVM1Z.slx*)
- AZVC-2, PWM with AZVC and two reverse active vectors (*Simulator_AZVC2.slx*).

The appropriate Simulink models are in *Chapter_2\Simulation_2_6_2*.

MATLAB®/Simulink model *Simulator_SVM.slx* is given in Figure 2.42.

The same style simulation file is for SVM1Z and AZVC-2 algorithms. The main simulation C file are *Main_SVM.c*, *Main_SVM1Z.c*, and *Main_AZVC2.c*. The appropriate PWM functions are defined in files: *PWM_SVM.c*, *PWM _SVM1Z.c*, and *PWM _AZVC2.c*.

The example results are presented in Figures 2.43–2.45.

The differences in voltage waveforms are easily noticed. It is worth noting that CMV *us0* has the smallest peak-to-peak value in case of AZVC-2 algorithm.

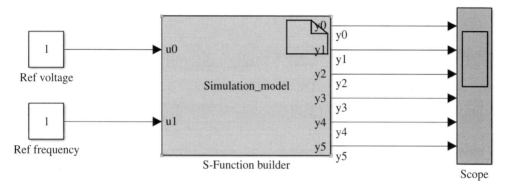

Figure 2.42 MATLAB®/Simulink model of induction motor drive with PWM inverter and CMV, Chapter_2\Simulation_2_6_1\Simulator_SVM.slx

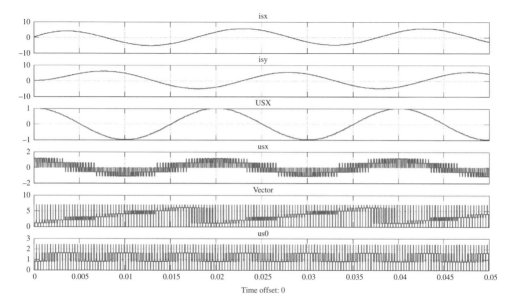

Figure 2.43 Results for SVM algorithm

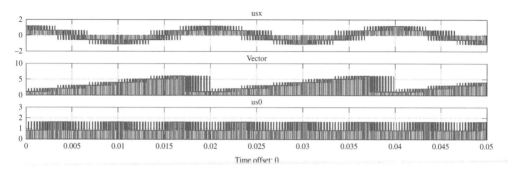

Figure 2.44 Results for SVM1Z algorithm

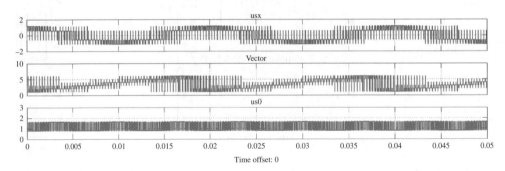

Figure 2.45 Results for AZVC-2 algorithm

2.7 Summary

Voltage inverters operating with PWM are a source of CMV. This voltage forces the CM current to flow through the parasitic motor capacitances. A part of the current flows through the motor bearings, causing their accelerated degradation.

Nowadays, the most common type of induction motor failures is bearing faults. Therefore, the most reasonable course of action is to seek and implement solutions that will help prevent this damage.

CM current reduction can be done either by increasing the impedance of the proper electrical circuit or by reducing the CMV. The impedance could be increased by using the appropriate passive filter, the CM filter. Reduction of the CMV is done by active solutions, mainly by methods for modification of the PWM.

The solutions that use PWM modification are cheaper compared to solutions involving the use of filters. With PWM modifications no additional elements are required; only the appropriate changes and extensions in the control software are needed. However, these PWM changes will result in increases in the number of transistor switching, which will cause a decrease of the inverter efficiency, and a more efficient inverter cooling system could be required.

Because of the complexity of the motor CM electrical circuit and the non-linearities of some elements, the usefulness of different PWM controls needs to be tested in real conditions. Nowadays the analytical or simulation models presented are insufficient for estimation of the CM current.

References

[1] Busse D, Erdman J, Kerkman RJ, Schlegel D, Skibinski G. Bearing currents and their relationship to PWM drives. IEEE Twenty-First International Conference on Industrial Electronics, Control, and Instrumentation, IECON 1995. November 6–10, 1995.

[2] Conraths HJ, Giebler F, Heining HD. Shaft-voltages and bearing currents—New phenomena in inverter driven induction machines. Eighth European Conference on Power Electronics and Applications EPE'99, Lausanne, France. September 7–9, 1999.

[3] Strom JP, Koski M, Muittari H, Silventoinen P. Analysis and filtering of common mode and shaft voltages in adjustable speed AC drives. Twelfth European Conference on Power Electronics and Applications, EPE'07, Aalborg, Denmark. September 2–5, 2007.

[4] Akagi H. Prospects and expectations of power electronics in the 21st century. Power Conversion Conference, PCC'02, Osaka, Japan. April 2–5, 2002.

[5] Abu-Rub H, Iqbal A, Guzinski J. *High performance control of AC drives with Matlab/Simulink models*. Chichester, UK: John Wiley & Sons, Ltd; 2012.

[6] Muetze A, Binder A. Practical rules for assessment of inverter-induced bearing currents in inverter-fed AC motors up to 500 kW. *IEEE Transactions on Industrial Electronics*. 2007; **54** (3): 1614–1622.

[7] Kikuchi M, Naoyuki AZ, Kubota H, Aizawa N, Miki I, Matsuse K. Investigation of common-mode voltages of PWM inverter with a small capacity auxiliary inverter. Twelfth International Conference on Electrical Machines and Systems, ICEMS'09, Tokyo, Japan. November 15–18, 2009.

[8] Pairodamonchai P Sangwongwanich S. Exact common-mode and differential-mode equivalent circuits of inverters in motor drive systems taking into account input rectifiers. IEEE Eleventh International Conference on Power Electronics and Drive Systems (PEDS), Singapore. December 5–8, 2011.

[9] Muetze A Binder A. High frequency stator ground currents of inverter-fed squirrel-cage induction motors up to 500 kW. Tenth European Conference on Power Electronics and Applications, EPE'03, Toulouse, France. September 2–4, 2003.

[10] Bhushan B. *Introduction to tribology*. New York: John Wiley & Sons, Inc.; 2002.

[11] Muetze A, Binder A. Calculation of influence of insulated bearings and insulated inner bearing seats on circulating bearing currents in machines of inverter-based drive systems. *IEEE Transactions on Industry Applications*. 2006; **42** (4): 965–972.

[12] Binder A, Muetze A. Scaling effects of inverter-induced bearing currents in AC machines. *IEEE Transactions on Industry Applications*. 2008; **44** (3): 769–776.

[13] Dymond JH, Findlay RD: Comparison of techniques for measurement of shaft currents in rotating machines. *IEEE Transactions on Energy Conversion*. 1997; **12** (4): 363–367.

[14] ASEA Brown Boveri. (2011) ABB Drives: Bearing currents in modern AC drive systems. Technical Guide No. 5. Zurich, Switzerland: Asea Brown Broveri

[15] AEGIS SGR [Internet]. (2014) Bearing Protection Rings, Catalog No. 2010-1. [accessed March 30, 2015] Available at www.est-aegis.com.

[16] Busse DF, Erdman J, Kerkman R, Schlegel D, Skibinski G. The effects of PWM voltage source inverters on the mechanical performance of rolling bearings. Eleventh Annual Applied Power Electronics Conference, APEC'96, San Jose, CA. March 3–7, 1996.

[17] Busse DF, Erdman JM, Kerkman RJ, Schlegel DW, Skibinski GL. An evaluation of the electrostatic shielded induction motor: a solution for rotor shaft voltage buildup and bearing current. *IEEE Transactions on Industry Applications*. 1997; **33** (6): 1563–1570.

[18] Ogasawara S, Akagi H. Modeling and damping of high-frequency leakage currents in PWM inverter-fed AC motor drive systems. *IEEE Transactions on Industry Applications*. 1996; **22** (5): 1105–1114.

[19] Muetze A, Binder A. Calculation of circulating bearing currents in machines of inverter-based drive systems. *IEEE Transactions on Industrial Electronics*. 2007; **54** (2): 932–938.

[20] Akagi H, Hasegawa H, Doumoto T. Design and performance of a passive EMI filter for use with a voltage-source PWM inverter having sinusoidal output voltage and zero common-mode voltage. *IEEE Transactions on Power Electronics*. 2004; **19** (4): 1069–1076.

[21] Lai R, Maillet Y, Wang F, Wang S, Burgos R, Boroyevich D. An integrated EMI choke for differential-mode and common-mode noise suppression. *IEEE Transactions on Power Electronics*. 2010; **25** (3): 539–544.

[22] Bin W, Rizzo S, Zargari N, Xiao Y. An integrated dc link choke for elimination of motor common-mode voltage in medium voltage drives. Thirty-Sixth IAS Annual Meeting Industry Applications Conference, Chicago, IL. September 30–October 4, 2001.

[23] Zhu N. (2013) Common-mode voltage mitigation by novel integrated chokes and modulation techniques in power converter systems. PhD thesis. Ryerson University, Toronto.

[24] Sun Y, Esmaeli A, Sun L. A new method to mitigate the adverse effects of PWM inverter. First IEEE Conference on Industrial Electronic and Applications, ICIEA'06. Singapore, May 24–26, 2006.

[25] Kempski A, Smolenski R, Kot E, Fedyczak Z. Active and passive series compensation of common mode voltage in adjustable speed drive system. Thirty-Ninth IEEE Industry Applications Conference, IAS 2004, Seattle. WA, October 3–7, 2004.

[26] Kikuchi M, Kubota H. A novel approach to eliminating common-mode voltages of PWM inverter with a small capacity auxiliary inverter. Thirteenth European Conference on Power Electronics and Applications EPE'09, Barcelona, Spain. September 8–10, 2009.

[27] Zitselsberger J, Hofmann W. Reduction of bearing currents by using asymmetric space-vector-based switching patterns. European Conference on Power Electronics and Applications, EPE'03, Toulouse, France. September 2–4, 2003.

[28] Zitzelsberger J, Hofmann W. Reduction of bearing currents in inverter fed drive applications by using sequentially positioned pulse modulation. *EPE Journal*; 2004; **14** (4): 19–25.

[29] Cacciato M, Consoli A, Scarcella G, Testa A. Reduction of common-mode currents in PWM inverter motor drives. *IEEE Transactions on Industry Applications*. 1999; **35** (2): 469–476.

[30] Cacciato M, Consoli A, Scarcella G, Testa A. Modified space-vector-modulation technique for common mode currents reduction and full utilization of the DC bus. Twenty-Fourth Annual IEEE Applied Power Electronics Conference and Exposition, APEC 2009, Singapore. November 14–15, 2009.

[31] Hofmann W, Zitzelsberger J. PWM-control methods for common mode voltage minimization—a survey. International Symposium on Power Electronics, Electrical Drives, Automation and Motion, SPEEDAM 2006, Taormina (Sicily). Italy, May 23–26, 2006.

[32] Lai YS, Shyu F-S. Optimal common-mode voltage reduction PWM technique for inverter control with consideration of the dead-time effects—part I: basic development. *IEEE Transactions on Industry Applications*. 2004; **40** (6): 1605–1612.

[33] Blasko V, Kaura V, Niewiadomski W. Sampling of discontinuous voltage and current signals in electrical drives—a system approach. *IEEE Transactions on Industry Applications*. 1998; **34** (5): 1123–1130.

[34] Briz F, Díaz-Reigosa D, Degner MW, Garcia P, Guerrero JM. Current sampling and measurement in PWM operated AC drives and power converters. The 2010 International Power Electronics Conference IPEC, Sapporo, Japan. June 21–24, 2010.

[35] Holtz J, Oikonomou N. Estimation of the fundamental current in low-switching-frequency high dynamic medium-voltage drives. *IEEE Transactions on Industrial Applications*. 2008; **44** (5): 1597–1605.

3

Model of AC Induction Machine

3.1 Introduction

In this chapter, the principal model of a three-phase induction machine is presented. The goal of the model presentation is to give a background to the simulation software and to understand how the model is implemented.

Synthesis of the control systems as well as estimation procedures for the induction motor drive system requires the definition of a mathematical model of the machine. Knowledge of the model is also necessary for simulation studies [1, 2].

In this book, the circuit model of the motor was used with the following simplifying assumptions [3, 4]:

- the machine is three-phase symmetrical
- an air gap is constant
- iron losses are negligible
- the magnetic circuit is linear, without a saturation effect
- motor parameters (resistances, inductances) are constant
- the conductor skin effect is neglected
- the stator windings are distributed spatially
- the power losses in the magnetic elements are negligible
- the motor parasitic capacitances (e.g., stator to frame) are negligible
- the stator windings are star connected without a neutral conductor.

Furthermore the motor common mode circuit was excluded in the machine model, which was applied in the analysis of the control and estimation system. This circuit model does not change the motor control properties because it does not influence the generated electromagnetic torque. The asymmetrical magnetic circuit that occurs in real motors, including the higher ordered harmonics of the magnetic flux that results from the nonsinusoidal motor supply, is the reason for noises and vibrations of the motor [5–7]. In the application of recent inverters with pulse width modulation, these effects do not influence the base electromagnetic torque component. Therefore, for a control task, the neglect of this effect—in addition to the common mode current effect—is allowed.

Based on the previous assumptions, it is possible to model an induction motor together with the dependencies of motor current, flux, and voltage. A natural method to represent a motor is the three phase *abc* reference frame. However, in practice it is more comfortable to assume the orthogonal reference frame system *xy0*, which rotates with a selectable speed of rotation ω_a.

In the literature it is common to introduce a determination for the orthogonal reference frame [3, 8–11]:

- if $\omega_a = 0$ then the orthogonal reference frame is motionless relative to the stator and the axes are determined as $\alpha\beta 0$.
- if ω_a is equal to the rotor flux pulse $\omega_a = \omega\psi_r$, then the orthogonal system is rotating synchronously with the matched flux rotor vector and the rotating axis system is described as dq.

The orthogonal reference frame applied in this book rotates synchronously with the stator voltage vector $\mathbf{u}_s$. In this case, the ω_a pulsation is equal to the stator voltage pulsation $\omega_a = \omega_u$.

For induction machine modeling the widely used equivalent circuit is the T-form model and in most cases this is used in the book. However, the T-form model is in fact more complicated than needed because of the separate stator and rotor leakage inductances used. These inductances cannot be identified separately during no-load and locked rotor tests [12]. During the test, only the common leakage inductance is identified. With that given parameter, the Inverse-Γ model induction motor [13] is more suitable. Both models are presented in the next subsections.

3.1.1   T-Model of Induction Machine

The structure of the T-model of the induction motor is presented in Figure 3.1. The equivalent circuit of the squirrel cage motor, presented in Figure 3.1, consists of a resistance, an inductance, and a voltage source. The quantities $l\sigma_s$, $l\sigma_r$ are the leakage inductances of the stator and rotor windings. L_m is the stator and rotor mutual inductance, where R_s and R_r are the stator and rotor winding resistances. In the applied model the rotor parameters were brought into the stator side. The voltage source $j\omega_r\psi_r$ models the electromotive force of the rotating motor.

Based on the simplified assumptions and furthermore by applying the per-unit system (see Section 3.4), the squirrel cage induction motor model can be described by the following dependencies [9]:

$$\mathbf{u}_s = R_s \mathbf{i}_s + \frac{d\psi_s}{d\tau} + j\omega_a \psi_s \tag{3.1}$$

$$0 = R_r \mathbf{i}_r + \frac{d\psi_r}{d\tau} + j(\omega_a - \omega_r)\psi_r \tag{3.2}$$

Model of AC Induction Machine

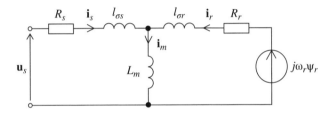

Figure 3.1 Equivalent circuit of the squirrel-cage induction motor

$$J\frac{d\omega_r}{d\tau} = \text{Im}\left|\psi_s^* i_s\right| - t_L \tag{3.3}$$

where:

u_s, i_s, ψ_s, ψ_r are the vectors of the voltage, current, and flux matched with the stator and rotor
ψ_s^* is the conjugate vector of ψ_s
ω_r is the angular rotor speed
J is the moment of inertia
t_L is the load torque on the motor shaft
τ is the relative time.

Main inductance of the stator and rotor:

$$L_s = l_{\sigma s} + L_m = L_m\left(1+\sigma_s\right) \tag{3.4}$$

$$L_r = l_{\sigma r} + L_m = L_m\left(1+\sigma_r\right) \tag{3.5}$$

where σ_s and σ_r are the leakage coefficients of the stator and rotor windings.
The dependencies between the current and flux in the circuit model of the induction motor are as follows:

$$i_s = \frac{1}{L_s}\psi_s - \frac{L_m}{L_s}i_r \tag{3.6}$$

$$i_r = \frac{1}{L_r}\psi_r - \frac{L_m}{L_r}i_s \tag{3.7}$$

The introduction of Equations 3.6 and 3.7 into the model equations [3.1 to 3.3] makes it possible to eliminate two vector values: the rotor current, i_r, and moreover the stator flux, ψ_s. This leads to describing the motor model with five differential equations noted in the rotating xy reference frame with the angular speed ω_a.

$$\frac{di_{sx}}{d\tau} = -\frac{R_s L_r^2 + R_r L_m^2}{L_r w_\sigma} i_{sx} + \frac{R_r L_m}{L_r w_\sigma} \psi_{rx} + \omega_a i_{sy} + \omega_r \frac{L_m}{w_\sigma} \psi_{ry} + \frac{L_r}{w_\sigma} u_{sx} \quad (3.8)$$

$$\frac{di_{sy}}{d\tau} = -\frac{R_s L_r^2 + R_r L_m^2}{L_r w_\sigma} i_{sy} + \frac{R_r L_m}{L_r w_\sigma} \psi_{ry} - \omega_a i_{sx} - \omega_r \frac{L_m}{w_\sigma} \psi_{rx} + \frac{L_r}{w_\sigma} u_{sy} \quad (3.9)$$

$$\frac{d\psi_{rx}}{d\tau} = -\frac{R_r}{L_r} \psi_{rx} + (\omega_a - \omega_r) \psi_{ry} + R_r \frac{L_m}{L_r} i_{sx} \quad (3.10)$$

$$\frac{d\psi_{ry}}{d\tau} = -\frac{R_r}{L_r} \psi_{ry} - (\omega_a - \omega_r) \psi_{rx} + R_r \frac{L_m}{L_r} i_{sy} \quad (3.11)$$

$$\frac{d\omega_r}{d\tau} = \frac{L_m}{L_r J} (\psi_{rx} i_{sy} - \psi_{ry} i_{sx}) - \frac{1}{J} t_L \quad (3.12)$$

where the coefficient $w\sigma$:

$$w_\sigma = \sigma L_r L_s = L_r L_s - L_m^2 \quad (3.13)$$

thus, σ is the total leakage coefficient of the motor.

$$\sigma = 1 - \frac{L_m^2}{L_s L_r} = 1 - \frac{1}{(1+\sigma_s)(1+\sigma_r)} \quad (3.14)$$

The total leakage inductance of the induction motor, $L\sigma$, is as follows:

$$L_\sigma = L_s - \frac{L_m^2}{L_r} = \frac{w_\sigma}{L_r} \quad (3.15)$$

For a more comfortable notation, the following coefficients will be introduced:

$$a_1 = -\frac{R_s L_r^2 + R_r L_m^2}{L_r w_\sigma} \quad (3.16)$$

$$a_2 = \frac{R_r L_m}{L_r w_\sigma} \quad (3.17)$$

$$a_3 = \frac{L_m}{w_\sigma} \quad (3.18)$$

Model of AC Induction Machine

$$a_4 = \frac{L_r}{w_\sigma} \tag{3.19}$$

$$a_5 = -\frac{R_r}{L_r} \tag{3.20}$$

$$a_6 = R_r \frac{L_m}{L_r} \tag{3.21}$$

For the fixed $\alpha\beta$ reference frame, the motor model equations [3.8 to 3.12] may be expressed in the following form:

$$\frac{di_{s\alpha}}{d\tau} = a_1 i_{s\alpha} + a_2 \psi_{r\alpha} + a_3 \omega_r \psi_{r\beta} + a_4 u_{s\alpha} \tag{3.22}$$

$$\frac{di_{s\beta}}{d\tau} = a_1 i_{s\beta} + a_2 \psi_{r\beta} - a_3 \omega_r \psi_{r\alpha} + a_4 u_{s\beta} \tag{3.23}$$

$$\frac{d\psi_{r\alpha}}{d\tau} = a_5 \psi_{r\alpha} - \omega_r \psi_{r\beta} + a_6 i_{s\alpha} \tag{3.24}$$

$$\frac{d\psi_{r\beta}}{d\tau} = a_5 \psi_{r\beta} + \omega_r \psi_{r\alpha} + a_6 i_{s\beta} \tag{3.25}$$

$$\frac{d\omega_r}{d\tau} = \frac{L_m}{L_r J}\left(\psi_{r\alpha} i_{s\beta} - \psi_{r\beta} i_{s\alpha}\right) - \frac{1}{J} t_L \tag{3.26}$$

The equations of the induction motor model are presented for the per-unit system given in Section 3.4.

3.2 Inverse-Γ Model of Induction Machine

The structure of the inverse-Γ model of the induction motor is presented in Figure 3.2 [13]. The model equations in the per unit system [12, 13] are:

$$\frac{d\mathbf{i}_s}{d\tau} = -\frac{R_s + R_r}{L_\sigma} \mathbf{i}_s + \frac{R_r}{L_m L_\sigma} \mathbf{\psi}_r - j\frac{1}{L_\sigma}\omega_r \mathbf{\psi}_r + \frac{1}{L_\sigma}\mathbf{u}_s \tag{3.27}$$

$$\frac{d\mathbf{\psi}_r}{d\tau} = -\frac{R_r}{L_m}\mathbf{\psi}_r + j\omega_r \mathbf{\psi}_r + R_r \mathbf{i}_s \tag{3.28}$$

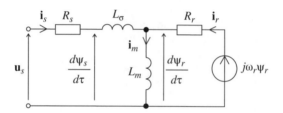

Figure 3.2 Induction motor inverse-Γ equivalent circuit model

$$\frac{d\omega_r}{d\tau} = \frac{1}{J_M}\left(\mathrm{Im}\left|\boldsymbol{\psi}_r^*\mathbf{i}_s\right| - t_L\right) \tag{3.29}$$

$$\mathbf{i}_s = \begin{bmatrix} i_{s\alpha} & i_{s\beta} \end{bmatrix}^T, \boldsymbol{\psi}_r = \begin{bmatrix} \psi_{r\alpha} & \psi_{r\beta} \end{bmatrix}^T, \mathbf{u}_s = \begin{bmatrix} u_{s\alpha} & u_{s\beta} \end{bmatrix}^T \tag{3.30}$$

where $\mathbf{u}_s, \mathbf{i}_s,$ and $\boldsymbol{\psi}_r$ denote the stator voltage, stator current, and rotor flux vectors; t_L is the motor load torque; J_M is the motor inertia; and L_m, $L\sigma$, R_s, and R_r are the motor parameters.

3.3 Per-Unit System

For modeling and simulation purposes the convenient solution is to recalculate all the variables into per-unit system. The different quantities are kept in the very close level. Moreover, the application of the per-unit system makes full use of the potential of the numbers range in computational calculations because the value dimensions are in a closer range. This will result in a better accuracy of computations. These solutions are widely used in the literature. Therefore, dimensionless units were used in this book [3, 9, 14, 15].

The classical way is to choose motor voltage and power as primary base quantities [15]. The *base voltage* is RMS value of phase-to-phase voltage (line voltage):

$$U_b = \sqrt{3}U_{n(ph)} \tag{3.31}$$

The motor *base power* is:

$$S_b = 3U_{n(ph)}I_b \tag{3.32}$$

so the *base current* has to be:

$$I_b = \sqrt{3}I_n \tag{3.33}$$

The next derived base values are *base impedance* and *resistance* as well:

$$Z_b = \frac{U_b}{I_b} \tag{3.34}$$

Model of AC Induction Machine

base torque is:
$$T_b = \frac{U_b I_b p}{\Omega_0} \tag{3.35}$$

base flux is:
$$\Psi_b = \frac{U_b}{\Omega_0} \tag{3.36}$$

base mechanical speed is:
$$\Omega_b = \frac{\Omega_0}{p} \tag{3.37}$$

where

p is number of motor poles pairs.
Ω_0 is rated voltage pulse of the supply:
$$\Omega_0 = 2 \cdot \pi \cdot f_n \tag{3.38}$$

and f_n is nominal frequency of motor stator voltage.

The next base value is *base inductance*:
$$L_b = \frac{\Psi_b}{I_b} \tag{3.39}$$

base capacitance is:
$$C_b = \frac{1}{\Omega_0 Z_b} \tag{3.40}$$

and *base moment of inertia* is:
$$J_b = \frac{T_b}{\Omega_b \Omega_0} \tag{3.41}$$

Also the time used in simulation is in per-unit according to relation:
$$\tau = \Omega_0 t \tag{3.42}$$

where:

t is real time
τ is relative time.

Every value in real units can be converted into an equivalent per-unit value with an appropriate base value:
$$X[pu] = \frac{X[real\ unit]}{Base\ value} \tag{3.43}$$

where

t_{br} is blocked-rotor torque in per-unit of the motor rated torque
i_{sbr} is blocked-rotor stator current in per-unit of the rated stator current
I_n is rated stator current.

Stator and rotor inductances are:

$$L_s = L'_r \cong \frac{1}{2\pi f_n} \frac{\frac{U_n}{\sqrt{3}}}{I_n\left(\sqrt{1-\cos^2\varphi_n} - \frac{S_n}{S_{bd}}\cos\varphi_n\right)} [\text{H}] \tag{3.49}$$

where

f_n is the nominal stator voltage frequency
$cos\varphi n$ is rated power factor.

Stator and rotor leakage inductances are:

$$l_s = l'_r \cong \frac{1}{4\pi f_n} \sqrt{\left(\frac{\frac{U_n}{\sqrt{3}}}{i_{sbr}I_n}\right)^2 - (R_s + R'_r)^2} \;[\text{H}] \tag{3.50}$$

where

f_n is the nominal stator voltage frequency
$cos\varphi_n$ is rated power factor.

Mutual inductances are:

$$L_m = L_s - l_s \;[\text{H}] \tag{3.51}$$

For verification purposes the coefficient c_1 is calculated

$$c_1 = 1 + \frac{l_s}{L_m} \tag{3.52}$$

Value of c_1 that is calculated according to Equation 3.52 have to be compared with the previous value assumed a priori as $c_1 = 1.03$. If the value differs, a new value for c_1 has to be chosen, the motor parameters have to be recalculated, and maybe some iterations have to be done.

The comparison of exemplary motor parameters measured in no-load and short-circuit test with the calculated values is presented in Table 3.1.

Model of AC Induction Machine

Table 3.1 Parameters of the induction motor equivalent circuit

P_n (kW)	p (–)	Measured				Calculated			
		R_s (Ω)	R_r' (Ω)	$L\sigma$ (mH)	L_m (H)	R_s (Ω)	R_r' (Ω)	$L\sigma$ (mH)	L_m (H)
0.75	1	9.6	11.6	34	0.819	8.8	8.9	53	1.055
0.75	4	5.2	3.8	20	0.253	2.8	3.6	39	0.332
1.5	2	4.7	4.7	17	0.303	4	4	30	0.422
3	2	1.9	2.7	12.3	0.257	1.7	1.8	14	0.237
5.5	2	0.98	0.97	5.9	0.143	0.72	0.73	8.5	0.162

It can be seen that a good convergence is obtained, which is enough for closed-loop induction motor drives as presented in the following chapters.

3.5 Simulation Examples

The induction motor model (T-model) for MATLAB/Simulink (*Chapter_3\IM_model\ Simulator.slx*) is shown in Figure 3.4.

The *Simulator.slx* model consists of a three-phase inverter with space vector pulse width modulation algorithm. The whole simulation is prepared in C language and done as S-Function *Simulation_model* of Simulink. Before starting the simulation, the S-Functions have to be compiled first what is done by [Build] button (Figure 3.5).

The main C language simulation file is *mainSIM.c*, which is added to the libraries of *Simulation_model* S-Function (directive #include <mainSIM.c> is added in [Libraries] tab).

The simulated drive is working according to the V/f = const principle. No ramp is added for stator voltage pulsation, so when the simulation starts, the motor direct start-up is performed. The model is for a four-pole, P_N = 1.5-kW, induction motor. The motor parameters are given in the *mainSIM.c* file:

```
/*------------------------------------------------------------*/
/* Motor parameters  -  Pn = 1.5 kW (Motor type 2Sg90L4) */
/*------------------------------------------------------------*/
double Rs=0.08,Rr=0.067,Lm=2.381,Ls=2.381+0.073,Lr=2.381+0.073,
   JJ=38.4*2.5;
```

All variables and parameters are given in per unit system as explained in Section 3.4. The examples of simulation results for reference speed ω_r^* = 1 pu (i.e., 50 Hz) are given in Figure 3.6.

Main function in *mainSIM.c* is *void SIMULATION(void)*. That function is called from S-function in each integration time. The T-model induction motor equations are placed in function:

```
/*------------------------------------------------------------*/
/* Differential equation of asynchronous motor model */
/* Motor model in stationary rectangular frame of references
   alpha-beta */
/* (noted in file as x and y) */
```

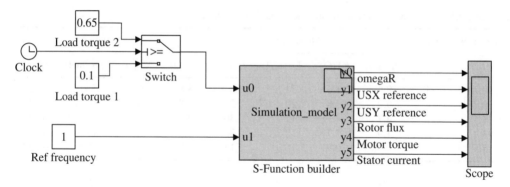

Figure 3.4 MATLAB/Simulink model of induction motor, *hapter_3\IM_model\Simulator.slx*

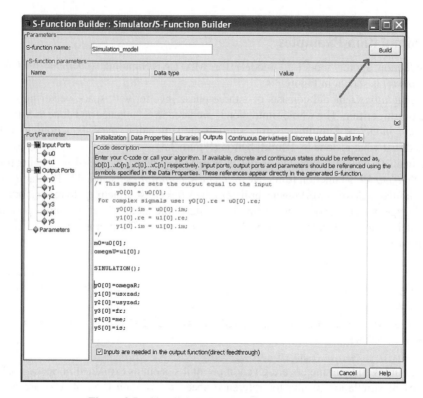

Figure 3.5 Simulink S-Function, *Simulation_model*

```
/*-------------------------------------------------------------*/
void F4DERY(double DV[32], double V[32])
{
tau=V[1];
isx=V[2]; isy=V[3];
frx=V[4]; fry=V[5];
omegaR=V[6];
```

Model of AC Induction Machine

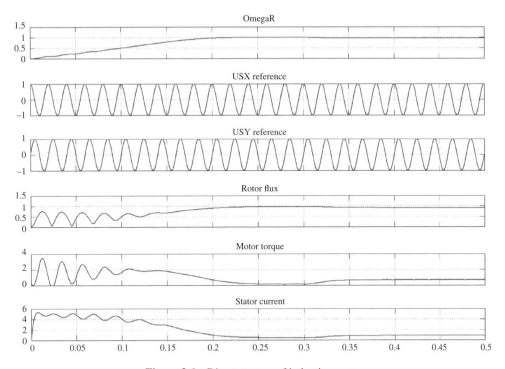

Figure 3.6 Direct start up of induction motor

```
DV[1]=1; /* time */
/* motor differential equations */
DV[2]=a1*isx+a2*frx+omegaR*a3*fry+a4*usx;
DV[3]=a1*isy+a2*fry-omegaR*a3*frx+a4*usy;
DV[4]=a5*frx+a6*isx-omegaR*fry;
DV[5]=a5*fry+a6*isy+omegaR*frx;
DV[6]=((frx*isy-fry*isx)*Lm/Lr-m0)/JJ;
}
```

Each of the differential equations is recognized as, for example, *DV[2]=a1*isx+a2*frx+omegaR*a3*fry+a4*usx*. On the left side of the differential is the derivative of an appropriate state variable, for example, *isx=V[2]*. The set of the differential equations given in the C file is equivalent to relations form Equations 3.22 to 3.26.

Additional functions are written in separate files, which are in the *LIBFILES* subdirectory:

```
/*---------------------------------------------------------------*/
/* additional functions included in LIBFILES subdirectory */
/*---------------------------------------------------------------*/
  #include ".\LIBFILES\VPWM1.c"  // PWM
  #include ".\LIBFILES\LIMIT.c"  // limiter
  #include ".\LIBFILES\ANGLE.c"  // angle calculation
  #include ".\LIBFILES\NZERO.c"  // no zero division
```

Each included file contains the definition of some auxiliary functions:

- *VPWM1.c*, space vector pulse width modulation SVPWM
- *LIMIT.c*, amplitude limiter
- *ANGLE.c*, calculation of the vector position
- *NZERO.c*, zero limiter to protect division by zero error.

SVPWM function arguments are:

```
void VPWM1(double usxzad_VPWM1,  /* alpha component of
reference voltage */
        double usyzad_VPWM1,   /* beta component of reference
        voltage */
        double ud_VPWM1,       /* DC link voltage */
        double Timp_VPWM1,     /* inverter switching period */
        double *t0_VPWM1,      /* passive vector time*/
        double *t1_VPWM1,      /* first active vector time */
        double *t2_VPWM1,      /* second active vector time */
        double *impuls_VPWM1,  /* internal counter */
        int *cycle_VPWM1,      /* active vectors cycle t1..t2 or
        t2..t1 */
        int *idu1_VPWM1,       /* number of first active vector */
        int *idu2_VPWM1)       /* number of second active vector */
```

In the simulation file *Simulator.slx*, the reference values for *Simulation_model S-Function* are:

- *omegaU*, inverter output voltage pulsation
- *m0*, motor load torque

as it is given in the [Outputs] tab:

```
m0=u0[0];
omegaU=u1[0];
```

The six-channel waveform scope is used to visualize selected global variables declared in *mainSIM*.c. The investigated variables are declared in [Outputs] tab of *Simulation_model*:

```
y0[0]=omegaR;
y1[0]=usxref;
y2[0]=usyref;
y3[0]=fr;
y4[0]=me;
y5[0]=is;
```

All of the global variables, used in C simulation, could be presented as waveforms in Simulink scope.

Integration of the motor differential equations is done with classical Runge–Kutta (RK4) method. The numerical procedure is defined in function void F4(int MM,double

HF,double Z[32]. The function is added to main simulation file *MainSim.c* using directive #include ".\LIBFILES\F4.C". The RK4 procedure is as follows:

```
void F4(int MM,double HF,double Z[32])
{
static double DZ[32];
static double Q[5][32];
int I;
F4DERY(DZ,Z);
for (I=0;I<MM;I+=1)
{
    Q[0][I]=Z[I];
    Q[1][I]=HF*DZ[I];
    Z[I]=Q[0][I]+0.5*Q[1][I];
}
F4DERY(DZ,Z);
for (I=0;I<MM;I+=1)
{
    Q[2][I]=HF*DZ[I];
    Z[I]=Q[0][I]+0.5*Q[2][I];
}
F4DERY(DZ,Z);
for (I=0;I<MM;I+=1)
{
    Q[3][I]=HF*DZ[I];
    Z[I]=Q[0][I]+Q[3][I];
}
F4DERY(DZ,Z);
for (I=0;I<MM;I+=1)
{
    Q[4][I]=HF*DZ[I];
    Z[I]=(Q[1][I]+Q[4][I])/6+(Q[2][I]+Q[3][I])/3+Q[0][I];
}
}
```

In the RK4 function, there are few of subroutine F4DERY(DZ,Z), in which the differential equations of the motor model are given.

In the next simulation examples, the same RK4 methods are used also for some observers calculations. In case of required integration for proportional-integral (PI) controllers, a simpler method is used (Euler).

References

[1] Ronkowski M, Szczęsny R. Circuit-oriented models of AC machines for converter-machine systems simulation. Proceedings of the Fourth European Conference on Power Electronics and Applications EPE'95, Sevilla, Spain. September 19–21, 1995.

[2] Szczęsny R, Ronkowski M. Modeling and simulation of converter systems. Part 1. Methods, models and techniques. *Journal on Circuits, Systems and Computers*. 1995; **5** (4): 635–668.

[3] Krzeminski Z. Nonlinear control of induction motor. In: *Proceedings of the Tenth IFAC World Congress*. Munich, Germany, July 1987. pp.394–354

[4] Orlowska-Kowalska T, Dybkowski M, Stator-current-based MRAS estimator for a wide range speed-sensorless induction-motor drive. *IEEE Transactions on Industrial Electronics*. 2010; **54** (4): 1296–1308.

[5] Karkosinski D, Ronkowski M, Moson I. An evaluation of acoustical noise level scattering for mass-produced induction motors fed by inverter. European Power Electronics Chapter Symposium on Electric Drive Design and Applications, Lausanne, Suisse. October 19–20, 1994.

[6] Kłapyta G, Kluszczyński K. Squirrel-cage induction motor with suppressed parasitic torques. Fourth International Symposium on Automatic Control AUTSYM 2005, Wismar, Germany. September 22–23, 2005.

[7] Krzemiński Z. A new speed observer for control system of induction motor. IEEE International Conference on Power Electronics and Drive Systems, PEDS'99, Hong-Kong, China. July 27–29, 1999.

[8] Kaźmierkowski MP, Krishnan R, Blaabjerg F: *Control in power electronics*. San Diego, CA: Academic Press; 2002.

[9] Krzeminski Z. Observer of induction motor speed based on exact disturbance model. Thirteenth Power Electronics and Motion Control Conference, EPE-PEMC 2008, Poznan, Poland. September 1–3, 2008.

[10] Krause P, Wasynczuk O, Sudhoff S, Pekarek S. *Analysis of electrical machinery and drive systems*. New York: John Wiley & Sons, Inc.; 2013.

[11] Dybkowski M, Orlowska-Kowalska T, Tarchała G. Sensorless direct torque control of the induction motor drive with sliding mode and MRAS estimators. Seventeenth International Conference on Electrical Drives and Power Electronics, EDPE'2011, High Tatras, Slovakia. September 28–30, 2011.

[12] Harnefors L, Hinkkanen M. Stabilization methods for sensorless induction motor drives—A survey. *IEEE Journal on Emerging and Selected Topics in Power Electronics*. 2014; **2** (2): 132–142.

[13] Slemon GR. Modelling of induction machines for electric drives. *IEEE Transactions on Industrial Applications*. 1989; **25** (6): 1126–1131.

[14] Abu-Rub H, Iqbal A, Guzinski J. *High performance control of AC drives with Matlab/Simulink models*. Chichester, UK: John Wiley & Sons, Ltd; 2012.

[15] Institute of Electrical and Electronics Engineers. *IEEE Standard 112-1996: IEEE Standard Test Procedure for Polyphase Induction Motors and Generators*. New York: Electric Machines Committee of the IEEE Power Engineering Society; 1997.

[16] Nieznanski J, Iwan K, Szczęsny R, Ronkowski M. *TCad for Windows. High performance power electronics simulation software*. Gdansk, Poland; Softech/University of Technology; 1996.

[17] Kozłowski H, Turowski J. *Induction motors: Design, construction and manufacture*. Warsaw, Poland: WNT; 1961.

[18] Dąbrowski M. *Design of AC electric machines*. Warsaw, Poland: WNT; 1988.

[19] Dabal K. Analysis of mechanical losses in three-phase squirrel-cage induction motors. Fifth International Conference on Electrical Machines and Systems, ICEMS 2001, Shenyang, China. August 18–20, 2001.

4

Inverter Output Filters

4.1 Structures and Fundamentals of Operations

An improvement of operation conditions for an induction motor in an inverter drive system is possible if the stator voltage waveform is close to a sinusoidal course.

Such a supply has proven advantages because the motor shows better efficiency in terms of minimization of additional losses in the motor. When motor is supplied with a pulsing voltage, the additional losses in motor will appear. The additional losses are results of high-frequency Eddy current (hysteresis losses). If compare with cooper losses it will dominate in motor [1, 2].

The application of inverter output filters also reduces the level of current and voltage disturbances [3].

Motor filters can be divided into three base categories:

- sinusoidal filter
- common mode filter
- dV/dt filter.

The aforementioned types of filters can be found in a single application or can be connected in different combinations, for example, a connection of a sinusoidal filter with a common mode filter. This book considers several different filter configurations.

Figure 4.1 presents a complex filter structure [4, 5]. The filter presented in Figure 4.1 shows a combination of two filters: a sinusoidal filter and a common mode filter. Such a connection makes it possible to achieve a sinusoidal current and voltage at the filter output, and moreover a limitation of the common mode current. This will allow a relatively long cable to be applied between the filter and the motor. The length of the cable is only limited by the accepted voltage drop on the cable.

Variable Speed AC Drives with Inverter Output Filters, First Edition. Jaroslaw Guzinski, Haitham Abu-Rub and Patryk Strankowski.
© 2015 John Wiley & Sons, Ltd. Published 2015 by John Wiley & Sons, Ltd.

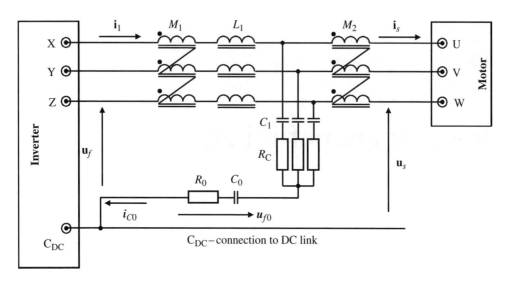

Figure 4.1 Voltage inverter output filter: combination of sinusoidal filter and common mode filter

The elements L_1, C_1, and R_C represent the sinusoidal filter, whereas M_1, M_2, R_0, and C_0 are the common mode filter elements. The coupled inductor M_2 limits the common mode current in the outside circuit considering the filter and inverter. The C_{DC} terminal connects the common mode filter with the inverter supply capacitor. For this reason a part of the common mode current of the motor flows to the inverter, bypassing the motor (Figure 4.2).

The connection of the filter with C_{DC} can be realized through the additional connection to the DC link of the frequency converter. In industrial inverters this terminal, named P, is used to connect an outside braking resistor.

The inductors L_1 are single-phase inductors, whereas the inductors M_1 and M_2 are common mode inductors, like those presented in Figure 2.12.

The resistors R_C and R_0 are damping elements of the filter that secure the system against dangerous oscillations.

For the high-switching frequency of the inverter, a significant current can flow through the branches R_0 and C_0. It requires the high power resistor R_0. Moreover it is necessary to use a capacitor C_0 with a significant component of the permissible alternating current.

An exemplary voltage and current waveforms in the drive of the inverter operation of the filter are shown in Figures 4.1 and 4.3 to 4.5. Figures 4.3 to 4.5 show a considerable smoothing of the motor voltage and current waveforms compared to the voltage and current of the inverter. The quality of the waveforms is dependent on the assumed filter parameter and on the level of the motor load. The selection of the filter elements is done by choosing the nominal load in such a way as to get the lowest value of the voltage and current harmonics and simultaneously the lowest voltage drop on the filter. Correct selection of the filter elements also ensures the minimization of the phase shift of the currents and voltages at the filter input and output (Figures. 4.4 and 4.5).

The filter structure shown in Figure 4.1 without inductor M_2 was presented for the first time in Akagi and colleagues [6, 7]. Figure 4.6 presents the structure of a sinusoidal filter with a common mode choke, patented by Schaffner [8].

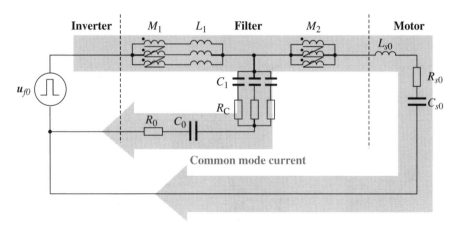

Figure 4.2 Equivalent circuit of the common mode current for the inverter, motor, and filter system shown in Figure 4.1

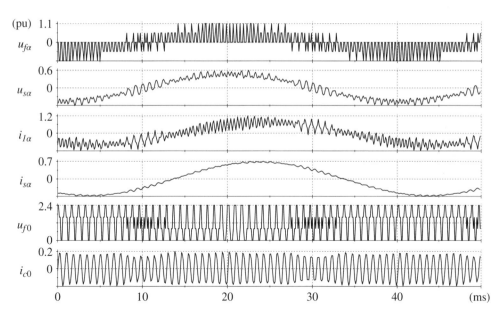

Figure 4.3 Waveforms of the voltages and currents in the inverter drive system with the filter shown in Figure 4.1: simulation

In the filter presented in Figure 4.6, the inductor M_2 and the structure comprising R_0 and C_0 were removed. A further elimination of the C_{DC} connection increases the common mode current in the motor compared to the filter in Figure 4.1. Nevertheless, the high-power resistor R_0 was simultaneously removed, which is beneficial for the filter efficiency and dimensions.

Figure 4.7 presents the sinusoidal filter of the voltage inverter. In the sinusoidal filter presented in Figure 4.7a, damping resistances R_C occur. The application of these resistors is

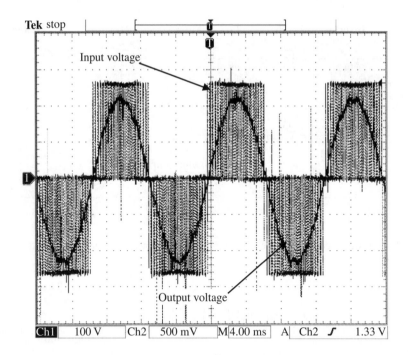

Figure 4.4 Waveforms of the voltages on the input and output of the filter shown in Figure 4.1: experiment

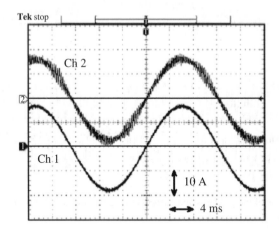

Figure 4.5 Current waveforms at Figure 4.1 filter input (CH2) and output (CH1): experiment

necessary for protection of the system in the case of inaccurate voltage generation and furthermore in the case of incomplete or absent death-time compensation. In such a situation it is possible for common mode voltages to appear in a frequency range that is close to the resonant frequency of the filter. Neglecting R_C allows a less complicated version of the sinusoidal filter to be implemented (Figure 4.7**b**).

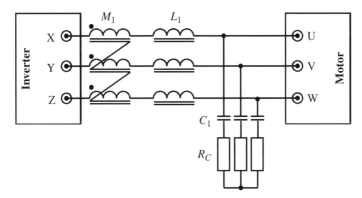

Figure 4.6 Output voltage inverter filter: a connection of a sinusoidal filter and common mode choke, patented by Schaffner [8]

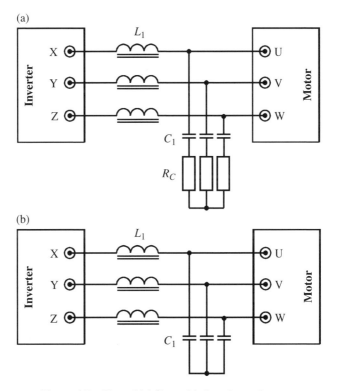

Figure 4.7 Sinusoidal filter with damping resistances

Figure 4.8 presents the filter suggested in Witkowski [3].

The filter structure of Figure 4.8 includes two common mode chokes, M_1 and M_2, and furthermore the capacitors, C_1 and C_2, connected to the positive and negative terminals of the frequency converter DC link. The task of the additional choke M_2 is to limit the common mode current in the outer circuit relative to the filter and inverter [3]. A single-phase choke L_3 was applied connected with the protection conductor PE. As shown in Witkowski [3], the aim of introducing L_3 is to protect the system against an uncontrolled voltage increase on the filter capacitors, which can occur in the case of asymmetrical voltage generation in each output phase of the inverter. The connection to the PE ground should eliminate the common mode current that flows to the motor. However, because of the significant inductance, L_3, this connection does not have a great influence on the motor common mode current.

The advantage of the filter presented in Figure 4.8 is the practical omission of the voltage and current phase-shift at the filter inputs and outputs and the minimization of the filter voltage

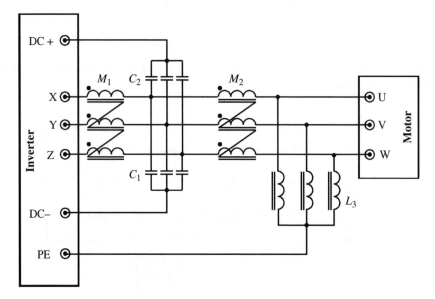

Figure 4.8 Voltage inverter with output filter suggested in Witkowski [3]

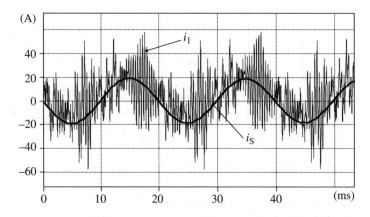

Figure 4.9 Current waveforms of the same circuit phase: i_1, current on filter input, and is, current on filter output

drop. This results from the small inductance of the M_1 and M_2 chokes for differential mode components of different currents and voltages. It is possible to replace the chokes M_1 and M_2 for these different components with a circuit that contains a low leakage resistance and inductance. Nevertheless, the disadvantage of this filter is the need for an oversized (because of the selection of high-current transistors) frequency converter resulting from the high current pulses, which are a result of the switching transistors on the circuit that contains practically only capacitances.

It can be seen in Figure 4.9 that there are significant momentary inverter current values i_1 that exceed the inverter current amplitude more than twice. To avoid the activation of the converter short-circuit protection, it will be necessary, in the case of the filter presented in Figure 5.8, to oversize the transistors rated current, which is unbeneficial for the system costs.

4.2 Output Filter Model

The analysis and simulation investigations of the drive systems require the use of a mathematical model of the filter. Figure 4.10 presents the circuit of the filter shown in Figure 4.1 with current and voltage notation.

It was assumed that the common mode chokes were ideal elements; that is, the leakage resistance and inductance were neglected. The same notation was chosen for the mutual and self-inductance of both chokes, M_1 and M_2.

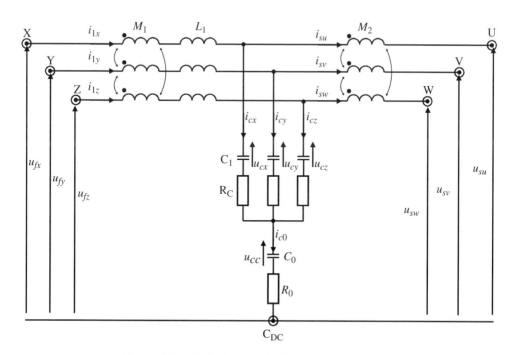

Figure 4.10 Equivalent circuit of the inverter output filter

In the circuit diagram shown in Figure 4.10, the winding resistance of the L_1 chokes was neglected. Under these assumptions made for the filter model, the equations considering C_{DC} as the reference point and the natural ABC frame are expressed as follows:

$$\begin{bmatrix} u_{fx} \\ u_{fy} \\ u_{fz} \end{bmatrix} = \frac{d}{d\tau}\begin{bmatrix} M_1 & M_1 & M_1 \\ M_1 & M_1 & M_1 \\ M_1 & M_1 & M_1 \end{bmatrix}\begin{bmatrix} i_{1x} \\ i_{1y} \\ i_{1z} \end{bmatrix} + L_1\frac{d}{d\tau}\begin{bmatrix} i_{1x} \\ i_{1y} \\ i_{1z} \end{bmatrix} + \frac{d}{d\tau}\begin{bmatrix} M_2 & M_2 & M_2 \\ M_2 & M_2 & M_2 \\ M_2 & M_2 & M_2 \end{bmatrix}\begin{bmatrix} i_{su} \\ i_{sv} \\ i_{sw} \end{bmatrix} + \begin{bmatrix} u_{su} \\ u_{sv} \\ u_{sw} \end{bmatrix} \quad (4.1)$$

$$\begin{bmatrix} u_{su} \\ u_{sv} \\ u_{sw} \end{bmatrix} = \begin{bmatrix} u_{cx} \\ u_{cy} \\ u_{cz} \end{bmatrix} + R_C\begin{bmatrix} i_{cx} \\ i_{cy} \\ i_{cz} \end{bmatrix} + \begin{bmatrix} u_{cc} \\ u_{cc} \\ u_{cc} \end{bmatrix} + R_0\begin{bmatrix} i_{cx}+i_{cy}+i_{cz} \\ i_{cx}+i_{cy}+i_{cz} \\ i_{cx}+i_{cy}+i_{cz} \end{bmatrix} - \frac{d}{d\tau}\begin{bmatrix} M_2 & M_2 & M_2 \\ M_2 & M_2 & M_2 \\ M_2 & M_2 & M_2 \end{bmatrix}\begin{bmatrix} i_{su} \\ i_{sv} \\ i_{sw} \end{bmatrix} \quad (4.2)$$

$$C_1\frac{d}{d\tau}\begin{bmatrix} u_{cx} \\ u_{cy} \\ u_{cz} \end{bmatrix} = \begin{bmatrix} i_{1x} \\ i_{1y} \\ i_{1z} \end{bmatrix} - \begin{bmatrix} i_{su} \\ i_{sv} \\ i_{sw} \end{bmatrix} \quad (4.3)$$

$$\begin{bmatrix} i_{1x} \\ i_{1y} \\ i_{1z} \end{bmatrix} = \begin{bmatrix} i_{su} \\ i_{sv} \\ i_{sw} \end{bmatrix} + \begin{bmatrix} i_{cx} \\ i_{cy} \\ i_{cz} \end{bmatrix} \quad (4.4)$$

$$C_0\frac{du_{cc}}{d\tau} = i_{cx}+i_{cy}+i_{cz} \quad (4.5)$$

$$i_{c0} = i_{cx}+i_{cy}+i_{cz} \quad (4.6)$$

The systems of equations of the filter model were changed to the orthogonal coordinates. The following equations describe the filter model for the $\alpha\beta0$ frame [3, 5, 9]:

$$\begin{bmatrix} u_{f0} \\ u_{f\alpha} \\ u_{f\beta} \end{bmatrix} = \frac{d}{d\tau}\begin{bmatrix} 3M_1 & & \\ & 0 & \\ & & 0 \end{bmatrix}\begin{bmatrix} i_{10} \\ i_{1\alpha} \\ i_{1\beta} \end{bmatrix} + L_1\frac{d}{d\tau}\begin{bmatrix} i_{10} \\ i_{1\alpha} \\ i_{1\beta} \end{bmatrix} + \frac{d}{d\tau}\begin{bmatrix} 3M_2 & & \\ & 0 & \\ & & 0 \end{bmatrix}\begin{bmatrix} i_{s0} \\ i_{s\alpha} \\ i_{s\beta} \end{bmatrix} + \begin{bmatrix} u_{s0} \\ u_{s\alpha} \\ u_{s\beta} \end{bmatrix} \quad (4.7)$$

$$\begin{bmatrix} u_{s0} \\ u_{s\alpha} \\ u_{s\beta} \end{bmatrix} = \begin{bmatrix} u_{c0} \\ u_{c\alpha} \\ u_{c\beta} \end{bmatrix} + R_c\begin{bmatrix} i_{c0} \\ i_{c\alpha} \\ i_{c\beta} \end{bmatrix} + \begin{bmatrix} u_{cc} \\ 0 \\ 0 \end{bmatrix} + R_0\begin{bmatrix} i_{c0} \\ 0 \\ 0 \end{bmatrix} - \frac{d}{d\tau}\begin{bmatrix} 3M_2 & & \\ & 0 & \\ & & 0 \end{bmatrix}\begin{bmatrix} i_{s0} \\ i_{s\alpha} \\ i_{s\beta} \end{bmatrix} \quad (4.8)$$

$$C_1 \frac{d}{d\tau} \begin{bmatrix} u_{c0} \\ u_{c\alpha} \\ u_{c\beta} \end{bmatrix} = \begin{bmatrix} i_{10} \\ i_{1\alpha} \\ i_{1\beta} \end{bmatrix} - \begin{bmatrix} i_{s0} \\ i_{s\alpha} \\ i_{s\beta} \end{bmatrix} \quad (4.9)$$

$$\begin{bmatrix} i_{10} \\ i_{1\alpha} \\ i_{1\beta} \end{bmatrix} = \begin{bmatrix} i_{s0} \\ i_{s\alpha} \\ i_{s\beta} \end{bmatrix} + \begin{bmatrix} i_{c0} \\ i_{c\alpha} \\ i_{c\beta} \end{bmatrix} \quad (4.10)$$

$$C_0 \frac{du_{cc}}{d\tau} = i_{c0} \quad (4.11)$$

The output filter structure for the orthogonal $\alpha\beta 0$ frame is illustrated in Figure 4.11. It can be seen that for the α and β components, only the elements of the sinusoidal filter appear. However, the elements for the common mode filter appear only for coordinate 0.

The presented filter model was used for simulation investigations in the following sections of this book and furthermore for the analysis of the influence of the filter on control and variable estimation in drive systems with induction motors.

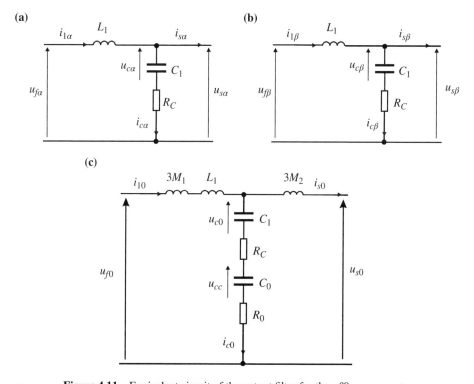

Figure 4.11 Equivalent circuit of the output filter for the $\alpha\beta 0$ components

4.3 Design of Inverter Output Filters

The choice of elements for the inverter output filter is a complex task. The following subsection presents the selection for the filter structure shown in Figure 4.1 [4, 5, 7, 9]. The complex structure of the filter can make the choice of parameters difficult. However, by considering the assumptions that $M_1 \gg L_1$, $C_0 \ll C_1$, and $M_1 = M_2$, it is possible to select the elements for the sinusoidal and common mode filters independently.

4.3.1 Sinusoidal Filter

The first element selection was done for the sinusoidal filter. The choice of the sinusoidal filter is determined by:

- the defined tolerance of the voltage deformation level of the output filter
- the maximum voltage drop on the filter
- the switching frequency of the inverter transistors.

The selection of the filter elements requires a compromise between the value of the higher ordered harmonics in the output voltage, which is described by the total harmonic distortion of the voltage (THD_U), and the filter size, costs, and current parameters of the inverter.

$$THD_U = \sqrt{\frac{\sum_{n=2}^{\infty} U_n^2}{U_1^2}} \cdot 100\% \tag{4.12}$$

where n is the harmonic order of the voltage and the index 1 is the fundamental frequency of the voltage.

The THD_U value of the uninterruptible power supply (UPS) systems should fulfill the following conditions [10]:

- $THD_U \leq 3\%$ in the case of linear load
- $THD_U \leq 5\%$ in the case of nonlinear load.

The choice of the sinusoidal filter differs from the selection of the UPS system because the THD value does not determine the filter parameters. In UPS systems, the THD value does not exceed 5% in full-load conditions. However, in drive systems, higher values of THD are allowed, even up to 20%. This follows from the fact that the THD is not the critical filter parameter. A higher or lower THD value in the motor supply voltage mainly influences the motor efficiency. Therefore, it should be remembered that the lower the THD_U value, the bigger the filter choke, voltage drop, and the limitation of the maximum motor torque at high speeds. A further essential reason for a filter application in a drive system is to avoid wave reflections on the motor terminals. A filter system that ensures a THD value greater than 5% can meet this need while satisfying the economic requirements. Moreover, it should be remembered that the inductance and capacitance values are influenced by the transistor switching frequency, which is significantly lower in a drive system than in UPS drive systems. In practice, the THD$_U$ value is a compromise between the voltage quality on the motor supply, the voltage drop on the filter choke, the distortion coefficient of the inverter output current, and the filter elements costs.

The first chosen filter element of the inverter output is the choke L_1. The inductance of choke L_1 is determined based on the inductor reactance X_1, which is calculated for the defined harmonic motor current tolerance, ΔI_{motor}, for the defined transistor switching frequency f_{imp} [7]. For transistors, the switching frequencies of a voltage inverter with a sinusoidal filter and squirrel cage motor can be represented by the equivalent circuit in Figure 4.12.

The circuit structure shown in Figure 4.12 is obtained though the omission of zero vectors in pulse-width modulation, where one voltage terminal of the DC link is connected to one output phase and the second pole is parallel to both output phases. This inverter operation complies with the maximum output voltage generation. The winding resistance and the leakage inductance of the motor were neglected in the schematic because these values create a path of a reactance circuit that is significantly higher than the filter reactance for frequency modulation. A reduced variant of the equivalent circuit is presented in Figure 4.13.

In Figures 4.12 and 4.13, the inverter was replaced by a sinusoidal voltage source $U_{P\,max}$ with modulation frequency f_{imp} and the maximum effective value of the first harmonic of the phase-to-phase voltage.

$$U_{P\,max} = \frac{U_d}{\sqrt{2}} \tag{4.13}$$

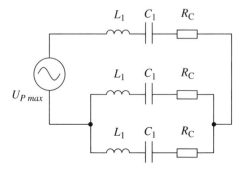

Figure 4.12 Equivalent circuit of inverter and sinusoidal filter

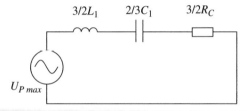

Figure 4.13 Reduced equivalent circuit of inverter and sinusoidal filter

In the case of supplying the inverter from the three-phase grid though the six-pulse rectifier (bridge) circuit, the average value of the inverter output voltage is described as follows:

$$U_d = \sqrt{2}U_n \frac{6}{\pi}\sin\frac{\pi}{6} \qquad (4.14)$$

where U_n is the effective value of the phase-to-phase voltage of the rectifier supply.

According to Equations 4.13 and 4.14, the value of the voltage $U_{P\,max}$ can be taken into account for the three-phase supply [11]:

$$U_{P\,max} = U_n \frac{6}{\pi}\sin\frac{\pi}{6} \approx 0.95 U_n \qquad (4.15)$$

In the circuit shown in Figure 4.13, the element that smoothes the current is mainly the choke L_1. For the modulation frequency, the reactance of L_1 is much larger than the reactance of the capacitor, C_1. Therefore, it is possible to have [7]:

$$2\cdot\pi\cdot f_{imp} \cdot \frac{3L_1}{2} \cdot \Delta I_{motor} = \frac{U_d}{\sqrt{2}} \qquad (4.16)$$

where ΔI_{motor} is the motor current pulsation for the modulation frequency (determined during the design).

Based on Equation 4.16, the choke inductance L_1 can be described as follows:

$$L_1 = 2\frac{U_d}{\sqrt{2}\cdot 2\cdot\pi\cdot f_{imp}\cdot 3\cdot \Delta I_{motor}} \qquad (4.17)$$

The quality of the current smoothing depends on the calculated inductance value L_1, which depends on the motor current pulsation ΔI_{motor} determined during the filter design process. Considering the size, weight, and costs of the filter, the inductance L_1 should be small as possible. A small inductance value L_1 simultaneously causes a small voltage drop on the filter and therefore smaller losses in the drive system.

Another method of determination of the inductance L_1 is presented in Xiyou and colleagues [9] and Danfoss [12], where the calculation of the value L_1 is based on the defined voltage drop ΔU_1 on the filter and further on the maximum of the first harmonic value of the output voltage $U_{out\,1har}$. The voltage drop has to be defined with nominal motor load on the reactance of choke L_1 for the frequency $f_{out\,1har}$. By accepting the allowed voltage drop ΔU_1, the choke can be determined as follows:

$$L_1 = \frac{\Delta U_1}{2\pi f_{out\,1har} I_n} \qquad (4.18)$$

In practice, the allowed voltage drop is $\Delta U_1 = 3.5\%$ [9, 12].

The value of the capacitor C_1 results from the resonant frequency of the system:

$$f_{res} = \frac{1}{2\pi\sqrt{L_1 C_1}} \qquad (4.19)$$

And thus:

$$C_1 = \frac{1}{4\pi^2 f_{res}^2 L_1} \qquad (4.20)$$

where f_{res} is the filter resonant frequency.

To ensure a good filter smoothing property, f_{res} has to be lower than the transistor switching frequency f_{imp}. Furthermore, to avoid the occurrence of resonance, the frequency f_{res} has to be higher than the maximum frequency of the first harmonic of the inverter output voltage $f_{out\ 1har}$. A secure determination is provided by [9]:

$$10 \cdot f_{out\ 1har} < f_{res} < 1/2 \cdot f_{imp} \qquad (4.21)$$

The upper range of the C_1 capacitance defines the maximum value of the current through the capacitor C_1 in nominal drive system operation. Moreover, the value of this current for the first harmonic must not exceed 10% of the nominal motor current.

There is a danger that a resonance will occur in the sinusoidal filter in the case of overmodulation operation in the drive system. The current, which flows through the C_1, has the first harmonic component and also the higher harmonics.

The filter damping resistor should be chosen in such a way that in the case of occurrence of a resonance, the current that flows through the capacitor will not be higher than the current defined value in the filter design process. Furthermore, it is necessary for the C_1 calculation to know the value of the voltage component with the frequency close to the filter resonant frequency. The corresponding equations are given elsewhere [13]. In practice, because of the complexity of these analytical calculations, the harmonic component with the resonant frequency is determined by using simulation tools.

Another method of selecting the filter damping resistance is to choose the value of R_C in such a way that the resulting power losses on the filter damping resistors were within a tolerance of 0.1% of the nominal power of the inverter [7]. After choosing the resistance R_C in this way, the filter quality factor has to be determined:

$$Q = \frac{Z_0}{R_C} \qquad (4.22)$$

The value Z_0 in Equation 4.22 is the characteristic impedance of the filter:

$$Z_0 = \sqrt{\frac{L_1}{C_1}} \qquad (4.23)$$

To ensure a sufficient attenuation of the filter while simultaneously minimizing the filter power losses, the quality factor Q should be within the range of 5 to 8 [7].

Knowledge of the parameters across the path of the sinusoidal filter, C_1 and R_C, and the assumed sinusoidal waveform of the filter output voltage makes it possible to calculate the power of the element R_C for the first harmonic of the current and voltage of the output filter:

$$P_{R1\ 1har} = \frac{U_{s\ phase}^2}{R_C} \tag{4.24}$$

where $U_{s\ phase}$ is the phase voltage on the filter output.

During the selection of the resistors R_C, it has to be considered that the filter output voltage includes a component of the inverter transistor switching frequency. If the filter is correctly designed, the value of this component is small. However, concerning high frequency, the impedance of the C_1, R_C path is low, which can cause a significant current flow with high frequency in this path. The power of the R_C element for this voltage component with the pulse width modulation frequency is described as follows:

$$P_{R1\ imp} = \frac{U_{s\ imp}^2}{R_C} \tag{4.25}$$

where $U_{s\ imp}$ is the filter output voltage component for the modulation frequency of the inverter.

The value $U_{s\ imp}$ corresponds to the resulting value because of the distortion coefficient THD of the filter output voltage determined by the design process.

It is possible to determine the frequency characteristics for the chosen elements of the sinusoidal filter by representing the sinusoidal filter as a passive four-terminal network (Figure 4.14).

The elements L_f, C_f, and R_f respond the values $3/2L_1$, $2/3C_1$, and $3/2R_C$ for the reduced filter circuit shown in Figure 4.13. Figure 4.15 presents the sinusoidal filter as a four-terminal network with impedances Z_1 and Z_2.

The impedances Z_1 and Z_2 are described as follows:

$$Z_1(j\omega) = j\omega L_f \tag{4.26}$$

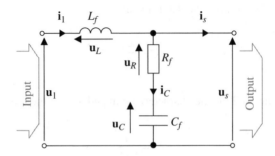

Figure 4.14 Equivalent circuit of filter for the α component

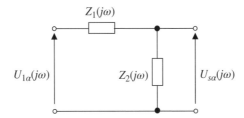

Figure 4.15 Equivalent circuit of the four-terminal sinusoidal filter for the α component

$$Z_2(j\omega) = R_f + \frac{1}{j\omega C_f} \tag{4.27}$$

or in the Laplace form:

$$Z_1(s) = sL_f \tag{4.28}$$

$$Z_2(s) = R_f + \frac{1}{sC_f} \tag{4.29}$$

The filter Laplace transfer function for the voltage is expressed as follows:

$$G_{fu}(s) = \frac{U_{s\alpha}}{U_{1\alpha}} \tag{4.30}$$

That is,

$$G_{fu}(s) = \frac{U_{s\alpha}}{U_{1\alpha}} = \frac{sR_f C_f + 1}{s^2 L_f C_f + sR_f C_1 + 1} \tag{4.31}$$

The filter resonant frequency is defined as follows:

$$f_{res} = \frac{1}{2\pi\sqrt{L_f C_f}} \tag{4.32}$$

while the damping coefficient ξ_d and the quality factor Q are described as follows:

$$\xi_d = \frac{R_f}{2\sqrt{L_f/C_f}} = \frac{1}{2Q} \tag{4.33}$$

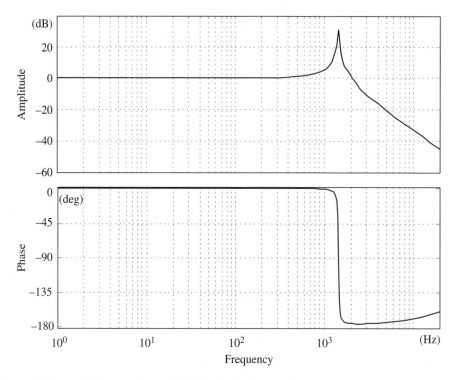

Figure 4.16 Bode diagram of sinusoidal filter (motor filter: 1.5 kW; $L_f = 4$ mH; $C_f = 3\ \mu$F; $R_f = 1\ \Omega$)

$$Q = \frac{\sqrt{L_f/C_f}}{R_f} \qquad (4.34)$$

Exemplary frequency characteristics—the amplitude and phase—of the sinusoidal filter are presented in Figure 4.16.

Knowledge of the sinusoidal filter elements properties and their characteristic parameters is necessary to make a correct consideration of the filter in the control process of the electrical drive.

4.3.2 Common Mode Filter

The selection of the common mode filter elements starts with the choice of the common mode inductors M_1 and M_2 (see Figure 4.10). It will be assumed that both chokes are identical. The common mode choke selection is carried out as shown in Chapter 2; L_{CM} is determined from Equation 2.5.

The capacitance C_0 will be defined from the known inductance values M_1 and M_2 and the resonant frequency $f_{res\ CM}$:

$$f_{res\ CM} = \frac{1}{2\pi\sqrt{L_{CM}C_0}} \qquad (4.35)$$

$$C_0 = \frac{1}{4\pi^2 f_{res\ CM}^2 L_0} \qquad (4.36)$$

while the resistance R_0 is:

$$R_0 = \frac{\sqrt{L_{CM}/C_0}}{Q_0} \qquad (4.37)$$

where Q_0 is the filter quality factor CM, which is assumed to be in the range of 5 to 8 [7].

The power P_{R0} can by determined with Kirchhoff's second law for the circuit in Figure 4.11 based on the equation:

$$P_{R0} = I_{C0(RMS)}^2 \cdot R_0 \qquad (4.38)$$

The current $I_{C0(RMS)}$ can be determined from the known effective value of the common mode voltage and the circuit parameters in Figure 4.11c. The calculation can be simplified if the elements L_1, C_1, are R_1 are neglected (error which causes a small oversize of the filter is accepted). From these results, for the inverter switching frequency, the total impedance of these elements is significantly lower compared to the impedance of a common mode filter. Through this simplification the power of the resistor R_0 can be described as follows:

$$P_{R0} = \frac{U_0^2}{R_0} \qquad (4.39)$$

where U_0 is the common mode voltage fulfilling the inequality

$$U_0 \leq \frac{U_d}{\sqrt{2}} \qquad (4.40)$$

The common mode filter function in Figure 4.1 is shown in Figures 4.17 and 4.18.

In the system without a common mode filter, the current of the PE line of a 5.5-kW motor (Figure 4.17) reaches its maximum close to 0.6 A, whereas after the implementation of the filter the current pulses do not exceed 0.2 A (Figure 4.18). As can be seen, thanks to the application of the common mode filter, the current pulses in the PE earth were significantly limited.

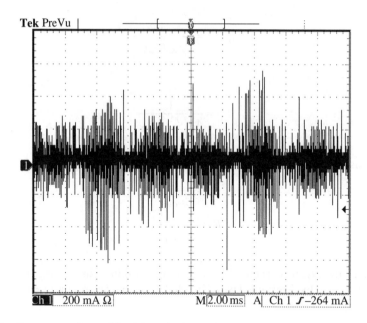

Figure 4.17 Current waveform of the PE-earth of a 5.5-kW motor supplied by an inverter without an output filter (scale: 200 mA/div)

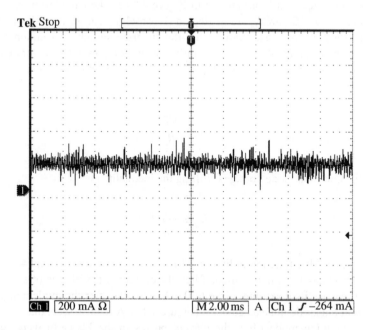

Figure 4.18 Current waveform of the PE earth of a 5.5-kW motor supplied by an inverter with a common mode filter (scale: 200 mA/div)

4.4 dV/dt Filter

For power electronics drives the long and reliable motor operation is guarantees only if dV/dt on motor terminals is limited.

Fast voltage rises in the switching transistors of the inverter output, that is, high dV/dt values, cause wave effects in the long motor supply cables that are characteristic for transmission lines [14]. Because the impedance is not fitted to the line and motor, wave reflections occur at the end of the cable that reach double value of the cable input voltage. This means that the momentary voltage value on the motor terminal can reach double the voltage value of the inverter DC link. For this reason, the degradation of the motor winding insulation will accelerate, which means that the fault-free operation time will be shortened. A solution to this issue is given by the implementation of adequate winding insulation or the application of a dV/dt filter. The typical dV/dt filter structure is presented in Figure 4.19.

The filter inductance L_f and capacitance C_f have small values compared to the analogous sinusoidal filter parameters. For this reason, the dV/dt filter does not smooth the motor supply voltage as much as a sinusoidal filter does, but only reduces the rise speed. Because of the small value of the dV/dt filter elements L_f and C_f, the filter size is significantly smaller if compare with appropriate sinusoidal filter.

The selection of elements of the dV/dt filter depends on the cable length, l_{cable}, connecting the inverter with the motor and moreover on the tolerance of the rise time, t_r, of the voltage on the motor terminal.

The filter capacitance is described as follows [15]:

$$C_f \geq l_{cable} \cdot 150 \cdot 10^{-12} \qquad (4.41)$$

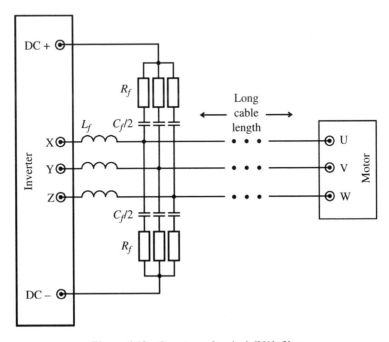

Figure 4.19 Structure of typical dV/dt filter

If l_{cable} is expressed in meters, the capacitance C_f is calculated in Farad. The filter inductance L_f is defined by the following condition:

$$L_f \geq \frac{t_r^2}{C_f} \qquad (4.42)$$

The damping resistance of dV/dt filter is defined as follows:

$$R_f \geq \sqrt{\frac{4L_f}{C_f}} \qquad (4.43)$$

The integrated sinusoidal filter with the common mode filter in Figure 4.1 also ensures motor protection against high dV/dt values. Figure 4.20 illustrates exemplary oscillograms of the motor supply voltage with a filter system and without filters.

Figure 4.20 presents the voltage waveforms of one measured motor terminal considering the (−) terminal of the inverter rectifier circuit. The measurement was done for a drive system with a 30-m motor supply cable. The inverter supply voltage was about 540 V. It can be seen that the voltage on the motor terminals rises to a value close to 800 V in the drive system without a filter. In contrast to this, the voltage of the drive system with an integrated filter was limited to approximately 600 V.

Except for the configuration of the dV/dt filter in Figure 4.19, the system can be realized with additional diodes (Figure 4.21) [16, 17].

The diodes in the filter shown in Figure 4.12 limit the voltage on the motor terminal to the level of the rectifier circuit of the frequency converter. However, this connection increases the current stress of the inverter switches, which have to be considered in the converter design.

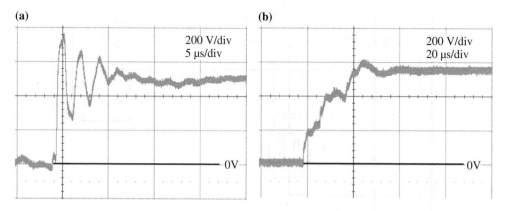

Figure 4.20 Influence of the output filter on the voltage rise on the terminals of the induction motor with an output power of $P_n = 1.5$ kW: **(a)** system without filter and **(b)** system with filter from Figure 4.1 ($L_1 = 5.6$ mH, $C_1 = 3$ µF, $R_C = 1$ Ω)

Inverter Output Filters

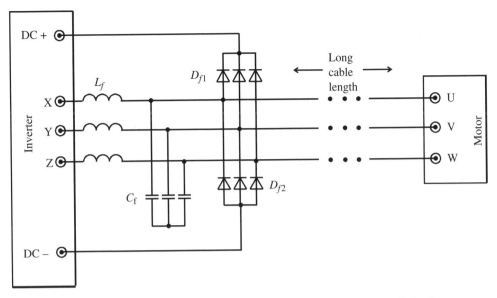

Figure 4.21 Structure of dV/dt filter with diodes of overvoltage cut-off circuit

4.5 Motor Choke

Motor chokes are applied to many industrial drive systems with frequency converters. That is the simplest form of inverter output filter. In three-phase systems, it is common to apply motor chokes as a three-phase choke or to use three single-phase chokes connected between the inverter output terminal and the motor input terminal (Figure 4.22).

The main task of motor chokes in AC drive systems is the limitation of voltage rise on the motor terminals, which reduces the danger of motor insulation damage [12, 18]. Motor chokes are also applied to reduce current shorting before the internal short-circuit protection of the frequency converter reacts. In the case of a long connection between the inverter and the motor, a motor choke also compensates the capacitance of the supply line. By neglecting the winding resistance, the motor choke can be modeled as a line induction (Figure 4.23).

The motor choke smoothes the motor current waveform and furthermore limits the speed of voltage rise on the motor terminals. Nevertheless, the influence of the motor choke on the form of the supply voltage, compared to the sinusoidal filter, is not of great importance (Figure 4.24). However, it significantly limits the steepness of motor voltage rise (reduces dV/dt) (Figure 4.25).

The inductance of the inductor L_1 can be calculated assuming that the voltage drop ΔU on the choke for the first harmonic with frequency $f_{out\ 1h}$ is smaller than the tolerance in the nominal motor current load I_n:

$$L_1 = \frac{\Delta U_1}{2\pi f_{out\ 1har} I_n} \qquad (4.44)$$

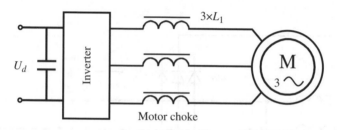

Figure 4.22 Drive system with squirrel cage motor, voltage inverter, and motor choke

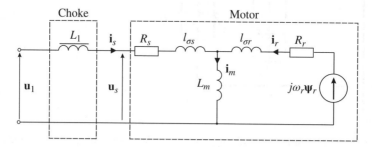

Figure 4.23 Equivalent circuit of choke: induction motor

It will be assumed that the accepted voltage drop on L_1 is within the range of 3% to 5% of the nominal motor voltage.

The choice of the choke considering the allowed voltage drop does not ensure the elimination of voltage wave reflections. Thus, the motor choke creates a resonant circuit with small damping in the connection with the cable capacitance. In the case of disadvantageous conditions, wave reflections can occur as presented in Figure 4.26.

Significant motor voltage oscillations can be seen in the waveform in Figure 4.26. Despite the implementation of the motor choke, the voltage reflections reach about 80%.

A correctly chosen motor choke reduces the voltage wave reflections in the motor supply cable. Figure 4.27 presents an exemplary waveform of the phase-to-phase voltage on the motor terminal with a long cable and correctly chosen motor choke.

It can be seen in Figure 4.27 that the application of a motor choke significantly reduces the unbeneficial effects of a long supply line. The maximum of the overvoltage spikes does not exceed 20%.

A considerable simplification of the selection process of the choke L_1 is achieved by the use of simulation tools. However, this requires knowledge of the circuit parameters, which allows a correct system to be modeled.

4.6 Simulation Examples

4.6.1 Inverter with LC Filter

The model of the inverter, LC filter, and induction motor (*Chapter_4\MotorLC\Simulator.slx*) is shown in Figure 4.28.

Inverter Output Filters

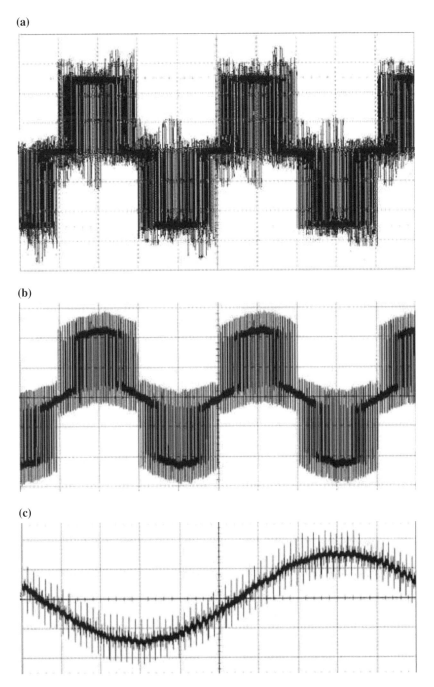

Figure 4.24 Waveforms of the experimental drive system with an inverter supplied induction motor and motor choke application: (**a**) inverter output voltage; (**b**) motor supply voltage; and (**c**) motor current (**a, b**: 200 V/div, 5 ms/div; **c**: 5 A/div, 2 ms/div, motor 300 V, 1.5 kW, choke L1 = 11 mH, motor with nominal load, the recorded waveforms are not synchronized)

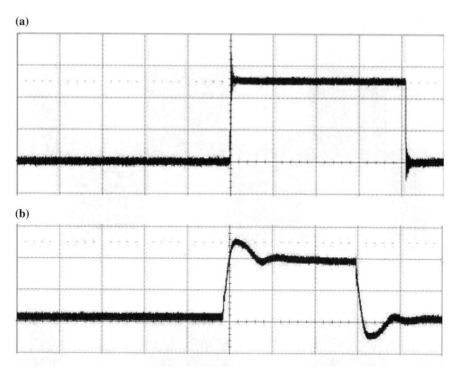

Figure 4.25 Waveforms: (a) inverter output voltage and (b) motor terminal voltage for drive system with motor choke (scale: 10 μs/div, 200 V/div; nonsynchronized waveforms)

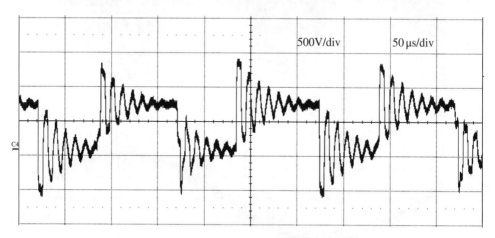

Figure 4.26 Phase-to-phase voltage on the motor terminal with inadequate choke chosen for the cable capacitances

The *Simulator.slx* model consists of a three-phase inverter with a space vector pulse width modulation algorithm. The sine wave filter (LC filter) is connected to the inverter output. The induction motor T model is considered as a load for the LC filter. The whole simulation system is found in *MotorLC.c* file.

Inverter Output Filters

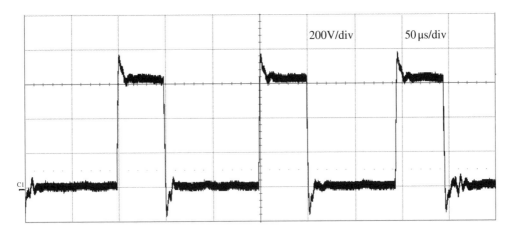

Figure 4.27 Phase-to-phase voltage of the motor terminals connected to the inverter with a 6-m cable, applying a motor inductor with 2.5 mH (inverter supply voltage: 650 V)

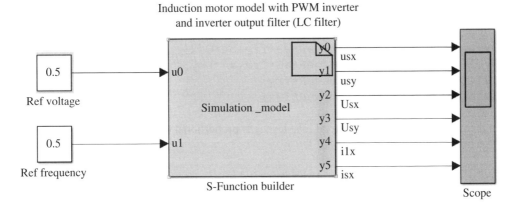

Figure 4.28 MATLAB/Simulink model of inverter and LC filter, *Chapter_4\MotorLC\Simulator.slx*

LC filter is the structure presented in Figure 4.27a, with a model for αβ components, while zero component is neglected. The filter equations are given as:

```
// LC filter model
ucx=V[7];   ucy=V[8];
i1x=V[9];   i1y=V[10];

ufx=usx;
ufy=usy;
icx=i1x-isx;
icy=i1y-isy;
Usx=Rc*(i1x-isx)+ucx;
Usy=Rc*(i1y-isy)+ucy;
```

```
DV[7]=icx/Cxy;
DV[8]=icy/Cxy;
DV[9]=(ufx-i1x*R1-icx*Rc-ucx)/L1;
DV[10]=(ufy-i1y*R1-icy*Rc-ucy)/L1;
```

The input signals (voltage) of the LC filter are *ufx* and *ufy*. The output are *Usx* and *Usy*, which are the sum of capacitor *Cxy* voltage and *Rc* damping resistance voltage drop. With:

```
ufx=usx;
ufy=usy;
```

LC filter is connected to the inverter output (*usx* and *usy* are inverter output voltages). The filter parameters are given as:

```
// ------------- LC filter with damping resistance ----------------
double L1=0.034,Cxy=0.0347,Rc=0.0271,R1=0.001357;
```

Filter output voltage appears as stator voltage in motor model:

```
DV[2]=a1*isx+a2*frx+omegaR*a3*fry+a4*Usx;
DV[3]=a1*isy+a2*fry-omegaR*a3*frx+a4*Usy;
DV[4]=a5*frx+a6*isx-omegaR*fry;
DV[5]=a5*fry+a6*isy+omegaR*frx;
```

In the mechanical equation the total load torque is proportional to square of motor speed (fan type load torque):

```
DV[6]=((frx*isy-fry*isx)*Lm/Lr-(m0+mMech*omegaR*omegaR))/JJ;
```

The simulated drive is working according to the *V/f* = const principle. Both voltage and frequency can be set separately by Simulink "constant block." No ramp is added for reference signals.

The example of simulation results for reference pulsation ω_u^* = 0.5 pu and reference voltage U_s^* = 0.5 pu is given in Figure 4.29.

It can be seen that inverter output voltages are smoothed to nearly sinusoidal shape. Also the motor current has less harmonic contents compared with the case when using filter input current.

4.6.2 Inverter with Common Mode and Differential Mode Filter

The model of inverter with common mode and differential mode filter and induction motor (*Chapter_4\Simulation_4_6_2\Simulator.slx*) is shown in Figure 4.30.

Inverter Output Filters

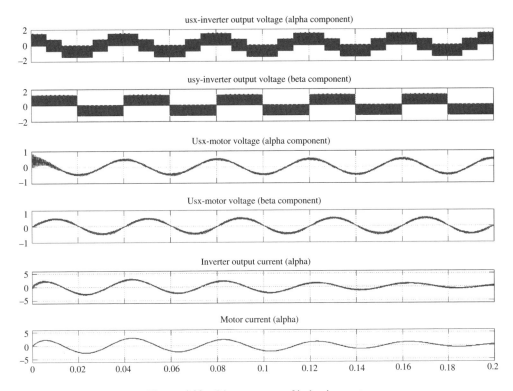

Figure 4.29 Direct start up of induction motor

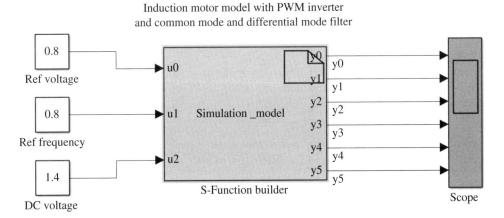

Figure 4.30 MATLAB/Simulink model of inverter with common mode and differential mode filter, *Chapter_4\Simulation_4_6_2\Simulator.slx*

The *Simulator.slx* model consists of a three-phase inverter with a space vector pulse width modulation algorithm. The main simulation file is Main.c, whereas space vector pulse width modulation (SVPWM) function is in PWM_0W.c file. The SVPWM function is:

```
void PWM_0W(double Timp, double Td, double US, double roU,
double ud, double h);
```

and includes generation of common mode voltage (variable *U0*). Common mode voltage is related to DC link negative potential as presented in Figure 2.3. It is possible to change it co DC link midpoint by change of comments in PWM_0W.c file:

```
// CM voltage related to (-) of DC link
u0[0]=0;        //000
u0[1]=Jp3;      //001
u0[2]=DJp3;     //010
u0[3]=Jp3;      //011
u0[4]=DJp3;     //100
u0[5]=Jp3;      //101
u0[6]=DJp3;     //110
u0[7]=p3;       //111

// CM voltage related to midpoint of DC link
/*u0[0]=-p3J2;  //000
u0[1]=-J2p3;    //001
u0[2]=J2p3;     //010
u0[3]=-J2p3;    //011
u0[4]=J2p3;     //100
u0[5]=-J2p3;    //101
u0[6]=J2p3;     //110
u0[7]=p3J2;     //111*/
```

and recompilation of the S-function.

The model of common mode and differential mode has a structure presented in Figure 4.1. The filter equations are given in function:

```
void F4DERY(double DV[32], double V[32])
```

together with the model of induction motor. Motor state variables are numbered from 1 to 6, whereas filter variables from 7 to 14 for differential mode:

```
// Differential mode filter model
ukx=V[7];   uky=V[8];
i1x=V[9];   i1y=V[10];
i2x=V[11];  i2y=V[12];
i3x=V[13];  i3y=V[14];

ufx=usx;
ufy=usy;
```

Inverter Output Filters

```
icx=i1x-i2x;
icy=i1y-i2y;
ucx=ukx+Rc*icx;
ucy=uky+Rc*icy;

x23=L2/L3;

Usx=(-L2*(a1*isx+a2*frx+omegaR*a3*fry)+i3x*R3*x23+ucx-i2x*R2)/
   (1+L2*a4+x23);
Usy=(-L2*(a1*isy+a2*fry-omegaR*a3*frx)+i3y*R3*x23+ucy-i2y*R2)/
   (1+L2*a4+x23);

DV[7] = (icx)/Cxy;
DV[8] = (icy)/Cxy;
DV[9] = (ufx-ucx-i1x*R1)/L1;
DV[10] = (ufy-ucy-i1y*R1)/L1;
DV[11] = (ucx-Usx-i2x*R2)/L2;
DV[12] = (ucy-Usy-i2y*R2)/L2;
DV[13] = (Usx-i3x*R3)/L3;
DV[14] = (Usy-i3y*R3)/L3;
```

and from 15 to 19 for common mode:

```
// Common mode filter model
I10=V[15];
i20=V[16];
i30=V[17];
uk0=V[18];
uks0=V[19];
uc0=uk0+ic0*R0;
ics0=i20-i30;
ucs0=uks0+ics0*R0s;
ic0=I10-i20;
UM1=U0-uc0-I10*Rm1;
DV[15] = (U0-uc0-I10*Rm1)/M1;
DV[16] = (uc0-ucs0-i20*Rm2)/M2;
DV[17] = (ucs0-i30*R3)/L3;
DV[18] = (I10-i20)/C0;
DV[19] = (i20-i30)/Cs0;
```

The simulation model includes also inverter input circuit with DC link capacitor *Cpp*, and model grid side with inductance *Lpp*, rectifier, and voltage source, *E*. The input circuit model is also given in function F4DERY for Equations numbered 4.20 and 4.21:

```
// Model of DC link circuit
ud=V[20];      // DC voltage
ie=V[21];      // rectifier output current
```

```
if (ie<0) Rdiod=500;   // rectifier off
   else Rdiod=6.0e-5;  // rectifier on
DV[20]=JCpp*(ie-id);
DV[21]=JLpp*(E-ud-ie*Rdiod);
```

The inverse Clarke transformation is used to evaluate inverter output and motor input currents, correspondingly by the functions:

```
// Inverse Clarke transformation - inverter currents
void tran2_3_f(double Xt,double Yt)
{
iaf=(p2/p3)*Xt;
ibf=(-1/p6)*Xt+(1/p2)*Yt;
icf=(-1/p6)*Xt-(1/p2)*Yt;
}

// Inverse Clarke transformation - motor currents
void tran2_3(double Xt,double Yt)
{
ia=(p2/p3)*Xt;
ib=(-1/p6)*Xt+(1/p2)*Yt;
ic=(-1/p6)*Xt-(1/p2)*Yt;
}
```

The inverter input current is calculated in `function double fun(int vector)`:

```
double fun(int vector)
{
double current;
switch(vector)
    {
    case 4:  current=icf;         break;
    case 6:  current=icf+ibf;     break;
    case 2:  current=ibf;         break;
    case 3:  current=ibf+iaf;     break;
    case 1:  current=iaf;         break;
    case 5:  current=icf+iaf;     break;
    case 0:  current=0;           break;
    case 7:  current=0;           break;
    }
return(current);
}
```

The example of simulation results for reference pulsation $\omega_u^* = 0.8$ pu, reference voltage $U_s^* = 0.8$ pu, and DC voltage E = 1.4 pu is given in Figure 4.31.

The system start, which is shown in Figure 4.31 has some oscillations noticeable in U_{sx} waveform. This is a result of filter capacitors charging. The DC link initial condition $ud = E$

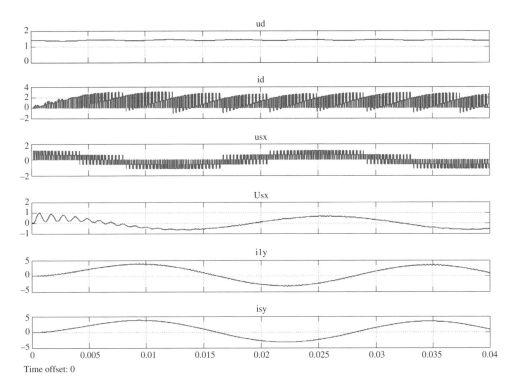

Figure 4.31 Waveforms of inverter input and output voltage and current in the system with a combination of sinusoidal filter and common mode filter (structure presented in Figure 4.1)

is noticeable in *ud* waveform. One can see that inverter output voltage and current waveforms are nearly sinusoidal. It is interesting to notice that inverter input current *id* has pulses and positive average value which is a result of motor mode operation of induction machine.

4.7 Summary

Different structures of inverter output filters are applied in drive systems: sinusoidal, common mode, and dV/dt filters. To reach a good filtration property, complex filter structures are designed which taking into account the filter frequency characteristics.

The design of a complex filter, considering the amount of elements used, requires an adequate approach in the development process. This process is difficult and demands experience by the developer.

Optimal selection of the filter elements requires numerous experimental investigations of, for example, the common mode voltage and current measurement in the drive system.

A correctly chosen complex inverter output filter ensures both smoothing of the motor voltage and current and limitation of the common mode current of the motor.

The manufacturers of inductors and capacitor offer some elements dedicated to use in filter structures. The elements parameters are related to driver power and inverter switching frequency.

References

[1] Ruifangi L, Mi CC, Gao DW. Modeling of iron losses of electrical machines and transformers fed by PWM inverters IEEE Transactions on Magnetics. 2008; **44** (8): 1–7.

[2] Yamazaki K, Fukushima N. (2009) Experimental validation of iron loss model for rotating machines based on direct eddy current analysis of electrical steel sheets. Electrical Machines and Drives Conference, IEMDC'09, Miami, FL. May 3–6, 2009.

[3] Witkowski S. Induction motor control with inverter output filter. PhD thesis. Gdansk, Poland: Gdansk University of Technology, Faculty of Electrical and Control Engineering; 2005.

[4] Adamowicz M, Guziński J. Control of sensorless electric drive with inverter output filter. Fourth International Symposium on Automatic Control AUTSYM 2005, Wismar, Germany. September 22–23, 2005.

[5] Krzeminski Z, Guziński J. Output filter for voltage source inverter supplying induction motor. International Conference on Power Electronics, Intelligent Motions and Power Quality, PCIM'05, Nuremburg, Germany. June 7–9, 2005.

[6] Akagi H. Prospects and expectations of power electronics in the 21st century. Power Conversion Conference, PCC'02, Osaka, Japan. April 2–5, 2002.

[7] Akagi H, Hasegawa H, Doumoto T. Design and performance of a passive EMI filter for use with a voltage–source PWM inverter having sinusoidal output voltage and zero common-mode voltage. *IEEE Transactions on Power Electronics*. 2002; **19** (5): 1069–1076.

[8] Schaffner Elektronik AG (1995) Einrichtung zur Begrenzung der Änderungsgeschwindigkeit der Ausgangsgrössen eines über einen Gleichspannungszwischenkreis selbstgeführten Umrichters. Patentschrift EP 0 682 402 B1, Anmeldetag April 18 1995, Veröffentlichungstag der Anmeldung November 15 1995.

[9] Xiyou C, Bin Y, Tu G. The engineering design and optimization of inverter output RLC filter in ac motor drive system. Twenty-Eighth Annual Conference of the IEEE Industrial Electronics Society, IECON'02, Sevilla, Spain. November 5–8, 2002.

[10] Institute of Electrical and Electronics Engineers. IEEE Standard 519-1992: *IEEE Recommended Practices and Requirements for Harmonic Control in Electrical Power Systems*. New York: IEEE Industry Applications Society/Power Engineering Society, Institute of Electrical and Electronics Engineers; 1993.

[11] Sodermanns P. Methods of generating pulse patterns in voltage source PWM inverters and implementation in an industrial drive. Third European Conference on Power Electronics and Applications, EPE '89, Aachen, Germany. October 9–12, 1989.

[12] Danfoss [Internet]. Output filters design guide. [Accessed May 17, 2005] Available at http://www.danfoss.com/nr/rdonlyres/27f81e71-3779-4406-8ea0-849044873f59/0/output_filters_design_guide.pdf.

[13] Franzo G, Mazzucchelli M, Puglisi L, Sciutto G. Analysis of PWM techniques using uniform sampling in variable-speed electrical drives with large speed range. *IEEE Transactions on Industry Applications*. 1985; **IA-21** (4): 966–974.

[14] Raghu Ram V, Kama Raju V, Durga Prasad, GS. Improvement of overvoltages of PWM inverter fed induction motor drives. *IOSR Journal of Electrical and Electronics Engineering (IOSR-JEEE)*. 2014; **9** (5): 1–7.

[15] Palma L, Enjeti P. An inverter output filter to mitigate dv/dt effects in PWM drive systems. Seventeenth Annual IEEE Applied Power Electronics Conference and Exposition, APEC'02, Dallas, TX. March 10–14, 2002.

[16] Kim S-J, Sul S-K. A novel filter design for suppression of high voltage gradient in voltage-fed PWM inverter. Twelfth Annual Applied Power Electronics Conference and Exposition, 1997, APEC'97, Atlanta, GA. February 23–27, 1997.

[17] Habetler TG, Naik R, Nondahl TA. Design and implementation of an inverter output LC filter used for dv/dt reduction. *IEEE Transactions on Power Electronics*. 2002; **17** (3): 327–331.

[18] ASEA Brown Boveri. Effects of AC drives on motor insulation—Knocking down the standing wave. A guide to understanding installation concerns and analyzing the use of motors with AC Drives below 600VAC. Technical Guide No. 102. Zurich, Switzerland: ASEA Brown Boveri; 1997.

5

Estimation of the State Variables in the Drive with LC Filter

5.1 Introduction

For control of electric drives operating in a closed loop system it is necessary to know the actual values of the regulated variables. The following are the most common: the angular velocity of the motor, the magnetic flux, and electromagnetic torque. Because of the difficulty in measuring torque and flux, these values are reproduced in the drive control system using different estimation methods. In modern drive systems the motor angular velocity must also be estimated, instead of its measurement. Fortunately, on the basis of the measured motor currents and motor reference voltage, all the controlled variables can be estimated if the system model is known with enough accuracy. Such drives, which do not require measurement of the angular position or speed of the motor, are commonly known as *sensorless drives*. For sensorless drives, only the use of the inverter output current sensors and inverter voltage supply sensor is required. The general structure of the sensorless drive is shown in Figure 5.1.

In the literature [1, 2], a lot of variable estimation methods for sensorless systems are presented; however, the methods do not take the inverter output filter into account.

Estimation of state variables becomes more complex in cases of the use of the inverter output filter. For the drive with LC filter, the motor current and voltages sensors can be used directly [3–7]. Unfortunately this simple solution is not practically accepted because it requires additional sensors installed outside the inverter housing. This arises from complications of the design, higher drive costs, and increased sensitivity of the drive to noises. A more preferred solution is to use the state estimation, which can estimate motor and filter state variables simultaneously. In this case, it is possible to use the same current and voltage sensors as in the drive without the filter. The general structure of the induction motor (IM) speed sensorless drive with inverter output filter is presented in Figure 5.2.

Variable Speed AC Drives with Inverter Output Filters, First Edition. Jaroslaw Guzinski, Haitham Abu-Rub and Patryk Strankowski.
© 2015 John Wiley & Sons, Ltd. Published 2015 by John Wiley & Sons, Ltd.

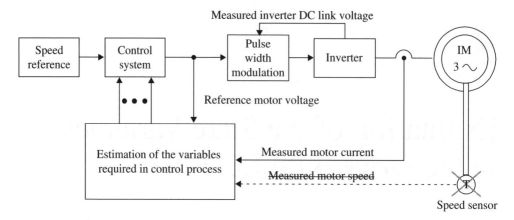

Figure 5.1 The general structure of speed sensorless induction motor drive

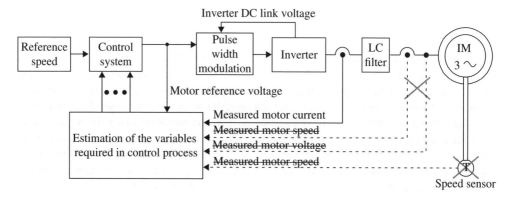

Figure 5.2 The general structure of speed sensorless induction motor drive with inverter output filter

As shown in Chapter 4, the control of the drive system, and hence the estimation of the state variables, are affected only by sine-wave filter elements. Effect of filter parameters for zero component is negligible.

Up to now, only a few works have given attention to the issues of state variables estimation in the drive with inverter output sine-wave filter. These methods were presented mainly in the works of the team: Salomäki, Luomi, and Hikkanen. In their works, the induction motor drive with LC filter, with [8] and without a speed sensor [9], were presented where the estimation system is based on the Kubota adaptive observer [10, 11]. The estimators are extended adequately with LC filter model dependencies and the observer feedback terms are properly changed.

In this book, the other observer solutions for drive systems with squirrel-cage motor and inverter output filter voltage are presented.

A new structure for a sensorless drive system was first proposed in Adamowicz and Guzinski [12]. The solution is based on use of a filter simulator working without feedback correction, which is incorporated into the structure of the disturbance state observer interference.

In other estimations solutions presented in the book the idea of the extended state observer with equations of filter model was used. Except additional equations of LC filter, the observer feedbacks were changed where the error between the measured and estimated inverter output current were used.

Another proposal of the estimation method is based on the concept of the speed observer which was first shown in Krzemiński [13]. Such an observer for the filter was shown for the first time in Guzinski and Abu-Rub [14].

The speed of the observer presented in Krzemiński [13] is characterized by high accuracy of estimation in a speed range up to rated speed. For the operation with field weakening its accuracy is significantly deteriorated. Therefore, the new observer was presented [12, 15, 16]. The observer was designed to work properly also for a high-speed motor. The extended speed range operation was possible because of more accuracy in the model in respect to the disturbances [17].

For solutions with the disturbance observer, the LC filter model was taken into account because of the concept presented in Salomaki and Luomi [8], which is based on observer extensions with additional differential equations for LC filter state variables and with changes of observer feedback terms modifications. Finally, the complete observer for complex objects, that is, an induction motor with LC filter, was derived.

The book also presents the solution of the estimation of the state variables for the induction motor drive with LC filter, based on IM stator circuit. Such an estimator for a drive without a filter is known from Abu-Rub [18]. The estimators have been modified according to the requirements of the filter. Also, the adaptive version of the state observer could be found in this book.

Except from the use of state observer in regulation it is possible to adopt it for diagnostic purposes, for example, for drive mechanical faults such as misalignment or unbalance. Such diagnostics could be based on analysis of the load torque or motor torque calculated by observer. For diagnostic purposes, it is important to have the observer cut-off frequency as high as possible [19].

5.2 The State Observer with LC Filter Simulator

The sensorless drive system with induction motor, with the state observer and the simulator filter, was presented in Adamowicz and Guzinski [12]. In the drive, the motor voltages and currents are calculated in the filter simulator operating in an open system, that is, without any correction terms. The structure of the observer and whole sensorless drive is shown in Figure 5.3.

According to the nomenclature presented in Figure 5.3, the filter simulator equations are as follows:

$$\hat{u}^F_{L\alpha} = L_1 \frac{di_{1\alpha}}{d\tau}, \qquad (5.1)$$

$$\hat{u}^F_{L\beta} = L_1 \frac{di_{1\beta}}{d\tau}, \qquad (5.2)$$

$$\hat{u}^F_{s\alpha} = u^*_{1\alpha} - \hat{u}^F_{L\alpha}, \qquad (5.3)$$

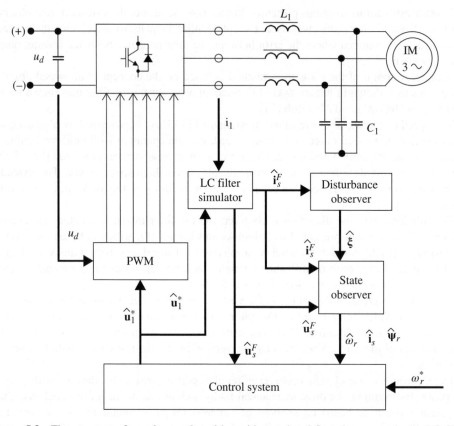

Figure 5.3 The structure of speed sensorless drive with speed and flux observer and with LC filter simulator

$$\hat{u}^F_{s\beta} = u^*_{1\beta} - \hat{u}^F_{L\beta}, \tag{5.4}$$

$$\hat{i}^F_{c\alpha} = C_1 \frac{d\hat{u}^F_{s\alpha}}{d\tau}, \tag{5.5}$$

$$\hat{i}^F_{c\beta} = C_1 \frac{d\hat{u}^F_{s\beta}}{d\tau}, \tag{5.6}$$

$$\hat{i}^F_{s\alpha} = i_{1\alpha} - \hat{i}^F_{c\alpha}, \tag{5.7}$$

$$\hat{i}^F_{s\beta} = i_{1\beta} - \hat{i}^F_{c\beta}, \tag{5.8}$$

where $\hat{\mathbf{u}}^F_s = \begin{bmatrix} \hat{u}^F_{s\alpha} & \hat{u}^F_{s\beta} \end{bmatrix}^T$, $\hat{\mathbf{i}}^F_s = \begin{bmatrix} \hat{i}^F_{s\alpha} & \hat{i}^F_{s\beta} \end{bmatrix}^T$ are motor voltage and current vectors, which are estimated in LC filter simulator block, whereas $\hat{\mathbf{u}}^F_L = \begin{bmatrix} \hat{u}^F_{L\alpha} & \hat{u}^F_{L\beta} \end{bmatrix}^T$ is filter inductor voltage drop vector.

On the basis of inverter output reference voltage, $\boldsymbol{u}_1^*$, and measured inverter output current, $\boldsymbol{i}_1$, the simulator block estimates actual values of motor voltage, $\hat{\boldsymbol{u}}_s^F$, and current, $\hat{\boldsymbol{i}}_s^F$, which are next used to estimate the speed, $\hat{\omega}_r$, and the motor flux, $\hat{\boldsymbol{\psi}}_r$ [12]:

$$\frac{d\hat{i}_{s\alpha}}{d\tau} = a_1 \hat{i}_{s\alpha} + a_2 \hat{\psi}_{r\alpha} + a_3 \hat{\xi}_\beta + a_4 \hat{u}_{s\alpha}^F + k_1 e_{is\alpha} - k_2 e_{is\beta}, \tag{5.9}$$

$$\frac{d\hat{i}_{s\beta}}{d\tau} = a_1 \hat{i}_{s\beta} + a_2 \hat{\psi}_{r\beta} + a_3 \hat{\xi}_\alpha + a_4 \hat{u}_{s\beta}^F + k_1 e_{is\beta} + k_2 e_{is\alpha}, \tag{5.10}$$

$$\frac{d\hat{\psi}_{r\alpha}}{d\tau} = a_5 \hat{\psi}_{r\alpha} + a_6 \hat{i}_{s\alpha} - \hat{\xi}_\beta + k_3 \left(\hat{\xi}_\alpha - \hat{\omega}_r \hat{\psi}_{r\alpha} \right), \tag{5.11}$$

$$\frac{d\hat{\psi}_{r\beta}}{d\tau} = a_5 \hat{\psi}_{r\beta} + a_6 \hat{i}_{s\beta} + \hat{\xi}_\alpha + k_3 \left(\hat{\xi}_\beta - \hat{\omega}_r \hat{\psi}_{r\beta} \right), \tag{5.12}$$

$$\frac{d\hat{\xi}_\alpha}{d\tau} = -\hat{\omega}_{\psi r} \hat{\xi}_\beta - k_4 e_{is\beta} + k_5 e_{is\alpha}, \tag{5.13}$$

$$\frac{d\hat{\xi}_\beta}{d\tau} = \hat{\omega}_{\psi r} \hat{\xi}_\alpha + k_4 \left(\hat{i}_{s\alpha}^F - \hat{i}_{s\alpha} \right) + k_5 \left(\hat{i}_{s\beta}^F - \hat{i}_{s\beta} \right), \tag{5.14}$$

$$\frac{d\hat{\omega}_{\psi r}}{d\tau} = k_6 \left(\left| \hat{\mathbf{i}}_{s\beta}^F \right| - \left| \hat{\mathbf{i}}_{s\beta} \right| \right), \tag{5.15}$$

$$\hat{\omega}_r = S \sqrt{\frac{\hat{\xi}_\alpha^2 + \hat{\xi}_\beta^2}{\hat{\psi}_{r\alpha}^2 + \hat{\psi}_{r\beta}^2}}, \tag{5.16}$$

$$S = \begin{cases} 1 & \text{if } \left(\hat{\xi}_\alpha \hat{\psi}_{r\alpha} + \hat{\xi}_\beta \hat{\psi}_{r\beta} \right) > 0 \\ -1 & \text{if } \left(\hat{\xi}_\alpha \hat{\psi}_{r\alpha} + \hat{\xi}_\beta \hat{\psi}_{r\beta} \right) \leq 0 \end{cases}, \tag{5.17}$$

where $e_{is\alpha} = \hat{i}_{s\alpha}^F - \hat{i}_{s\alpha}$, $e_{is\beta} = \hat{i}_{s\beta}^F - \hat{i}_{s\beta}$, and $\xi = \begin{bmatrix} \xi_\alpha & \xi_\beta \end{bmatrix}^T$ is a disturbance value represented as vector with components $\xi_\alpha = \omega_r \psi_{r\alpha}$, $\xi_\beta = \omega_r \psi_{r\beta}$.

Relations (5.9) to (5.17) are known as equations of Krzeminski observer [13]. The idea of Krzeminski observer is presented in more detail in Sections 5.3 to 5.5.

In the observer feedback terms, which appear in Equations 5.9, 5.10, 5.13, and 5.14, the estimated current, $\hat{\mathbf{i}}_s^F$, is used (estimated in the LC filter simulator block) instead of measured motor stator current, $\mathbf{i}_s$.

The waveforms showing the sensorless drive operation with structure presented in Figure 5.3 are presented in Figures 5.4 and 5.5.

Based on results [12], the motor voltages and current estimation errors have not exceeded 10%, which is noticeable in Figure 5.5.

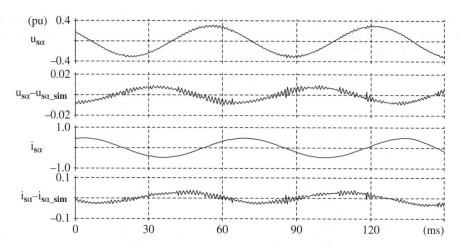

Figure 5.4 The waveforms for the drive with LC filter simulator: measured stator voltage and current and compared with estimated values

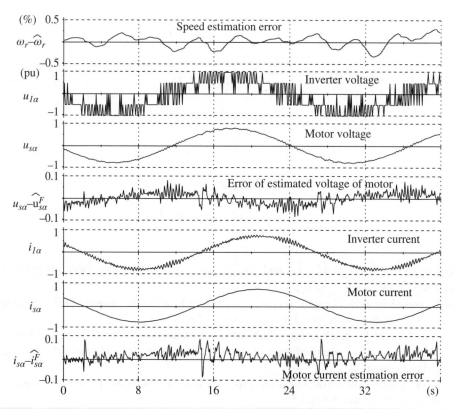

Figure 5.5 The operation of the drive with LC filter simulator (induction motor $P_n = 5.5$ kW, $n_n = 950$ rpm, filter $L_1 = 4$ mH, $C_1 = 10$ μF)

5.3 Speed Observer with Simplified Model of Disturbances

The observer presented in Section 5.2 has an internal filter simulator operating without any corrections feedback. It can lead to inaccurate estimation of variables in the case of changes of system parameters and inaccurate generation of inverter output voltage. Such adverse properties can be reduced by using a state observer for one dynamic system, which includes an IM motor and LC filter. For this, complex systems have to be formulated simultaneously. This solution was presented for a sensorless drive with an adaptive observer speed [20].

This concept can be extended also to other kinds of state observer, for example, for the Krzeminski observer [13, 21], where the concept similar to that for a synchronous motor with permanent magnets applied for induction motor [22, 23]. As shown [13, 21], the Krzeminski observer has high accuracy of variable estimation in a wide speed operating range.

In equations of the Krzeminski observer, the motor electromotive force (EMF) is treated as a disturbance. This disturbance is estimated by pure integrators [13]. For the drive with an LC filter, the observer equations have to be changed properly.

The disturbance, which is equivalent to motor EMF, is formulated as $\xi = \begin{bmatrix} \xi_\alpha & \xi_\beta \end{bmatrix}^T$ and is defined as follows:

$$\xi_\alpha = \omega_r \psi_{r\alpha}, \tag{5.18}$$

$$\xi_\beta = \omega_r \psi_{r\beta}. \tag{5.19}$$

For a short time interval, it can be assumed that the EMF of the motor does not change and therefore we can write:

$$\frac{d\xi_\alpha}{d\tau} = 0, \tag{5.20}$$

$$\frac{d\xi_\beta}{d\tau} = 0. \tag{5.21}$$

Equations of the state observer for both sine-wave filter and induction motor have the following form for $\alpha\beta$ [14, 15]:

$$\frac{d\hat{i}_{s\alpha}}{d\tau} = a_1 \hat{i}_{s\alpha} + a_2 \hat{\psi}_{r\alpha} + a_3 \hat{\xi}_\beta + a_4 \hat{u}_{s\alpha} + k_3 \left(k_1 \left(i_{1\alpha} - \hat{i}_{1\alpha} \right) - \hat{\omega}_r \xi_\alpha \right), \tag{5.22}$$

$$\frac{d\hat{i}_{s\beta}}{d\tau} = a_1 \hat{i}_{s\beta} + a_2 \hat{\psi}_{r\beta} - a_3 \hat{\xi}_\alpha + a_4 \hat{u}_{s\beta} + k_3 \left(k_1 \left(i_{1\beta} - \hat{i}_{1\beta} \right) - \hat{\omega}_r \xi_\beta \right), \tag{5.23}$$

$$\frac{d\hat{\psi}_{r\alpha}}{d\tau} = a_5 \hat{\psi}_{r\alpha} - \hat{\xi}_\beta + a_6 \hat{i}_{s\alpha} - k_2 \left(\hat{\omega}_r \hat{\psi}_{r\beta} - \hat{\xi}_\beta \right), \tag{5.24}$$

$$\frac{d\hat{\psi}_{r\beta}}{d\tau} = a_5 \hat{\psi}_{r\beta} + \hat{\xi}_\alpha + a_6 \hat{i}_{s\beta} + k_2 \left(\hat{\omega}_r \hat{\psi}_{r\alpha} - \hat{\xi}_\alpha \right), \tag{5.25}$$

$$\frac{d\hat{\xi}_\alpha}{d\tau} = k_1\left(i_{1\beta} - \hat{i}_{1\beta}\right), \tag{5.26}$$

$$\frac{d\hat{\xi}_\beta}{d\tau} = -k_1\left(i_{1\alpha} - \hat{i}_{1\alpha}\right), \tag{5.27}$$

$$\frac{d\hat{u}_{c\alpha}}{d\tau} = \frac{i_{1\alpha} - \hat{i}_{s\alpha}}{C_1}, \tag{5.28}$$

$$\frac{d\hat{u}_{c\beta}}{d\tau} = \frac{i_{1\beta} - \hat{i}_{s\beta}}{C_1}, \tag{5.29}$$

$$\frac{d\hat{i}_{1\alpha}}{d\tau} = \frac{u_{1\alpha} - \hat{u}_{c\alpha} - R_c\left(\hat{i}_{1\alpha} - \hat{i}_{s\alpha}\right)}{L_1} + k_4\left(i_{1\alpha} - \hat{i}_{1\alpha}\right) - k_5\left(i_{1\beta} - \hat{i}_{1\beta}\right), \tag{5.30}$$

$$\frac{d\hat{i}_{1\beta}}{d\tau} = \frac{u_{1\beta} - \hat{u}_{c\beta} - R_c\left(\hat{i}_{1\beta} - \hat{i}_{c\beta}\right)}{L_1} + k_4\left(i_{1\beta} - \hat{i}_{1\beta}\right) + k_5\left(i_{1\alpha} - \hat{i}_{1\alpha}\right), \tag{5.31}$$

$$\hat{u}_{s\alpha} = R_c\left(\hat{i}_{1\alpha} - \hat{i}_{s\alpha}\right) + \hat{u}_{c\alpha}, \tag{5.32}$$

$$\hat{u}_{s\beta} = R_c\left(\hat{i}_{1\beta} - \hat{i}_{s\beta}\right) + \hat{u}_{c\beta}, \tag{5.33}$$

$$\hat{i}_{cd} = \hat{i}_{1d} - \hat{i}_{sd}, \tag{5.34}$$

$$\hat{i}_{cq} = \hat{i}_{1q} - \hat{i}_{sq}, \tag{5.35}$$

$$\frac{dS_{bF}}{d\tau} = \frac{1}{T_{Sb}}\left(S_b - S_{bF}\right), \tag{5.36}$$

$$\frac{d\hat{\omega}_{rF}}{d\tau} = \frac{1}{T_{KT}}\left(\hat{\omega}_r - \hat{\omega}_{rF}\right), \tag{5.37}$$

$$S = \begin{cases} 1 & \text{if } \left(\hat{\xi}_\alpha \hat{\psi}_{r\alpha} + \hat{\xi}_\beta \hat{\psi}_{r\beta}\right) > 0 \\ -1 & \text{if } \left(\hat{\xi}_\alpha \hat{\psi}_{r\alpha} + \hat{\xi}_\beta \hat{\psi}_{r\beta}\right) \le 0 \end{cases}, \tag{5.38}$$

$$S_b = S\left(\hat{\xi}_\alpha \hat{\psi}_{r\beta} - \hat{\xi}_\beta \hat{\psi}_{r\alpha}\right), \tag{5.39}$$

$$\hat{\omega}_r = S\sqrt{\frac{\hat{\xi}_\alpha^2 + \hat{\xi}_\beta^2}{\hat{\psi}_{r\alpha}^2 + \hat{\psi}_{r\beta}^2}} - S_b - S_{bF}, \tag{5.40}$$

where:

S is the motor speed sign
S_b and S_{bF} are the the observer internal variables used to stabilize estimation
$\hat{\xi}_\alpha$, $\hat{\xi}_\beta$ are the variables equivalent to motor EMF and treated as disturbances
T_{KT} and T_{Sb} are the time constant for low-pass filters applied for variables: S_b and $\hat{\omega}_r$
k_1, k_2, k_3, k_4, k_5 are the observers gains.

Observer Equations 5.22 to 5.40 result from equations of the filter model, presented in Chapter 4. The filter model was introduced to the Krzeminski observer and feedback parts were changed to errors between the measured and the estimated output current of the inverter. So the operation of the observer requires knowledge of the current and voltage output of the inverter only.

Sample waveforms for sensorless control multiscalar control systems with the observer which use the simplified model of disturbance are shown in Figure 5.6.

As presented in the Figure 5.6 waveforms, the speed estimation error does not exceed 3%. Compared with the filter simulator (see Section 5.2 and Figure 5.5) less estimation error of the motor current is noticeable.

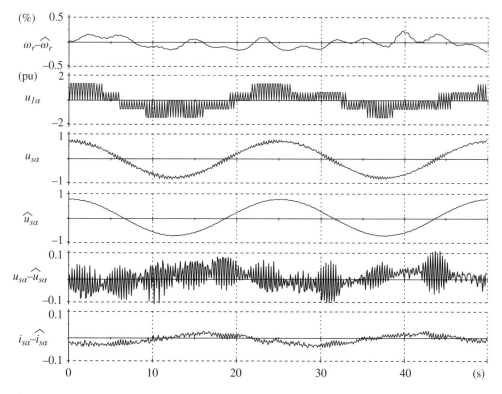

Figure 5.6 The operation of the sensorless drive with state observer with simplified disturbance model (Induction motor $P_n = 1.5$ kW, $U_n - 300$ V, filter $L_1 = 4$ mH, $C_1 = 3$ µF)

5.4 Speed Observer with Extended Model of Disturbances

More precise operation of the speed observer shown in Section 5.3 is possible by extending the disturbance model Equations 5.26 and 5.27 into the following [24]:

$$\frac{d\xi_\alpha}{d\tau} = -\hat{\omega}_{\psi r}\xi_\beta, \tag{5.41}$$

$$\frac{d\xi_\beta}{d\tau} = \hat{\omega}_{\psi r}\xi_\alpha. \tag{5.42}$$

In the observer model with extended disturbance model, it is assumed that for a short time the motor speed is not changed, and thus the derivative of speed, which should appear in Equations 5.41 and 5.42, is zero.

For a drive with sinusoidal filter the observer form [24] was modified, as shown in Section 5.3 for the observer with simplified model of disturbances. The equations of state observer were extended with filter model dependencies and feedback terms were turned to inverter output current. The final set of speed observer equations (a variant for LC drive with LC filter) has the form:

$$\frac{d\hat{i}_{s\alpha}}{d\tau} = a_1\hat{i}_{s\alpha} + a_2\hat{\psi}_{r\alpha} + a_3\xi_\beta + a_4\hat{u}_{s\alpha} + k_3\left(i_{1\alpha} - \hat{i}_{1\alpha}\right), \tag{5.43}$$

$$\frac{d\hat{i}_{s\beta}}{d\tau} = a_1\hat{i}_{s\beta} + a_2\hat{\psi}_{r\beta} - a_3\xi_\alpha + a_4\hat{u}_{s\beta} + k_3\left(i_{1\beta} - \hat{i}_{1\beta}\right), \tag{5.44}$$

$$\frac{d\hat{\psi}_{r\alpha}}{d\tau} = a_5\hat{\psi}_{r\alpha} - \xi_\beta + a_6\hat{i}_{s\alpha} - k_2 S_b\hat{\psi}_{r\alpha} + Sk_2 k_3\hat{\psi}_{r\beta}\left(S_b - S_{bF}\right)$$
$$+ Sk_5\left(\left(S_x - S_{xF}\right)\hat{\psi}_{r\alpha} - \left(S_b - S_{bF}\right)\hat{\psi}_{r\beta}\right), \tag{5.45}$$

$$\frac{d\hat{\psi}_{r\beta}}{d\tau} = a_5\hat{\psi}_{r\beta} + \xi_\alpha + a_6\hat{i}_{s\beta} - k_2 S_b\hat{\psi}_{r\beta} - Sk_2 k_3\hat{\psi}_{r\alpha}\left(S_b - S_{bF}\right)$$
$$+ Sk_5\left(-\left(S_x - S_{xF}\right)\hat{\psi}_{r\beta} - \left(S_b - S_{bF}\right)\hat{\psi}_{r\alpha}\right), \tag{5.46}$$

$$\frac{d\hat{\xi}_\alpha}{d\tau} = -\hat{\omega}_{\psi r}\hat{\xi}_\beta - k_1\left(i_{1\beta} - \hat{i}_{1\beta}\right), \tag{5.47}$$

$$\frac{d\hat{\xi}_\beta}{d\tau} = \hat{\omega}_{\psi r}\hat{\xi}_\alpha + k_1\left(i_{1\alpha} - \hat{i}_{1\alpha}\right), \tag{5.48}$$

$$\frac{d\hat{u}_{c\alpha}}{d\tau} = \frac{i_{1\alpha} - \hat{i}_{s\alpha}}{C_1}, \tag{5.49}$$

$$\frac{d\hat{u}_{c\beta}}{d\tau} = \frac{i_{1\beta} - \hat{i}_{s\beta}}{C_1}, \tag{5.50}$$

$$\frac{d\hat{i}_{1\alpha}}{d\tau} = \frac{u_{1\alpha}^* - \hat{u}_{s\alpha}}{L_1} + k_A\left(i_{1\alpha} - \hat{i}_{1\alpha}\right) - k_B\left(i_{1\beta} - \hat{i}_{1\beta}\right), \tag{5.51}$$

$$\frac{d\hat{i}_{1\beta}}{d\tau} = \frac{u^*_{1\beta} - \hat{u}_{s\beta}}{L_1} + k_A \left(i_{1\beta} - \hat{i}_{1\beta} \right) + k_B \left(i_{1\alpha} - \hat{i}_{1\alpha} \right), \tag{5.52}$$

$$\hat{u}_{s\alpha} = \hat{u}_{c\alpha} + \left(i_{1\alpha} - \hat{i}_{s\alpha} \right) R_c, \tag{5.53}$$

$$\hat{u}_{s\beta} = \hat{u}_{c\beta} + \left(i_{1\beta} - \hat{i}_{s\beta} \right) R_c, \tag{5.54}$$

$$\frac{dS_{bF}}{d\tau} = \frac{1}{T_{Sb}} (S_b - S_{bF}), \tag{5.55}$$

$$\frac{d\hat{\omega}_{rF}}{d\tau} = \frac{1}{T_{KT}} (\hat{\omega}_r - \hat{\omega}_{rF}), \tag{5.56}$$

$$\frac{dS_{xF}}{d\tau} = \frac{1}{T_{Sx}} (S_x - S_{xF}), \tag{5.57}$$

$$S = \begin{cases} 1 & \text{if } \hat{\omega}_{\psi r} > 0 \\ -1 & \text{if } \hat{\omega}_{\psi r} \leq 0 \end{cases}, \tag{5.58}$$

$$S_x = \hat{\xi}_\alpha \hat{\psi}_{r\alpha} + \hat{\xi}_\beta \hat{\psi}_{r\beta}, \tag{5.59}$$

$$S_b = \hat{\xi}_\alpha \hat{\psi}_{r\beta} - \hat{\xi}_\beta \hat{\psi}_{r\alpha}, \tag{5.60}$$

$$\hat{\omega}_{\psi r} = \hat{\omega}_{rF} + R_r \frac{L_m}{L_r} \left(\frac{\hat{\psi}_{r\alpha} \hat{i}_{s\beta} + \hat{\psi}_{r\beta} \hat{i}_{s\alpha}}{\hat{\psi}_{r\alpha}^2 + \hat{\psi}_{r\beta}^2} \right), \tag{5.61}$$

$$\hat{\omega}_r = \frac{\hat{\zeta}_\alpha \hat{\psi}_{r\alpha} + \hat{\zeta}_\beta \hat{\psi}_{r\beta}}{\hat{\psi}_{r\alpha}^2 + \hat{\psi}_{r\beta}^2}, \tag{5.62}$$

where k_A and k_B are observer gains.

The equations of the disturbances model in the form of dependencies Equations 5.47 and 5.48 have integrating elements and terms equivalent to rotational EMF.

5.5 Speed Observer with Complete Model of Disturbances

An improvement of the properties of the speed observers presented successively in Sections 5.3 and 5.4 could be achieved by further development of the disturbance model. This was obtained by introducing, instead of Equations 5.47 and 5.48, equations of the complete disturbance model [17]:

$$\frac{d\xi_\alpha}{d\tau} = a_6 \omega_r i_{s\alpha} + a_5 \xi_\alpha - \omega_r \xi_\beta + \psi_{r\alpha} \frac{d\omega_r}{d\tau}, \tag{5.63}$$

$$\frac{d\xi_\beta}{d\tau} = a_6 \omega_r i_{s\beta} + a_5 \xi_\beta + \omega_r \xi_\alpha + \psi_{r\beta} \frac{d\omega_r}{d\tau}. \tag{5.64}$$

The observer with complete disturbance model can be modified similarly, as shown in the previous subsections (i.e., by adding the equations of the filter model and modification of observer feedback terms). In this case, the observer dependencies take the following form:

$$\frac{d\hat{i}_{s\alpha}}{d\tau} = a_1\hat{i}_{s\alpha} + a_2\hat{\psi}_{r\alpha} + a_3\hat{\xi}_\beta + a_4\hat{u}_{s\alpha} + k_3\left(i_{1\alpha} - \hat{i}_{1\alpha}\right), \tag{5.65}$$

$$\frac{d\hat{i}_{s\beta}}{d\tau} = a_1\hat{i}_{s\beta} + a_2\hat{\psi}_{r\beta} - a_3\hat{\xi}_\alpha + a_4\hat{u}_{s\beta} + k_3\left(i_{1\beta} - \hat{i}_{1\beta}\right), \tag{5.66}$$

$$\frac{d\hat{\psi}_{r\alpha}}{d\tau} = a_5\hat{\psi}_{r\alpha} + a_6\hat{i}_{s\alpha} - \hat{\xi}_\beta - k_2 S_b \psi_{r\alpha} + Sk_2 k_3 \hat{\psi}_{r\beta}\left(S_b - S_{bF}\right)$$
$$+ Sk_5\left(\left(S_x - S_{xF}\right)\hat{\psi}_{r\alpha} - \left(S_b - S_{bF}\right)\hat{\psi}_{r\beta}\right), \tag{5.67}$$

$$\frac{d\hat{\psi}_{r\beta}}{d\tau} = a_5\hat{\psi}_{r\beta} + a_6\hat{i}_{s\beta} + \hat{\xi}_\alpha - k_2 S_b \psi_{r\beta} - Sk_2 k_3 \hat{\psi}_{r\alpha}\left(S_b - S_{bF}\right)$$
$$+ Sk_5\left(-\left(S_x - S_{xF}\right)\hat{\psi}_{r\beta} - \left(S_b - S_{bF}\right)\hat{\psi}_{r\alpha}\right), \tag{5.68}$$

$$\frac{d\hat{\xi}_\alpha}{d\tau} = a_6\hat{\omega}_r\hat{i}_{s\alpha} + a_5\hat{\xi}_\alpha - \hat{\omega}_r\hat{\xi}_\beta - k_1\left(i_{1\beta} - \hat{i}_{1\beta}\right), \tag{5.69}$$

$$\frac{d\hat{\xi}_\beta}{d\tau} = a_6\hat{\omega}_r\hat{i}_{s\beta} + a_5\hat{\xi}_\beta + \hat{\omega}_r\hat{\xi}_\beta + k_1\left(i_{1\alpha} - \hat{i}_{1\alpha}\right), \tag{5.70}$$

$$\frac{d\hat{u}_{c\alpha}}{d\tau} = \frac{i_{1\alpha} - \hat{i}_{s\alpha}}{C_1}, \tag{5.71}$$

$$\frac{d\hat{u}_{c\beta}}{d\tau} = \frac{i_{1\beta} - \hat{i}_{s\beta}}{C_1}, \tag{5.72}$$

$$\frac{d\hat{i}_{1\alpha}}{d\tau} = \frac{u_{1\alpha}^* - \hat{u}_{s\alpha}}{L_1} + k_A\left(i_{1\alpha} - \hat{i}_{1\alpha}\right) - k_B\left(i_{1\beta} - \hat{i}_{1\beta}\right), \tag{5.73}$$

$$\frac{d\hat{i}_{1\beta}}{d\tau} = \frac{u_{1\beta}^* - \hat{u}_{s\beta}}{L_1} + k_A\left(i_{1\beta} - \hat{i}_{1\beta}\right) + k_B\left(i_{1\alpha} - \hat{i}_{1\alpha}\right), \tag{5.74}$$

$$\hat{u}_{s\alpha} = \hat{u}_{c\alpha} + \left(i_{1\alpha} - \hat{i}_{s\alpha}\right) R_c, \tag{5.75}$$

$$\hat{u}_{s\beta} = \hat{u}_{c\beta} + \left(i_{1\beta} - \hat{i}_{s\beta}\right) R_c, \tag{5.76}$$

$$\frac{dS_{bF}}{dt} = \frac{1}{T_{Sb}}\left(S_b - S_{bF}\right), \tag{5.77}$$

$$\hat{\omega}_r = \frac{\hat{\xi}_\alpha \hat{\psi}_{r\alpha} + \hat{\xi}_\beta \hat{\psi}_{r\beta}}{\hat{\psi}_{r\alpha}^2 + \hat{\psi}_{r\beta}^2}. \tag{5.78}$$

In the equations of disturbance state variables Equations 5.69 and 5.70, the terms appearing in Equations 5.63 and 5.64 that are associated with derivative of the motor speed are eliminated. In this way, the observer calculations are protected from the possibility of an overlapping calculation error, resulting from the numerical counting of derivative of the estimated motor speed.

5.6 Speed Observer Operating for Rotating Coordinates

The primary method of comparison of state observers is in most cases the investigations of estimation error in steady state and determination of the minimum value of the estimated speed [2]. However, in some applications, it is important to analyze observer amplitude-frequency response. This characteristic is particularly important in the diagnostic applications of the observer. There are known solutions of the systems in which the speed observer is used as a detection system, for example, detection of the drive mechanical part faults [25–28]. Generally, in electric drives, mechanical damage faults can be detected based on analysis of mechanical vibrations measured or reconstructed in the observer [29]. Mechanical defects are characterized by the growth of harmonic amplitudes of motor torque and speed for the particular frequencies related to the motor mechanical speed, intermeshing gear wheels speed, or frequencies related to faulty motor bearings. The defects are characterized by different load torque harmonic frequencies in the range of hundreds of hertz to a few hertz. Hence, it is important that the observer used for diagnostic purposes should have a high-frequency response.

Such an approach to evaluation of the quality of the estimation variables is presented where the amplitude-frequency response is compared for three observers [30]. The widest frequency band was obtained for the speed observer with disturbance model [24], which was shown in Section 5.3 in the application of the sensorless drive with inverter output filter.

For the observer with disturbance estimation, the state variables are calculated in stationary reference frame $\alpha\beta$. Most of these estimated variables in $\alpha\beta$ coordinates are of sinusoidal shape (e.g., rotor and stator fluxes and currents). However it is possible to transform the observer equations into the other coordinates. It has been shown [31] that the acceptance of the rotating reference system, for example, dq fixed with rotor flux, could be more attractive than $\alpha\beta$ coordinates. With the rotating dq coordinates, the observer state variables do not change instantaneously. As a result, a determinant of observability matrix could be only occasionally and only temporarily set to zero. Thus, apart from such rare cases, the system observability is preserved [31]. The advantage of observer operation in the rotating coordinates was also noted elsewhere [32]. This publication indicates that for calculation of a continuous system, both the observers are mathematically identical independent of a reference system. A significant difference exists, however, if the observer is implemented in a discrete system. For discrete systems, the observer frequency response is heavily dependent on the adopted reference system. The variables that are estimated by the discrete observer in $\alpha\beta$ have larger errors than variables estimated by the observer calculated as a continuous system. Whereas, the same variables estimation errors are smaller for the dq observer and they are simultaneously less dependent on the calculation step.

The observer in the rotating dq coordinates is derived based on speed observer with full model of disturbances presented in Section 5.5 in the form of differential equations:

$$\frac{d\hat{i}_{sd}}{d\tau} = a_1\hat{i}_{sd} + a_2\hat{\psi}_{rd} + \omega_u\hat{i}_{sq} + a_3\hat{\xi}_q + a_4 u_{sd}^* + k_3\left(i_{sd} - \hat{i}_{sd}\right), \tag{5.79}$$

$$\frac{d\hat{i}_{sq}}{d\tau} = a_1\hat{i}_{sq} + a_2\hat{\psi}_{rq} - \omega_u\hat{i}_{sd} - a_3\hat{\xi}_d + a_4 u_{sq}^* + k_3\left(i_{sq} - \hat{i}_{sq}\right), \tag{5.80}$$

$$\frac{d\hat{\psi}_{rq}}{d\tau} = a_5\hat{\psi}_{rq} + a_6\hat{i}_{sq} - \left(\omega_{\psi r} - \hat{\omega}_r\right)\hat{\psi}_{rd} - \hat{\xi}_d - k_2 S_b\hat{\psi}_{rq}$$
$$+ S\left(k_2 k_3\hat{\psi}_{rd}\left(S_b - S_{bF}\right) + k_5\left(-\left(S_x - S_{xF}\right)\hat{\psi}_{rq} - \left(S_b - S_{bF}\right)\hat{\psi}_{rd}\right)\right) \tag{5.81}$$

$$\frac{d\hat{\psi}_{rq}}{d\tau} = a_5\hat{\psi}_{rq} + a_6\hat{i}_{sq} - \left(\omega_{\psi r} - \hat{\omega}_r\right)\hat{\psi}_{rd} - \hat{\xi}_d - k_2 S_b\hat{\psi}_{rq}$$
$$+ S\left(k_2 k_3\hat{\psi}_{rd}\left(S_b - S_{bF}\right) + k_5\left(-\left(S_x - S_{xF}\right)\hat{\psi}_{rq} - \left(S_b - S_{bF}\right)\hat{\psi}_{rd}\right)\right) \tag{5.82}$$

$$\frac{d\hat{\xi}_d}{d\tau} = -\hat{\omega}_{\psi r}\hat{\xi}_d + \omega_{\psi r}\hat{\omega}_r\hat{\psi}_{rd} + k_1\left(i_{sq} - \hat{i}_{sq}\right), \tag{5.83}$$

$$\frac{d\hat{\xi}_q}{d\tau} = \hat{\omega}_{\psi r}\hat{\xi}_d - \omega_{\psi r}\hat{\omega}_r\hat{\psi}_{rq} + k_1\left(i_{sd} - \hat{i}_{sd}\right), \tag{5.84}$$

$$\frac{dS_{bF}}{d\tau} = \frac{1}{T_{Sb}}\left(S_b - S_{bF}\right), \tag{5.85}$$

$$\frac{d\hat{\omega}_{rF}}{d\tau} = \frac{1}{T_{KT}}\left(\hat{\omega}_r - \hat{\omega}_{rF}\right), \tag{5.86}$$

$$\frac{dS_{xF}}{d\tau} = \frac{1}{T_{Sx}}\left(S_x - S_{xF}\right), \tag{5.87}$$

$$S = \begin{cases} 1 & \text{if } \hat{\omega}_{\psi r} > 0 \\ -1 & \text{if } \hat{\omega}_{\psi r} \leq 0 \end{cases}, \tag{5.88}$$

$$S_x = \hat{\xi}_d\hat{\psi}_{rd} + \hat{\xi}_q\hat{\psi}_{rq}, \tag{5.89}$$

$$S_b = \hat{\xi}_d\hat{\psi}_{rq} - \hat{\xi}_q\hat{\psi}_{rd}, \tag{5.90}$$

$$\hat{\omega}_{\psi r} = \hat{\omega}_{rF} + R_r\frac{L_m}{L_r}\left(\frac{\hat{\psi}_{rd}\hat{i}_{sq} + \hat{\psi}_{rq}\hat{i}_{sd}}{\hat{\psi}_{rd}^2 + \hat{\psi}_{rq}^2}\right), \tag{5.91}$$

$$\hat{\omega}_r = \frac{\hat{\xi}_d\hat{\psi}_{rd} + \hat{\xi}_q\hat{\psi}_{rq}}{\hat{\psi}_{rd}^2 + \hat{\psi}_{rq}^2}, \tag{5.92}$$

where:

k_1, k_2, k_3, k_4, k_5 are observer gains
S_x and S_{xF} are the observer internal variables used to improve stability of the system
T_{Sx} is the time constant for inertia of S_x variable.

The observer amplitude-frequency characteristics were determined experimentally using developed simulation software. In the simulation program, induction motor with voltage inverter was modeled. The drive was operating in open loop control because of the U/f = const. principle. In the simulation program, the discrete nature of the inverter operation and state observer were included. For the motor speed signal, the disturbance sinusoidal signal was added with amplitude much lower (about 10%) than actual motor speed. This solution was presentedfor the speed observer investigations [30, 33]. A similar approach [34] was used to determine the frequency characteristics of the whole control system. For the observer Equations 5.79 to 5.92, operating in open loop system, the actual value of the motor speed was calculated. The estimated speed has DC and AC components, where the AC component includes input function of disturbance signal. The structure of the test system is shown in Figure 5.7.

In this system (Figure 5.7), the frequency characteristics of the observer were derived for two cases: the observer operating in a stationary αβ and in a dq rotating system. The resulting frequency-amplitude characteristics are shown in Figure 5.8.

In Figure 5.8, it can be seen that the frequency response of the observer in the rotating dq system is higher than for an observer in the stationary system αβ. For the same IM motor and the same operating conditions, the frequency limit of the observer in the system dq is about 600 Hz, and for the observer in the system αβ, it is about 80 Hz. Both versions of the observer speed

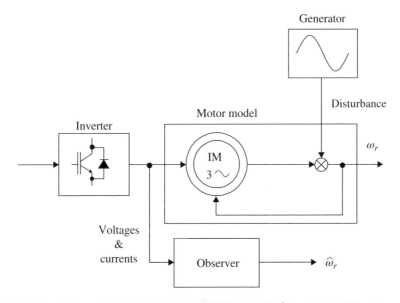

Figure 5.7 Structure of the system to determine frequency characteristic of the observer

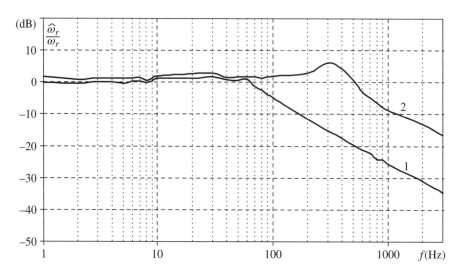

Figure 5.8 Frequency-amplitude characteristics of the speed observer in cases of: (1) stationary frame of references αβ and (2) rotating coordinates dq (motor P_n = 1.5 kW, n_n = 1420 rpm, average value of the motor speed ω_r (AV) = 1 pu)

are operating as low-pass filter. The frequency bandwidth is affected by both the observer gains as well the motor parameters. This can be observed, for example, where the results for observers αβ with a simplified and extended disturbance model without a filter was presented [30, 33].

For an observer in dq system, an increase in the ratio of signals $\hat{\omega}_r / \omega_r$ at a frequency of about 300 Hz is noticeable. This phenomenon was also observed for the observer in the stationary coordinates [30].

For the drive with inverter output filter, the rotating coordinates were fixed with stator voltage vector where the speed of the coordinated rotation ω_a is equal to ω_u. For that coordinate system the axes, are noted as *KL*. Additionally, the model of inverter LC filter is taken into account as shown in Sections 5.2 to 5.4.

For KL coordinates, the observer dependencies Equations 5.79 to 5.92 have the following form:

$$\frac{d\hat{i}_{sK}}{d\tau} = a_1 \hat{i}_{sK} + a_2 \hat{\psi}_{rK} + \omega_u i_{sL} + a_3 \hat{\xi}_L + a_4 u_{sK}^* + k_3 \left(i_{1K} - \hat{i}_{1K} \right), \tag{5.93}$$

$$\frac{d\hat{i}_{sL}}{d\tau} = a_1 \hat{i}_{sL} + a_2 \hat{\psi}_{rL} - \omega_u i_{sK} - a_3 \hat{\xi}_K + a_4 u_{sL}^* + k_3 \left(i_{1L} - \hat{i}_{1L} \right), \tag{5.94}$$

$$\frac{d\hat{\psi}_{rK}}{d\tau} = a_5 \hat{\psi}_{rK} + a_6 \hat{i}_{sK} + (\omega_u - \hat{\omega}_r) \hat{\psi}_{rL} - \hat{\xi}_L - k_2 S_b \hat{\psi}_{rK}$$
$$+ S \left(k_2 k_3 \hat{\psi}_{rL} (S_b - S_{bF}) + k_5 \left((S_x - S_{xF}) \hat{\psi}_{rK} - (S_b - S_{bF}) \hat{\psi}_{rL} \right) \right), \tag{5.95}$$

$$\frac{d\hat{\psi}_{rL}}{d\tau} = a_5 \hat{\psi}_{rL} + a_6 \hat{i}_{sL} + (\omega_u - \hat{\omega}_r) \hat{\psi}_{rK} - \hat{\xi}_K - k_2 S_b \hat{\psi}_{rL}$$
$$+ S \left(k_2 k_3 \hat{\psi}_{rK} (S_b - S_{bF}) + k_5 \left(-(S_x - S_{xF}) \hat{\psi}_{rL} - (S_b - S_{bF}) \hat{\psi}_{rK} \right) \right), \tag{5.96}$$

Estimation of the State Variables in the Drive with LC Filter

$$\frac{d\hat{\xi}_K}{d\tau} = -\hat{\omega}_{\psi r}\hat{\xi}_L + \omega_u\hat{\omega}_r\hat{\psi}_{rK} + k_1\left(i_{sL} - \hat{i}_{sL}\right), \tag{5.97}$$

$$\frac{d\hat{\xi}_L}{d\tau} = \hat{\omega}_{\psi r}\hat{\xi}_K - \omega_u\hat{\omega}_r\hat{\psi}_{rL} + k_1\left(i_{sK} - \hat{i}_{sK}\right), \tag{5.98}$$

$$\frac{d\hat{u}_{cK}}{d\tau} = \frac{i_{1K} - \hat{i}_{sK}}{C_1}, \tag{5.99}$$

$$\frac{d\hat{u}_{cL}}{d\tau} = \frac{i_{1L} - \hat{i}_{sL}}{C_1}, \tag{5.100}$$

$$\frac{d\hat{i}_{1K}}{d\tau} = \frac{u^*_{1K} - \hat{u}_{sK}}{L_1} + \omega_u i_{1L} + k_A\left(i_{1k} - \hat{i}_{1K}\right) - k_B\left(i_{1L} - \hat{i}_{1L}\right), \tag{5.101}$$

$$\frac{d\hat{i}_{1L}}{d\tau} = \frac{u^*_{1L} - \hat{u}_{sL}}{L_1} - \omega_u i_{1K} + k_A\left(i_{1L} - \hat{i}_{1L}\right) + k_B\left(i_{1K} - \hat{i}_{1K}\right), \tag{5.102}$$

$$\hat{u}_{sK} = \hat{u}_{cK} + \left(i_{1K} - \hat{i}_{sK}\right)R_c, \tag{5.103}$$

$$\hat{u}_{sL} = \hat{u}_{cL} + \left(i_{1L} - \hat{i}_{sL}\right)R_c, \tag{5.104}$$

$$\frac{dS_{bF}}{d\tau} = \frac{1}{T_{Sb}}\left(S_b - S_{bF}\right), \tag{5.105}$$

$$\frac{d\hat{\omega}_{rF}}{d\tau} = \frac{1}{T_{KT}}\left(\hat{\omega}_r - \hat{\omega}_{rF}\right), \tag{5.106}$$

$$\frac{dS_{xF}}{d\tau} = \frac{1}{T_{Sx}}\left(S_x - S_{xF}\right), \tag{5.107}$$

$$S = \begin{cases} 1 & \text{if } \hat{\omega}_{\psi r} > 0 \\ -1 & \text{if } \hat{\omega}_{\psi r} \leq 0 \end{cases}, \tag{5.108}$$

$$S_x = \hat{\xi}_K\hat{\psi}_{rK} + \hat{\xi}_L\hat{\psi}_{rL}, \tag{5.109}$$

$$S_b = \hat{\xi}_K\hat{\psi}_{rL} - \hat{\xi}_L\hat{\psi}_{rK}, \tag{5.110}$$

$$\hat{\omega}_{\psi r} = \hat{\omega}_{rF} + R_r\frac{L_m}{L_r}\left(\frac{\hat{\psi}_{rK}\hat{i}_{sL} + \hat{\psi}_{rL}\hat{i}_{sK}}{\hat{\psi}^2_{rK} + \hat{\psi}^2_{rL}}\right), \tag{5.111}$$

$$\hat{\omega}_r = \frac{\hat{\xi}_K\hat{\psi}_{rK} + \hat{\xi}_L\hat{\psi}_{rL}}{\hat{\psi}^2_{rK} + \hat{\psi}^2_{rL}}. \tag{5.112}$$

5.7 Speed Observer Based on Voltage Model of Induction Motor

One estimation method of the motor state variables is based on the stator circuit equations. The advantage of this type of implementation is that it is simply realized, but nevertheless this method is connected with problems, such as drifts [2, 35, 36]. The application and modification of state estimators based on to the stator model were previously presented for drive systems without filter.

A solution for a filterless drive system [37–40] depends on the voltage model equations of the motor for variables ψ_s and ψ_r presented in the following form:

$$\tau_s' \frac{d\psi_{s\alpha}}{d\tau} + \psi_{s\alpha} = k_r \psi_{r\alpha} + u_{s\alpha}, \tag{5.113}$$

$$\tau_s' \frac{d\psi_{s\beta}}{d\tau} + \psi_{s\beta} = k_r \psi_{r\beta} + u_{s\beta}, \tag{5.114}$$

where the time constant of the stator circuit is described as:

$$\tau_s' = \sigma \frac{L_s}{R_s}, \tag{5.115}$$

and the rotor coupling factor is noted as:

$$k_r = \frac{L_m}{L_r}. \tag{5.116}$$

To eliminate the drift issue, correction terms were introduced, which worked on the error of measured and generated currents of the motor. Furthermore, suitably limiters were introduced for the estimated stator flux. The filter equations taken into account enable the usage of an observer in a drive system with sinusoidal filter [18]. The following equations are based on the voltage stator model for a system with filter:

$$\frac{d\hat{\psi}_{s\alpha}}{d\tau} = \frac{-\hat{\psi}_{s\alpha} + k_r \hat{\psi}_{r\alpha} + \hat{u}_{s\alpha}}{\tau_s'} - k_a \left(i_{1\alpha} - \hat{i}_{1\alpha}\right) + k_b \left(i_{1\beta} - \hat{i}_{1\beta}\right), \tag{5.117}$$

$$\frac{d\hat{\psi}_{s\beta}}{d\tau} = \frac{-\hat{\psi}_{s\beta} + k_r \hat{\psi}_{r\beta} + \hat{u}_{s\beta}}{\tau_s'} - k_a \left(i_{1\beta} - \hat{i}_{1\beta}\right) - k_b \left(i_{1\alpha} - \hat{i}_{1\alpha}\right), \tag{5.118}$$

$$\hat{\psi}_{r\alpha} = \frac{\hat{\psi}_{s\alpha} - \sigma L_s \hat{i}_{s\alpha}}{k_r}, \tag{5.119}$$

$$\hat{\psi}_{r\beta} = \frac{\hat{\psi}_{s\beta} - \sigma L_s \hat{i}_{s\beta}}{k_r}, \tag{5.120}$$

$$\frac{d\hat{u}_{s\alpha}}{d\tau} = \frac{i_{1\alpha} - \hat{i}_{s\alpha}}{C_1}, \tag{5.121}$$

$$\frac{d\hat{u}_{s\beta}}{d\tau} = \frac{i_{1\beta} - \hat{i}_{s\beta}}{C_1}, \tag{5.122}$$

$$\frac{d\hat{i}_{1\alpha}}{d\tau} = \frac{u_{1\alpha}^* - \hat{u}_{s\alpha}}{L_1} + k_A \left(i_{1\alpha} - \hat{i}_{1\alpha} \right) - k_B \left(i_{1\beta} - \hat{i}_{1\beta} \right), \tag{5.123}$$

$$\frac{d\hat{i}_{1\beta}}{d\tau} = \frac{u_{1\beta}^* - \hat{u}_{s\beta}}{L_1} + k_A \left(i_{1\beta} - \hat{i}_{1\beta} \right) + k_B \left(i_{1\alpha} - \hat{i}_{1\alpha} \right), \tag{5.124}$$

where $\psi_{s\alpha}$ and $\psi_{s\beta}$ are the vector components of the stator linked flux.
The rotor linked flux vector amplitude and position are described as follows:

$$|\hat{\psi}_r| = \sqrt{\hat{\psi}_{r\alpha}^2 + \hat{\psi}_{r\beta}^2}, \tag{5.125}$$

$$\hat{\rho}_{\psi r} = \text{arc tg} \frac{\hat{\psi}_{r\beta}}{\hat{\psi}_{r\alpha}}. \tag{5.126}$$

Estimated vector components of the stator current are:

$$\hat{i}_{s\alpha} = \frac{\hat{\psi}_{s\alpha} - k_r \hat{\psi}_{r\alpha}}{\sigma L_s}, \tag{5.127}$$

$$\hat{i}_{s\beta} = \frac{\hat{\psi}_{s\beta} - k_r \hat{\psi}_{r\beta}}{\sigma L_s}. \tag{5.128}$$

The rotor speed can be written as a difference of pulsing rotor flux $\omega_{\psi r}$ and slip ω_2:

$$\hat{\omega}_r = \hat{\omega}_{\psi r} - \hat{\omega}_2, \tag{5.129}$$

where rotor flux pulsation:

$$\hat{\omega}_{\psi r} = \frac{d\hat{\rho}_{\psi r}}{d\tau}, \tag{5.130}$$

and slip pulsation:

$$\hat{\omega}_2 = \frac{L_m}{T_r} \frac{\hat{\psi}_{r\alpha} \hat{i}_{s\beta} - \hat{\psi}_{r\beta} \hat{i}_{s\alpha}}{|\hat{\psi}_r|^2}, \tag{5.131}$$

where T_r is the time constant magnitude of the rotor circuit:

$$T_r = \frac{L_r}{R_r}. \qquad (5.132)$$

The dependencies of the observer neglect the damping resistance of the filter. The structure of the speed observer is based on the stator voltage model for a drive system with sinusoidal filter, which is illustrated in Figure 5.9.

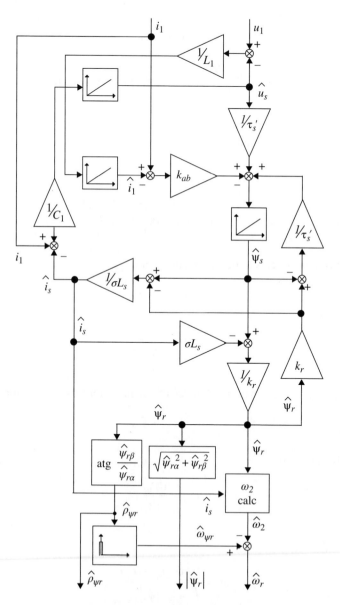

Figure 5.9 Structure of speed and flux observer based on the stator voltage model

The determination of the rotor flux pulsation, in the presented observer (Figure 5.9), requires an application of numerical differentiation in the estimated angular position. However, this can cause significant calculation errors, explained by the limited accuracy of the estimated rotor flux. Even a small error in the flux estimation can cumulate in the numerical differentiation into an improper calculation of flux pulsation. The elimination of this issue in the observer (Figure 5.9) requires a determination of the rotor flux vector position, based on filtered components of the estimated rotor flux. But nevertheless, a filtration will result in undesired delays.

It is possible to eliminate the differentiation block of the flux angle, for instance though the implementation of an adaptive mechanism to determine the rotor flux pulsation. Such a solution for other observers is known from the literature [10, 11, 39].

An application of the adaptive mechanism for the observer of Figure 5.9 is possible by an introduction of an additional algorithm for the rotor flux coordinates in the rotating dq system. In such a case, this furthermore enables the addition of the estimation procedure of a proportional-integral regulator, where the input signal is the zero value of the rotor flux component q:

$$\hat{\psi}^*_{rq} = 0. \tag{5.133}$$

The proportional-integral controller, which compares the desired value with the estimated stator flux q component, simultaneously estimates the pulsation:

$$\hat{\omega}_{\psi r} = K_{pq}\left(\hat{\psi}^*_{rq} - \hat{\psi}_{rq}\right) + \frac{1}{T_{iq}}\int\left(\hat{\psi}^*_{rq} - \hat{\psi}_{rq}\right)d\tau, \tag{5.134}$$

where K_{pq} and T_{iq} are amplification and reset time of the proponet-integral controller.

Considering Equation 5.133, the Equation 5.134 changes the form into:

$$\hat{\omega}_{\psi r} = -K_{pq}\hat{\psi}_{rq} - K_{iq}\int\hat{\psi}_{rq}d\tau. \tag{5.135}$$

Based on the pulsation $\hat{\omega}_{\psi r}$ it is possible to determine the position angle, $\hat{\rho}_{\psi r}$:

$$\hat{\rho}_{\psi r} = \int\hat{\omega}_{\psi r}d\tau. \tag{5.136}$$

The estimated stator flux component q, depending on the estimated $\alpha\beta$ component and angle $\hat{\rho}_{\psi r}$, is described as follows:

$$\hat{\psi}_{rq} = -\hat{\psi}_{r\alpha}\sin\left(\hat{\rho}_{\psi r}\right) + \hat{\psi}_{r\beta}\cos\left(\hat{\rho}_{\psi r}\right). \tag{5.137}$$

The state observer structure, based on the stator voltage model, designed for a drive system with sinusoidal filter, is presented in Figure 5.10.

Exemplary results obtained in a sensorless drive system with sinusoidal filter, with observer application of Figure 5.10, are illustrated in Figure 5.11.

The results of Figure 5.11 illustrate the characteristics of the closed-loop drive system, in whichthe desired values of drive speed and magnetic flux are controlled. A sinusoidal filter was applied in the drive system, which is considered in the field orientated control presented in Chapter 6. The obtained sensorless system works properly both in the lower speed range and in the change of rotation direction.

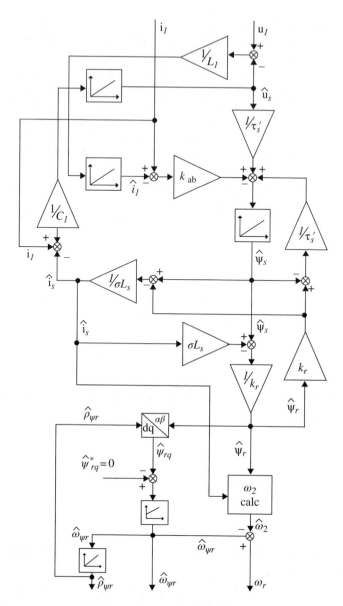

Figure 5.10 Speed and flux observer structure based on the stator voltage model with an adaptive mechanism for rotor flux pulsation

Furthermore, the observer should keep the highest resistance on natural changes of these parameters during motor operation and on inaccuracy identification on measurement of the equivalent circuit parameters. Besides the calculation accuracy, the quality of the observer depends on sensitivity of inaccuracy and changes of the motor parameters.

In the next subsection, the observer is investigated in case of sensitivity of parameter changes of the motor. The rotor resistance, R_r, stator resistance, R_s, and the mutually inductance, L_m,

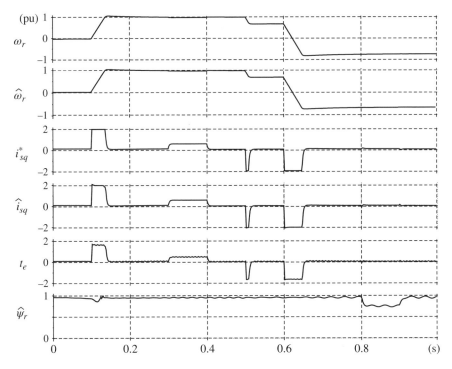

Figure 5.11 Functionality of sensorless drive system with induction motor, sinusoidal filter, and observer presented in Figure 5.10, investigation of simulations (motor 1.5 kW, filter $L_1 = 11.2$ mH, C1 = 10 µF)

were simultaneously changed in the simulation. Furthermore, the investigated drive system operated in closed-loop of field orientated control, assuming accessibility of all variables. Chosen state variables were reproduced in the independent operating observer, which were not used in the control process. The obtained results of the investigations are illustrated in Figure 5.12.

Figure 5.12 presents the dependence of estimation error in case of individual changes of motor parameters. For each parameter, calculation errors were assigned for three different mechanical speeds: 0.1, 0.5, and 1.0 pu. It can be recognized that the mutually inductance, L_m, and the rotor resistance, R_r, had the greatest influence on the observer calculation accuracy. A variation of the stator resistance, R_s, insignificantly affects the observer operation. Observer errors are higher for the lower speed range of $\omega_r = 0.1$ pu, whereas errors for higher speeds, ω_r = 0.5 and 1 pu, are nearly the same.

Figure 5.13 presents the simulations for a field orientation system operation with changed motor parameters: R_r, R_s, and L_m.

During the investigations all controlled values were reproduced in the observer. The changes of motor parameters were approximated to real variations. The rotor and stator resistance increase with rising temperature. The fastest temperature jump affects the rotor, which causes an increase of the parameter R_r up to 150% of the nominal value. This temperature increase results in the rising temperature of the stator, where the time constant is lower. The system sensitivity was tested by a variation of L_m, whereby the maximal range of L_m was determined

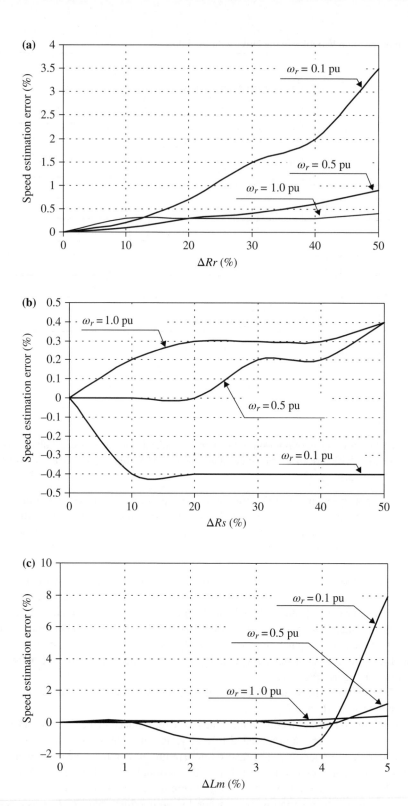

Figure 5.12 Estimation errors of motor speed in different range of speeds: **(a)** changes in R_r; **(b)** changes in R_s; and **(c)** changes in L_m, investigation of simulation (motor 1.5-kW, filter L1 = 11.2 mH, C1 = 10 μF)

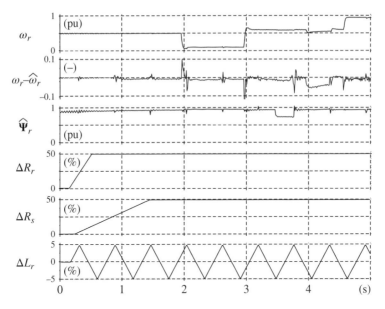

Figure 5.13 Operation of field orientated control system with state observer presented in Figure 5.10, with simultaneously changed motor

in the stable operation of the drive system. It was found that the proper function is ensured in an alternate rising and falling of L_m in a range of ±5% of the initial value.

Furthermore, it was proved that the system still controls the speed, flux, and load changes after a variation of the resistance. A proper system operation for a time range from 2 to 5 s is shown in Figure 5.13. Nevertheless, if the motor parameters are changed, the error of speed estimation does not rise too much to have a serious influence on the full closed-loop motor control.

Figure 5.14 illustrates the reproduced speed errors of an experimental system. In the presented method, the observer parameters were varied during the investigations.

The top three subplots in Figure 5.14 present the error of the estimated speed for different motor angular speed, whereas the bottom subplots show the individual changes of the motor parameter. During the realized test the system operated stably. The estimation error of the observer is insignificantly higher in the lower speed range and did not influence the system operation.

Moreover the system resistance with observer was experimentally tested on measurement errors of the inverter output currents.

In the experiment, illustrated in Figure 5.15, a disturbing signal *dist* was introduced and added to the value of one analog to digital (AD) converter channel. Signal *ia1* is the directly read current of the AD converter, whereas *dist* is the disturbance impact on *ia1*. The signals *ia1* and *dist* are shown in the natural units of the converter. The value *ia1* = 900 decimal corresponds to the rated motor current, whereas the peak value of the disturbance signal forms about 10% of this current. To consider the range and variation of the disturbance signal, the drive must operate stably. The only difference can be recognized in the insignificant increase of the reproduced motor speed error.

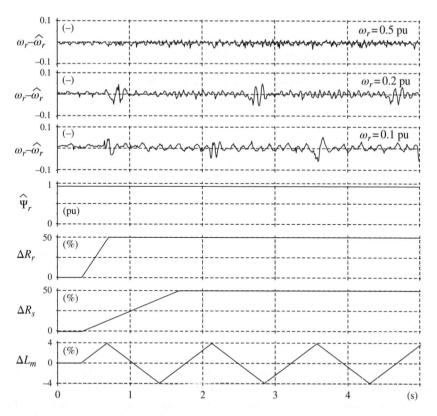

Figure 5.14 Functionality of field orientated control system with speed observer. Variation of motor parameters used depending of the observer for speeds 0.1, 0.2, and 0.5 pu, experiment (motor 1.5 kW, filter $L_1 = 11.2$ mH, $C_1 = 10$ μF)

5.8 Speed Observer with Dual Model of Stator Circuit

The disadvantage of the presented observer solution in Section 5.7 is the estimation structure. It appears as an algebraic loop resulting from the application of the same rotor flux estimation and stator current estimation dependencies: Equations 5.119 and 5.120, as well as Equations 5.128 and 5.129. Caused by this, it requires an implementation of an iterative start procedure of the observer or application of additional terms to delay the estimation process of the variables.

A possible elimination of this disadvantage can be realized by a simultaneous determination of the motor model stator flux for state variables ψ_s and ψ_r (model I) and for the model variables ψ_s and i_s (model II) [19]:

$$\frac{d\psi_{s\alpha}}{d\tau} = u_{s\alpha} - R_s i_{s\alpha}, \tag{5.138}$$

$$\frac{d\psi_{s\beta}}{d\tau} = u_{s\beta} - R_s i_{s\beta}. \tag{5.139}$$

Estimation of the State Variables in the Drive with LC Filter

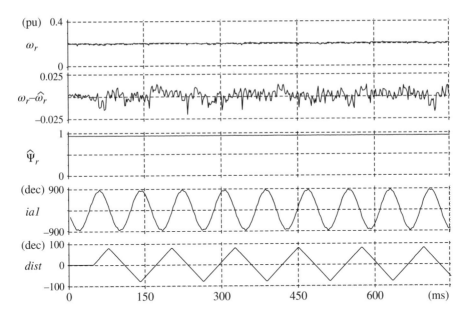

Figure 5.15 Functionality of field orientated control with speed observer in case of measure error of the inverter output current, experiment (motor 1.5 kW, filter $L_1 = 11.2$ mH, $C_1 = 10$ μF)

The stator flux is determined by the equations of model I, whereas the rotor flux is reproduced by the stator flux, which depends on model II and current-flux dependencies:

$$\psi_{r\alpha} = \frac{\psi_{s\alpha} - \sigma L_s i_{s\alpha}}{k_r}, \tag{5.140}$$

$$\psi_{r\beta} = \frac{\psi_{s\beta} - \sigma L_s i_{s\beta}}{k_r}. \tag{5.141}$$

Figure 5.16 presents the flux observer structure that takes advantage of two stator circuit models for a drive without filter.

For a drive system with sinusoidal filter the illustrated observer structure in Figure 5.16 had to be extended in such a way as to estimate simultaneously the current and voltage of the motor, based on the known inverter output current and voltage values. This approach is identical to the application of the observer in Section 5.7.

The flux observer equations that use the voltage stator models I and II for a drive with sinusoidal filter are described as follows:

$$\frac{d\hat{\psi}_{s\alpha}^{(1)}}{d\tau} = \frac{-\hat{\psi}_{s\alpha}^{(1)} + k_r \hat{\psi}_{r\alpha} + \hat{u}_{s\alpha}}{\tau'_s} - k_A\left(i_{1\alpha} - \hat{i}_{1\alpha}\right) + k_B\left(i_{1\beta} - \hat{i}_{1\beta}\right), \tag{5.142}$$

$$\frac{d\hat{\psi}_{s\beta}^{(1)}}{d\tau} = \frac{-\hat{\psi}_{s\beta}^{(1)} + k_r \hat{\psi}_{r\beta} + \hat{u}_{s\beta}}{\tau'_s} - k_A\left(i_{1\beta} - \hat{i}_{1\beta}\right) - k_B\left(i_{1\alpha} - \hat{i}_{1\alpha}\right), \tag{5.143}$$

$$\frac{d\hat{\psi}_{s\alpha}^{(\text{II})}}{d\tau} = \hat{u}_{s\alpha} - R_s \hat{i}_{s\alpha}, \quad (5.144)$$

$$\frac{d\hat{\psi}_{s\beta}^{(\text{II})}}{d\tau} = \hat{u}_{s\beta} - R_s \hat{i}_{s\beta}, \quad (5.145)$$

$$\hat{\psi}_{r\alpha} = \frac{\hat{\psi}_{s\alpha}^{(\text{II})} - \sigma L_s \hat{i}_{s\alpha}}{k_r}, \quad (5.146)$$

$$\hat{\psi}_{r\beta} = \frac{\hat{\psi}_{s\beta}^{(\text{II})} - \sigma L_s \hat{i}_{s\beta}}{k_r}, \quad (5.147)$$

$$\frac{d\hat{u}_{s\alpha}}{d\tau} = \frac{i_{1\alpha} - \hat{i}_{s\alpha}}{C_1}, \quad (5.148)$$

$$\frac{d\hat{u}_{s\beta}}{d\tau} = \frac{i_{1\beta} - \hat{i}_{s\beta}}{C_1}, \quad (5.149)$$

$$\frac{d\hat{i}_{1\alpha}}{d\tau} = \frac{u_{1\alpha}^* - \hat{u}_{s\alpha}}{L_1} + k_A \left(i_{1\alpha} - \hat{i}_{1\alpha} \right) - k_B \left(i_{1\beta} - \hat{i}_{1\beta} \right), \quad (5.150)$$

$$\frac{d\hat{i}_{1\beta}}{d\tau} = \frac{u_{1\beta}^* - \hat{u}_{s\beta}}{L_1} + k_A \left(i_{1\beta} - \hat{i}_{1\beta} \right) + k_B \left(i_{1\alpha} - \hat{i}_{1\alpha} \right), \quad (5.151)$$

where $\psi_{s\alpha}^{(\text{I})}, \psi_{s\beta}^{(\text{I})}, \psi_{s\alpha}^{(\text{II})}, \psi_{s\beta}^{(\text{II})}$ are the components of the flux linkage vector estimated in models I and II.

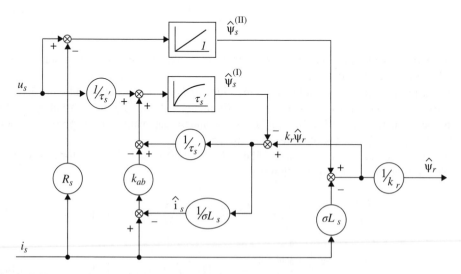

Figure 5.16 Flux observer based on the motor model for state variables ψ_s and ψ_r (model I) and ψ_s, is (model II).

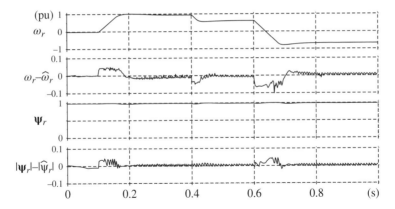

Figure 5.17 Estimation errors of the angular speed and the module of rotor flux in variation of motor speed, simulations (motor 1.5 kW, filter $L_1 = 11.2$ mH, $C_1 = 10$ μF)

The method of rotor speed estimation is applied in the same way as for the observer in Section 5.7. Figure 5.17 presents the functionality of the observer in motor speed variations. As shown in Figure 5.17, the reproduced errors were determined for the observer that operates in the field orientated control. The estimated observer values were not used for the control. In steady state, the error of the estimated speed does not exceed 3%, whereas in the transients this rises to about 6%. Only in cases of motor reverse does a momentary estimation error of 10% occur. Nevertheless, compared to the observer presented in Section 5.7 a lower estimation error was identified in transients.

5.9 Adaptive Speed Observer

Among the algebraic methods of the state variable estimation of the motor, there are further known adaption methods [1], one of which was applied in the observer in Section 5.7.

The following subsection presents an adaptive observer, which is modified for a drive system with filter application [39, 40]. The observer concept is based on the motor stator circuit equations and the current-flux relations.

The voltage model of motor equivalent circuit was used in the observer structure:

$$\frac{d\psi_{s\alpha}}{d\tau} = u_{s\alpha} - R_s i_{s\alpha}, \qquad (5.152)$$

$$\frac{d\psi_{s\beta}}{d\tau} = u_{s\beta} - R_s i_{s\beta}. \qquad (5.153)$$

Furthermore, the relations between the stator current, stator, and rotor flux are:

$$\sigma \frac{L_s L_r}{L_m} i_{s\alpha} = L_m \psi_{r\alpha} - L_r \psi_{s\alpha}, \qquad (5.154)$$

$$\sigma \frac{L_s L_r}{L_m} i_{s\beta} = L_m \psi_{r\beta} - L_r \psi_{s\beta}, \qquad (5.155)$$

After the consideration of the sinusoidal filter equation and the assumption of the dq coordinates, the dependencies of the adaptive flux and motor speed observer are described as follows:

$$\frac{d\hat{\psi}_{sd}}{d\tau} = \hat{u}_{sd} - R_s \hat{i}_{sd} + \hat{\omega}_{\psi r} \hat{\psi}_{sq} + k_1 \left(L_m \hat{i}_{sd} - \hat{\psi}_{rd} \right), \qquad (5.156)$$

$$\frac{d\hat{\psi}_{sq}}{d\tau} = \hat{u}_{sq} - R_s \hat{i}_{sq} - \hat{\omega}_{\psi r} \hat{\psi}_{sd} - k_1 \hat{\psi}_{rq}, \qquad (5.157)$$

$$\hat{\psi}_{rd} = \frac{L_r}{L_m} \hat{\psi}_{sd} + \sigma \frac{L_s L_r}{L_m} \hat{i}_{sd}, \qquad (5.158)$$

$$\hat{\psi}_{rq} = \frac{L_r}{L_m} \hat{\psi}_{sq} - \sigma \frac{L_s L_r}{L_m} \hat{i}_{sq}, \qquad (5.159)$$

$$\frac{d\hat{u}_{sd}}{d\tau} = \frac{i_{1d} - \hat{i}_{sd}}{C_1}, \qquad (5.160)$$

$$\frac{d\hat{u}_{sq}}{d\tau} = \frac{i_{1q} - \hat{i}_{sq}}{C_1}, \qquad (5.161)$$

$$\frac{d\hat{i}_{1d}}{d\tau} = \frac{u^*_{1d} - \hat{u}_{sd}}{L_1} + k_A \left(i_{1d} - \hat{i}_{1d} \right) - k_B \left(i_{1q} - \hat{i}_{1q} \right), \qquad (5.162)$$

$$\frac{d\hat{i}_{1q}}{d\tau} = \frac{u^*_{1q} - \hat{u}_{sq}}{L_1} + k_A \left(i_{1q} - \hat{i}_{1q} \right) + k_B \left(i_{1d} - \hat{i}_{1d} \right), \qquad (5.163)$$

$$\hat{\omega}_{\psi r} = -k_\omega \left(\hat{\psi}_{rq} + \frac{1}{T_\omega} \int_0^\tau \hat{\psi}_{rq} \right), \qquad (5.164)$$

$$\hat{\omega}_r = \hat{\omega}_{\psi r} - \frac{R_r}{L_r} \frac{\hat{i}_{sq}}{\hat{i}_{sd}}, \qquad (5.165)$$

$$\hat{\rho}_{\psi r} = \int_0^\tau \hat{\omega}_{\psi r}. \qquad (5.166)$$

A scheme representation of the angular speed observer system is illustrated in Figure 5.18.

The relation of the motor steady state was applied to the observer structure presented in Figure 5.18:

$$\psi_{sd} = L_m i_{sd}, \qquad (5.167)$$

which was implemented in Equation 5.156 as an element of the correction component $k_1 \left(L_m \hat{i}_{sd} - \hat{\psi}_{rd} \right)$.

However, this can cause calculation errors in transients, and therefore proper drive system functionality requires adequate selection of the dynamic control system. It is possible to

Estimation of the State Variables in the Drive with LC Filter

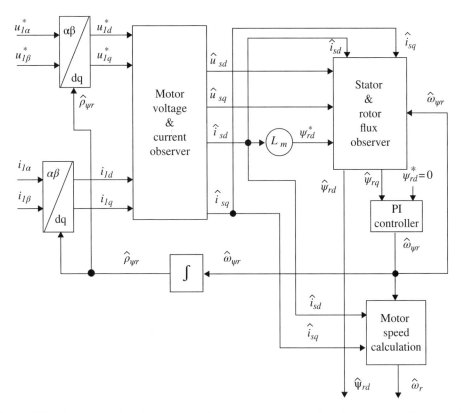

Figure 5.18 Structure of the adaptive flux and motor speed observer for a drive system with an inverter output filter

correct the observer properties by adding to the output of the proportional integral controller the rotor flux pulsation based on the determination of the temporary active and reactive power values, P_t and Q_t. In such a case, the Equation 5.164 should be replaced with the generated rotor flux pulsation according to:

$$\hat{\omega}_{\psi r} = -k_\omega \left(\hat{\psi}_{rq} + \frac{1}{T_\omega} \int_0^\tau \hat{\psi}_{rq} \right) \hat{\omega}_{\psi r} + \frac{1}{T_{\omega 1}} \int_0^\tau \left(\omega_{rm} + R_r \frac{L_m}{L_r} \frac{x_{12m}}{x_{21m}} \right), \quad (5.168)$$

where $T_{\omega 1}$ is the time constant of the inertial filter.

A further component of Equation 5.168 is the matching rotor pulsation, determined by the steady state relations [40]:

$$\omega_{rm} = \frac{1}{L_m x_{22}} \left(L_r Q_t - w_\sigma \left| \hat{i}_s \right|^2 \omega_i - \frac{R_r L_m}{L_r} x_{12m} \right), \quad (5.169)$$

$$x_{12m} = \frac{\left(L_m + \dfrac{a_1}{a_2}\right)|\hat{\mathbf{i}}_s|^2 + \dfrac{a_4}{a_2}P_t}{\omega_i \dfrac{L_r}{R_r}}, \qquad (5.170)$$

$$x_{22m} = \frac{a_4 Q - \omega_i |\hat{\mathbf{i}}_s|^2}{\omega_i \dfrac{L_m}{w_\sigma}}, \qquad (5.171)$$

$$x_{21m} = L_m x_{22m}. \qquad (5.172)$$

An example of such an observer modification was implemented in an induction drive system operating with load angle control in the case of a filterless drive system [40]. The results demonstrate a significant improvement of the observer functionality in transients [40]. Moreover, the outcomes justify the aim of the modifications depending on Equations 5.168 to 5.168.

Figures 5.19 and 5.20 present the errors of the estimated angular motor speed.

The results shown in Figures 5.19 and 5.20 were obtained for the observer that operates in the scalar control of the induction motor system. It can be observed that the estimation errors rise with lower, constant motor speed (Figure 5.19). The observed error contains a higher frequency component that should be filtered before further use in the control system. Figure 5.20

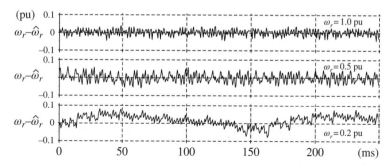

Figure 5.19 Errors of reproduced angular speed in steady state for different motor speeds, simulations (motor 1.5 kW, filter $L_1 = 11.2$ mH, $C_1 = 10$ μF)

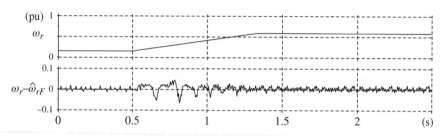

Figure 5.20 Error of reproduced angular speed with motor speed variation, simulations (motor 1.5 kW, filter $L_1 = 11.2$ mH, $C_1 = 10$ μF)

shows, for a higher time scale, the control error between the real value, the estimated, and filtered motor speed, $\hat{\omega}_{rF}$. Moreover an increase of the estimation error in the transition state is observable, which reaches a value of up to 7%.

5.10 Luenberger Flux Observer

Previously presented observers can precisely estimate both flux and speed so it could be used in speed sensorless drives. On other side, lots of existing electrical drives unnecessarily operate with speed sensors, and therefore, mechanical speed estimation is not required. Hence, simpler structures of the observers could be used. One of the well-known solutions is the classical Luenberger observer [41]. The full order Luenberger observer can estimate the speed in addition to the flux, but it requires simultaneous calculation of the load torque. If the flux is the only required variable, the mechanical equation is eliminated in Luenberger observer. The motor speed is considered as one of motor parameters, which can vary. The feedbacks of the observer can compensate variations to get an accurate flux estimation. The elimination of the mechanical equations is acceptable because of the high differences of time constants of mechanical and electromagnetical variables.

Equations of induction motor rotor flux of Luenberger observer have the form [42]:

$$\frac{d\hat{i}_{s\alpha}}{d\tau} = a_1 \hat{i}_{s\alpha} + a_2 \hat{\psi}_{r\alpha} + a_3 \omega_r \hat{\psi}_{r\beta} + a_4 \hat{u}_{s\alpha} + k_i \left(i_{s\alpha} - \hat{i}_{s\alpha} \right), \tag{5.173}$$

$$\frac{d\hat{i}_{s\beta}}{d\tau} = a_1 \hat{i}_{s\beta} + a_2 \hat{\psi}_{r\beta} - a_3 \omega_r \hat{\psi}_{r\alpha} + a_4 \hat{u}_{s\beta} + k_i \left(i_{s\beta} - \hat{i}_{s\beta} \right), \tag{5.174}$$

$$\frac{d\hat{\psi}_{r\alpha}}{d\tau} = a_5 \hat{\psi}_{r\alpha} - \omega_r \hat{\psi}_{r\beta} + a_6 \hat{i}_{s\alpha} + k_{f1} \left(i_{s\alpha} - \hat{i}_{s\alpha} \right) - k_{f2} \left(i_{s\beta} - \hat{i}_{s\beta} \right), \tag{5.175}$$

$$\frac{d\hat{\psi}_{r\beta}}{d\tau} = a_5 \hat{\psi}_{r\beta} + \omega_r \hat{\psi}_{r\alpha} + a_6 \hat{i}_{s\beta} + k_{f1} \left(i_{s\beta} - \hat{i}_{s\beta} \right) + k_{f2} \left(i_{s\alpha} - \hat{i}_{s\alpha} \right), \tag{5.176}$$

where k_i, k_{f1}, and k_{3f2} are observer gains.

Taking into account the LC filter model, it is possible to obtain the equations of Luenberger observer for rotor flux calculation:

$$\frac{d\hat{i}_{s\alpha}}{d\tau} = a_1 \hat{i}_{s\alpha} + a_2 \hat{\psi}_{r\alpha} + a_3 \omega_r \hat{\psi}_{r\beta} + a_4 \hat{u}_{s\alpha} + k_i \left(i_{1\alpha} - \hat{i}_{1\alpha} \right), \tag{5.173}$$

$$\frac{d\hat{i}_{s\beta}}{d\tau} = a_1 \hat{i}_{s\beta} + a_2 \hat{\psi}_{r\beta} - a_3 \omega_r \hat{\psi}_{r\alpha} + a_4 \hat{u}_{s\beta} + k_i \left(i_{1\beta} - \hat{i}_{1\beta} \right), \tag{5.174}$$

$$\frac{d\hat{\psi}_{r\alpha}}{d\tau} = a_5 \hat{\psi}_{r\alpha} - \omega_r \hat{\psi}_{r\beta} + a_6 \hat{i}_{s\alpha} + k_{f1} \left(i_{1\alpha} - \hat{i}_{1\alpha} \right) - k_{f2} \left(i_{1\beta} - \hat{i}_{1\beta} \right), \tag{5.175}$$

$$\frac{d\hat{\psi}_{r\beta}}{d\tau} = a_5\hat{\psi}_{r\beta} + \omega_r\hat{\psi}_{r\alpha} + a_6\hat{i}_{s\beta} + k_{f1}\left(i_{1\beta} - \hat{i}_{1\beta}\right) + k_{f2}\left(i_{1\alpha} - \hat{i}_{1\alpha}\right), \tag{5.176}$$

$$\frac{d\hat{u}_{c\alpha}}{d\tau} = \frac{i_{1\alpha} - \hat{i}_{s\alpha}}{C_1}, \tag{5.177}$$

$$\frac{d\hat{u}_{c\beta}}{d\tau} = \frac{i_{1\beta} - \hat{i}_{s\beta}}{C_1}, \tag{5.178}$$

$$\frac{d\hat{i}_{1\alpha}}{d\tau} = \frac{u_{1\alpha}^* - \hat{u}_{s\alpha}}{L_1} + k_A\left(i_{1\alpha} - \hat{i}_{1\alpha}\right) - k_B\left(i_{1\beta} - \hat{i}_{1\beta}\right), \tag{5.179}$$

$$\frac{d\hat{i}_{1\beta}}{d\tau} = \frac{u_{1\beta}^* - \hat{u}_{s\beta}}{L_1} + k_A\left(i_{1\beta} - \hat{i}_{1\beta}\right) + k_B\left(i_{1\alpha} - \hat{i}_{1\alpha}\right), \tag{5.180}$$

$$\hat{u}_{s\alpha} = \hat{u}_{c\alpha} + \left(i_{1\alpha} - \hat{i}_{s\alpha}\right)R_c, \tag{5.181}$$

$$\hat{u}_{s\beta} = \hat{u}_{c\beta} + \left(i_{1\beta} - \hat{i}_{s\beta}\right)R_c, \tag{5.182}$$

where k_A and k_B are observer gains.

The extension of the flux observer is the same as presented for speed and flux observers in Sections 5.3, 5.4, and 5.6.

5.11 Simulation Examples

5.11.1 Model of the State Observer with LC Filter Simulator

The model of LC filter simulator for the inverter with common mode and differential mode filter and induction motor *(Chapter_5\Simulation_1\Simulator.slx)* is shown in Figure 5.21.

The *Simulator.slx* model consists of a three-phase inverter with a space vector pulse width modulation (SVPWM) algorithm. The control principle is V/f without ramp. The main simulation file is FILTSIM.c, whereas SVPWM function is in PWM_OW.c file.

The exemplary waveforms for start up of the system is presented in Figure 5.22.
In Figure 5.22 the next variables are presented (from top to bottom):

- omegaR, real motor speed
- omega_sof, estimated motor speed
- fr, magnitude of real motor flux
- fr_so, magnitude of estimated motor flux
- Usx, alpha component of motor voltage
- is, stator current magnitude.

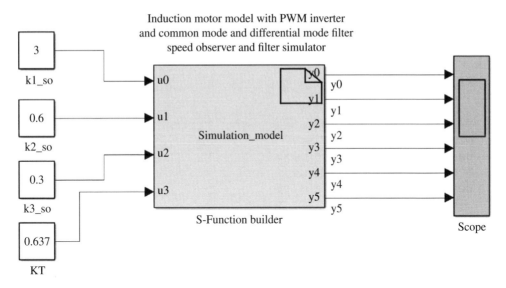

Figure 5.21 MATLAB®/Simulink model of LC filter simulator for inverter with common mode and differential mode filter and induction motor, *Chapter_5\Simulation_1\Simulator.slx*

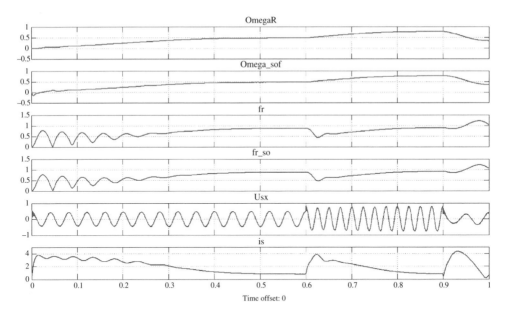

Figure 5.22 Example of the results for LC filter simulator (estimator structure presented in Section 5.2)

At the beginning the motor direct start up is done with $f^* = 25$ Hz (0.5 pu), next at 0.6 s the frequency is increased to 40 Hz (0.8 pu) and at 0.9 s is decreased to 20 Hz (0.4 pu).

The structure of the estimator has two parts: disturbance observer and LC filter simulator. The internal variables of filter simulator have subscript _sf_, whereas speed observer variables have suffix _so_. The LC filter estimator calculates motor input voltages and currents based on the knowledge of inverter output voltages and currents. The estimation is done within a step of the pulse width modulation (PWM) switching period:

```
if(impulse>Timp)
  {
  impulse=0;

  // L1, R1 voltage drop
  Di1x=(i1x-i1x_old)/Timp;
  ucx_sf=USX-i1x*R1-L1*Di1x;
  Di1y=(i1y-i1y_old)/Timp;
  ucy_sf=USY-i1y*R1-L1*Di1y;
  i1x_old=i1x; // previous value
  i1y_old=i1y; // previous value
  ukx_sf=ucx_sf;
  uky_sf=ucy_sf;
  Ducx_sf=(ucx_sf-ucx_sf_old)/Timp;
  icx_sf=Cxy*Ducx_sf;
  Ducy_sf=(ucy_sf-ucy_sf_old)/Timp;
  icy_sf=Cxy*Ducy_sf;
  ucx_sf_old=ucx_sf;
  ucy_sf_old=ucy_sf;
  // current of L2, R2 branch
  i2x_sf=i1x-icx_sf;
  i2y_sf=i1y-icy_sf;
  // L2, R2 voltage drop
  Di2x_sf=(i2x_sf-i2x_sf_old)/Timp;
  Usx_sf=ucx_sf-i2x_sf*R2-L2*Di2x_sf;   // motor voltage
  Di2y_sf=(i2y_sf-i2y_sf_old)/Timp;
  Usy_sf=ucy_sf-i2y_sf*R2-L2*Di2y_sf;   // motor voltage
  i2x_sf_old=i2x_sf;
  i2y_sf_old=i2y_sf;
  // current of L3, R3
  i3x_sf=((Usx_sf-R3*i3x_sf)/L3)*Timp+i3x_sf_pop;
  i3x_sf_pop=i3x_sf;
  i3y_sf=((Usy_sf-R3*i3y_sf)/L3)*Timp+i3y_sf_pop;
  i3y_sf_pop=i3y_sf;
  // motor current
  isx_sf=i2x_sf-i3x_sf;
  isy_sf=i2y_sf-i3y_sf;
  // Speed observer call
  usx_so=Usx_sf;
  usy_so=Usy_sf;
  isx0=isx_sf;
  isy0=isy_sf;

  SPEED_OBSERVER(Timp);
  }
```

The speed observer function is called subroutine *SPEED_OBSERVER()*. The subroutine contains the initialization of the observer coefficients, right side of observer differential equations, and numerical procedure of RK4. The right side differential equations are set in subroutine *p_SPEED_OBSERVER()*. The speed observer equations are:

```
Dspeed_obs[0]=a1_so*isx_so+a2_so*frx_so+zetaY*a3_so+a4_so*usx_so
              +k3_so*(k1_so*(isx0-isx_so)-omega_so*zetaX);
Dspeed_obs[1]=a1_so*isy_so+a2_so*fry_so-zetaX*a3_so+a4_so*usy_so
              +k3_so*(k1_so*(isy0-isy_so)-omega_so*zetaY);
Dspeed_obs[2]=a5_so*frx_so+a6_so*isx_so-zetaY-k2_so*(omega_
so*fry_so-zetaY);
Dspeed_obs[3]=a5_so*fry_so+a6_so*isy_so+zetaX+k2_so*(omega_
so*frx_so-zetaX);
Dspeed_obs[4]=k1_so*(isx0-isx_so);
Dspeed_obs[5]=-k1_so*(isy0-isy_so);Dspeed_obs[4]=k1_
so*(isx0-isx_so);
Dspeed_obs[5]=-k1_so*(isy0-isy_so);
```

In *p_SPEED_OBSERVER()* the algebraic equations of the observer could be found:

```
frfr_so=frx_so*frx_so+fry_so*fry_so; if(frfr_so<.001) frfr_
so=.001;
fr_so=sqrt(frfr_so);
omega_so=sqrt((zetaX*zetaX+zetaY*zetaY)/(frfr_so));
Sb=MINUS*(zetaX*fry_so-zetaY*frx_so); // the value foto
stabilize observer
omega_so=MINUS*omega_so-Sb+Sbf; // calculated motor speed
```

A value MINUS is the sign of estimated motor speed:

```
if(zetaX*frx_so+zetaY*fry_so>0) MINUS=1; else MINUS=-1;
//speed direction
```

The observer gains are variables *k1_so*, *k2_so*, and *k3_so* whereas *KT* is coefficient of estimated speed inert filter:

```
omega_sof=speed_obs[7];
Dspeed_obs[7]=(omega_so-omega_sof)*KT;
```

The observer gains are set by Simulink parameters blocks.

5.11.2 Model of Speed Observer with Simplified Model of Disturbances

The model of speed observer with simplified model of disturbances (*Chapter_5\Observer\Simulator.slx*) is shown in Figure 5.23.

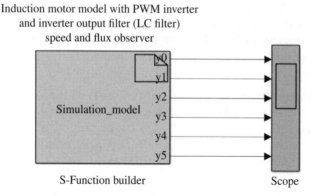

Figure 5.23 MATLAB®/Simulink model of speed observer with simplified model of disturbances, *Chapter_5\Observer\Simulator.slx*

The model *Simulator.slx* consists of a three-phase inverter with a SVPWM algorithm. The main simulation file is MainObs.c, whereas SVPWM function is in PWMU_2XY.c file. At the beginning the control principle is *V/f*, and after 200 ms the control is changed to closed loop. The switching from open loop to closed-loop control is imposed by variable *START_REG* (for *START_REG = 0* it is V/f, and for *START_REG = 1* it is closed-loop control). The speed observer is modeled according to the structure presented in Section 5.3. The differential equations of the observer are given in function *p_SPEED_OBSERVER()*:

```
Dspeed_obs[0]=a1_so*isx_so+a2_so*frx_so+zetaY*a3_so+a4_so*usx_so
              +k3_so*(k1_so*(isx0-i1x_so)-omega_so*zetaX);
Dspeed_obs[1]=a1_so*isy_so+a2_so*fry_so-zetaX*a3_so+a4_so*usy_so
              +k3_so*(k1_so*(isy0-i1y_so)-omega_so*zetaY);
Dspeed_obs[2]=a5_so*frx_so+a6_so*isx_so-zetaY-k2_so*(omega_
so*fry_so-zetaY);
Dspeed_obs[3]=a5_so*fry_so+a6_so*isy_so+zetaX+k2_so*(omega_
so*frx_so-zetaX);
Dspeed_obs[4]=k1_so*(isx0-i1x_so);
Dspeed_obs[5]=-k1_so*(isy0-i1y_so);
Dspeed_obs[6]=0.01*(Sb-Sbf);
Dspeed_obs[7]=(-omega_sof+omega_so)*KT;
Dspeed_obs[8]=(i1x_so-isx_so)/Cxy;
Dspeed_obs[9]=(i1y_so-isy_so)/Cxy;
Dspeed_obs[10]=(-R1/L1)*i1x_so+(usx0-usx_so)/L1+k_LC1*
(isx0-i1x_so)
              -k_LC2*(isy0-i1y_so);
Dspeed_obs[11]=(-R1/L1)*i1x_so+(usy0-usy_so)/L1+k_LC1*
(isy0-i1y_so)
              +k_LC2*(isx0-i1x_so);
```

Estimation of the State Variables in the Drive with LC Filter

where the integrated variables are:

```
isx_so=speed_obs[0];
isy_so=speed_obs[1];
frx_so=speed_obs[2];
fry_so=speed_obs[3];
zetaY=speed_obs[4];
zetaX=speed_obs[5];
Sbf=speed_obs[6];
omega_sof=speed_obs[7];
usx_so=speed_obs[8];
usy_so=speed_obs[9];
ilx_so=speed_obs[10];
ily_so=speed_obs[11];
```

The estimated variables are used in the control process to make the drive sensorless.
An example of the waveforms for start up of the system is presented in Figure 5.24.
In Figure 5.24 the next variables are presented (from top to bottom):

- omegaR, real motor speed,
- omega_sof, estimated motor speed,
- isx, real motor current (alpha component),
- isx_so, estimated motor current (alpha component),
- Usx, alpha component of motor voltage,
- Usx_so, estimated value of the alpha component of motor voltage.

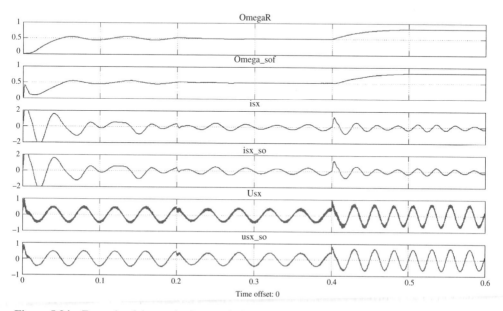

Figure 5.24 Example of the results for speed observer (estimator structure presented in Section 5.3)

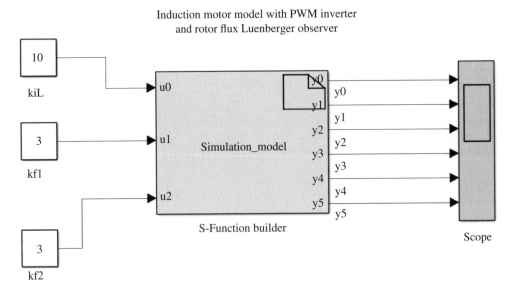

Figure 5.25 MATLAB®/Simulink model of speed observer with simplified model of disturbances, *Chapter_5\Luenberger\Simulator.slx*

With *V/f* principle the motor starts for *omegaU* = 0.5 (without ramp). After 200 ms the drive is switched to sensorless mode, and at instant 400 ms the referenced motor speed is increased to 0.8 pu. It is noticeable that real motor voltage *Usx* has high frequency component coming from PWM switching frequency. Such high frequency is not observed in the estimated voltage *usx_so* because of damping properties of the observer.

Implementing other versions of the speed observer (extended disturbance model and full disturbance model from Sections 5.4 and 5.5) is left to the reader.

5.11.3 Model of Rotor Flux Luenberger Observer

The model of Luenberger observer (*Chapter_5\Luenberger\Simulator.slx*) is shown in Figure 5.25.

The *Simulator.slx* model consists of a three-phase inverter with a SVPWM algorithm and closed-loop speed and flux control. The drive is without a filter. The main simulation file is Luenberg.c. The estimation procedure includes disturbance observer and Luenberger observer. The Luenberger observer equations are:

```
D_obs[9]=a1_so*isx_o+a2_so*frx_o+x11*a3_so*fry_o+a4_so*usx_
so+kiL*(isx-isx_o);
D_obs[10]=a1_so*isy_o+a2_so*fry_o -x11*a3_so*frx_o+a4_so*usy_
so+kiL*(isy-isy_o);
D_obs[11]=a5_so*frx_o+a6_so*isx_o-x11*fry_o+kf1*(isx-isx_o)-kf2*
x11*(isy-isy_o);
D_obs[12]=a5_so*fry_o+a6_so*isy_o+x11*frx_o+kf1*(isy-isy_o)+
kf2*x11*(isx-isx_o);
```

Estimation of the State Variables in the Drive with LC Filter

where the integrated variables are:

```
isx_o=obs[9]=limit(obs[9],8.0);
isy_o=obs[10]=limit(obs[10],8.0);
frx_o=obs[11]=limit(obs[11],8.0);
fry_o=obs[12]=limit(obs[12],8.0);
```

Each of the estimated variables is limited to prevent numerical problems in case of wrong operation of the observer, for example, improper observer gains. The Luenberger variables are noted with subscript _o, whereas the disturbance observer has subscript _so.

The closed-loop control system operates using real values such as speed, flux, and so on. The observers are not used in the control process, so the observer gains can be changed in wide range and will not influence the proper system operation.

The Luenberger gains are *kiL*, *kf1*, and *kf2* and could be changed in Simulink file using block parameters settings.

An example of the results is presented in Figure 5.25.

In Figure 5.26 the next variables are presented (from top to bottom):

- omegaR, real motor speed
- fr, magnitude of the real rotor flux
- fr_o, magnitude of the estimated rotor flux
- frx, alpha component of the real rotor flux
- frx_o, alpha component of the estimated rotor flux
- fry_o, beta component of the estimated rotor flux.

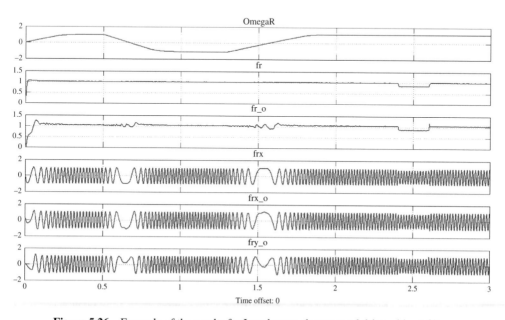

Figure 5.26 Example of the results for Luenberger observer and drive without filter

The drive starts with referenced speed 1 pu. At 0.5 and 1.3 s, the motor speed is reversed. Next at 2.4 s the flux is decreased and finally at 2.6 s is increased. During motor reverse at low speed a higher flux error is observed, which is usual in such structure of the Luenberger observer.

Adding the LC filter model and changing Luenberger observer to the form which includes the LC filter model is left to the reader.

5.12 Summary

The Sections 5.2 to 5.10 presented observer structures elaborated for drive systems with sinusoidal filter application.

The common aim of the presented solutions is the introduction of the filter model to the observer equations. The proposed equations were different in terms of several observers. In sensorless drive systems without motor filters, which can already be found, miscellaneous observer systems were applied. A presentation of all possible types and necessary modifications for a drive system with a motor filter would go beyond the scope of this book. As a result of a lot of observers existing in sensorless drives, the observer concepts presented in this chapter describe only a fragment of possible solutions. The intention of the book was not to compare the properties of and recommend several observers, but to point to concrete answers for several applications found in sensorless control systems.

The investigations led to the conclusion that both operating range and sensitivity of observers on variations or errors of motor parameter identification will not result in a significant difference between observers for filterless systems and their equivalent for filter systems. Furthermore, because of the knowledge of several estimation systems for filterless drive systems it should be noted that similar properties will be reached for their equivalents with filters.

Gain factors of correction components were contained in each of the observers. The right selection of this factor, which ensures low static and dynamic error and an adequate reserve of stability, proves to be a complex and difficult calculation task. Different solving approaches and suggestions can be found [2, 10, 11, 23, 31, 35, 39, 43].

The practical implementation of observer tuning is often realized by an empirical approach. Knowledge and practice of experts, supported by modern calculation technology, enable effective implementation of simulations, the results of which allow the choice of the right gain coefficients. Hence, multiple simulation experiments were performed with the same transition state, however with different parameters.

Such observer tuning was applied in most presented cases. The gain factors were searched for using elaborate simulation programs. Firstly, the coefficients were tuned for the observer in an open loop operation; this means that all values, with the control process, were measured. Next, the coefficients were corrected for the observer in closed-loop operation. After the simulation analysis completion, a further setting was performed in the experimental setup.

References

[1] Orłowska-Kowalska T, Wojsznis P, Kowalski Cz. Comparative study of different flux estimators for sensorless induction motor drive. *Archives of Electrical Engineering*. 2000; **49** (1): 49–63.

[2] Rajashekara K, Kawamura A, Matsuse K. *Sensorless control of AC motor drives*, New York: IEEE Press; 1996.

[3] Kawabata T, Miyashita T, Yamamoto Y. Digital control of three phase PWM inverter with LC filter. Nineteenth Annual IEEE Power Electronics Specialists Conference PESC'88, Kyoto, Japan. April 11–14, 1988.

[4] Kawabata T, Miyashita T, Yamamoto Y. Dead beat control of three phase PWM inverter, IEEE Transactions on Power Electronics. 1990; 5 (1): 21–28.
[5] Kawabata T, Miyashita T, Yamamoto Y. Digital control of three phase PWM inverter with LC filter. IEEE Transactions on Power Electronics. 1991; 6 (1): 62–72.
[6] Seliga R, Koczara W. Multiloop feedback control strategy in sine-wave voltage inverter for an adjustable speed cage induction motor drive system. European Conference on Power Electronics and Applications, EPE'2001, Graz, Austria. August 27–29, 2001.
[7] Seliga R, Koczara W. High quality sinusoidal voltage inverter for variable speed ac drive systems. Tenth International Power Electronics and Motion Control Conference, EPE–PEMC 2002, Dubrovnik, Croatia. September 9–11, 2002.
[8] Salomaki J, Luomi J. Vector control of an induction motor fed by a PWM inverter with output LC filter. Fourth Nordic Workshop on Power and Industrial Electronics, NORPIE'04, Trondheim, Norway. June 14–16, 2004.
[9] Salomaki J, Hikkanen M, Luomi, J. Sensorless control of induction motor drives equipped with inverter output filter. IEEE International Conference on Electric Machines and Drives, IEMDC'05, San Antonio, TX. May 15–18, 2005.
[10] Kubota H, Matsuse K, Nakano T. DSP-based speed adaptive flux observer of induction motor. IEEE Transactions on Industrial Applications. 1993; 29 (2): 344–348.
[11] Kubota H, Sato I, Tamura Y, Matsuse K, Ohta H, Hori Y. Regenerating-mode low-speed operation of sensorless induction motor drive with adaptive observer. IEEE Transactions on Industry Applications. 2002; 38 (4): 1081–1086
[12] Adamowicz M, Guzinski J. Control of sensorless electric drive with inverter output filter. Fouth International Symposium on Automatic Control AUTSYM 2005, Wismar, Germany. September 22–23, 2005.
[13] Krzemiński Z. Sensorless control of the induction motor based on new observer. International Conference on Power Electronics, Intelligent Motions and Power Quality PCIM'00, Nuremberg, Germany. June 6–8, 2000.
[14] Guzinski J, Abu-Rub H. Speed sensorless control of induction motors with inverter output filter. International Review of Electrical Engineering. 2008; 3 (2): 337–343.
[15] Guzinski J Abu-Rub H. Asynchronous motor nonlinear control with inverter output LC filter. Second Mediterranean Conference on Intelligent Systems and Automation CISA 2009, Zarzis, Tunisia. March 23–25, 2009.
[16] Guziński J. Sensorless AC drive control with LC filter. The Thirteenth European Conference on Power Electronics and Applications EPE'09, Barcelona, Spain. September 8–10, 2009.
[17] Krzeminski Z. Observer of induction motor speed based on exact disturbance model. Thirteenth Power Electronics and Motion Control Conference, EPE–PEMC 2008, Poznan, Poland. September 1–3, 2008.
[18] Guziński J, Abu-Rub H. Speed sensorless induction motor drive with motor choke and predictive control. COMPEL: The International Journal for Computation and Mathematics in Electrical and Electronic Engineering. 2011; 30 (3): 686–705.
[19] Abu-Rub H, Iqbal A, Guzinski J. High performance control of AC drives with MATLAB/Simulink models. New York: John Wiley & Sons, Ltd; 2012.
[20] Salomaki J, Luomi J. Vector control of an induction motor fed by a PWM inverter with output LC filter. EPE Journal. 2006; 16 (1): 37–43.
[21] Krzemiński Z. Estimation of rotor speed for nonlinear control of the induction motor. Tenth International Power Electronics and Motion Control Conference, EPE–PEMC 2002, Dubrovnik, Croatia. September 9–11, 2002.
[22] Parasaliti F, Petrella R, Tursini M. Speed sensorless control of a PM Motor by sliding mode observer. IEEE International Symposium on Industrial Electronics, ISIE'97, Guimaraes, Portugal. July 7–11, 1997.
[23] Urbanski K, Zawirski K. Observer based sensorless control of PMSM. International Conference in Electrical Dives and Power Electronics, EDPE 2003, The High Tatras, Slovakia. September 24–26, 2003.
[24] Krzemiński Z. Observer of induction motor speed based on exact disturbance model. Thirteenth Power Electronics and Motion Control Conference, EPE–PEMC 2008, Poznań, Poland. September 1–3, 2008.
[25] Guzinski J, Abu-Rub H, Diguet M, Krzeminski Z, Lewicki A. Speed and load torque observer application in high–speed train electric drive. *IEEE Transactions on Industrial Electronics*. 2007; **57** (2): 565–574.
[26] Guziński J, Abu-Rub H, Toliyat HA. Speed sensorless ac drive with inverter output filter and fault detection using load torque signal. ISIE 2010, Bari, Italy. July 4–7, 2010.
[27] Guzinski J, Diguet M, Krzeminski Z, Lewicki A, Abu-Rub H. Application of speed and load torque observers in high-speed train drive for diagnostic purposes. *IEEE Transactions on Industrial Electronics*. 2009; **56** (1): 248–256.

[28] Guziński J, Diguet M, Krzemiński Z, Lewicki A, Abu-Rub H. Application of speed and load torque observers in high speed train. Thirteenth International Power Electronics and Motion Conference EPE–PEMC 2008, Poznań, Poland. September 1–3, 2008.
[29] Kowalski CT, Orlowska-Kowalska T. Neural networks application for induction motor faults diagnosis. *Mathematics and Computers in Simulation, Modelling and Simulation of Electric Machines, Converters and Systems.* 2003; **63** (3–5): 435–448.
[30] Bogalecka E, Kolodziejek P. Frequency characteristics of induction machine speed observers. XX Symposium Electromagnetic Phenomena in Nonlinear Circuits, EPNC 2008, Lille, France. July 2–4, 2008.
[31] Brdys MA, Du T. Algorithms for joint state and parameter estimation in induction motor drive systems. IEE International Conference on Control, Control'91, Edinburgh, Great Britain. March 25–28, 1991.
[32] Rothenhagen K, Fuchs FW. Implementation of state and input observers for doubly-fed induction generators. International Conference on "Computer as a Tool," EUROCON 2007, Warsaw, Poland. September 9–12, 2007.
[33] Kołodziejek P. Non-invasive method for rotor fault diagnosis in inverter fed induction motor drive. The Eight International Conference and Exhibition on Ecological Vehicles and Renewable Energies, Monte Carlo, Monaco. March 27–30, 2013.
[34] Levine WS. *The control handbook control system applications.* 2nd ed. Boca Raton, FL: CRC Press; 2010.
[35] Holtz J. Sensorless control of induction motor drive. *Proceedings of IEEE.* 2002; **90** (8): 1359–1394.
[36] Holtz J. Sensorless control of induction machines—with or without signal injection? *IEEE Transactions on Industrial Electronics.* 2006; **53** (1): 7–30.
[37] Abu-Rub H, Guzinski J, Rodriguez J, Kennel R, Cortes P. Predictive current controller for sensorless induction motor drive. IEEE–ICIT 2010 International Conference on Industrial Technology, Viña del Mar, Chile. March 14–17, 2010.
[38] Abu-Rub H, Oikonomou N. Sensorless observer system for induction motor control. Thirty-Ninth IEEE Power Electronics Specialists Conference, PESC0'08, Rodos, Greece. June 15–19, 2008.
[39] Tsuji M, Chen S, Ohta T, Izumi K, Yamada E. A speed sensorless vector-controlled method for induction motor using q-axis flux. Second International Power Electronics and Motion Control Conference, IPEMC'97, Hangzhou, China. November 3–6, 1997.
[40] Abu-Rub H, Guzinski J, Krzeminski Z, Toliyat HA. Advanced control of induction motor based on load angle estimation. *IEEE Transactions on Industrial Electronics.* 2004; **51** (1): 5–14.
[41] Luenberger DG. Observers for multivariable systems. *IEEE Transactions on Automatic Control.* 1966; **AC-11** (2): 190–197.
[42] Krzeminski Z. Observer system for the induction motor. Ninth International Conference on Electrical Drives and Power Electronics, the High Tatras, Czechoslovakia. 1990.
[43] Dębowski A, Lewandowski D. Application of the induction motor torque—observer to the control of turbo—machines. Thirteenth Power Electronics and Motion Control Conference, EPE–PEMC 2008, Poznan, Poland. September 1–3, 2008.

6

Control of Induction Motor Drives with LC Filters

6.1 Introduction

As shown in previous parts of this book, current control is affected by sinusoidal filter parameters. The use of such filters enables a voltage drop to the motor feeding circuit and a phase-shift between the filter input and output voltages and currents. As an example, a sinusoidal filter for a 5.5-kW induction motor (3 × 400 V) with a rated load and 50 Hz supply frequency introduces around a 5% voltage drop and a 5-degree phase-shift between the filter input and output voltages.

Figure 6.1 shows an example of such waveforms: voltage reference (commanded) at the filter input and the real voltage at the filter output.

Regardless of the small amount of filter voltage drop and the small phase shift, some advanced drives will not work properly or will require limited dynamics in the system [1]. Hence, the filter used should be taken into account both in the control algorithm as well as in the state variables estimation [2–9].

This chapter presents various solutions to the control problems of electric drives with sinusoidal filters. The solutions are presented for the field oriented control (FOC) and for nonlinear control schemes (multiscalar control). A method for FOC with load angle control is also described.

6.2 A Sinusoidal Filter as the Control Object

A sinusoidal filter is a two-dimensional linear stationary control object. The control variables of the filter are: motor supply voltage, u_s, the inverter output current, i_I, and the control variable is the inverter output voltage u_I. The current i_I is an internal variable.

The main structure of the automatic filter control system is presented in Figure 6.2.

Variable Speed AC Drives with Inverter Output Filters, First Edition. Jaroslaw Guzinski, Haitham Abu-Rub and Patryk Strankowski.
© 2015 John Wiley & Sons, Ltd. Published 2015 by John Wiley & Sons, Ltd.

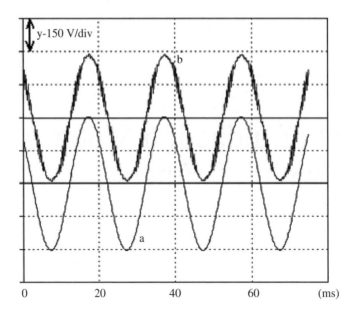

Figure 6.1 Waveforms: **(a)** reference inverter output voltage and **(b)** actual filter output voltage, experimental results

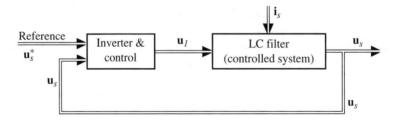

Figure 6.2 Automatic filter control structure

In Figure 6.2, the commanded filter output voltage is u_s^*, which is the reference value of the motor supply voltage. The command voltage value is compared with the real motor voltage u_s. In the controller block, the desired value of the inverter output voltage u_I, which directly acts on the control object, is identified as a filter. The motor current i_s is assumed to be a disturbance. A decrease in the disturbance i_s effect is possible by considering it in the control process [9]. This requires introducing compensation of the disturbance i_s to the control structure, as seen in Figure 6.3.

In the control schemes in Figures 6.2 and 6.3, the inverter output current i_I is not controlled. Leaving the current i_I uncontrolled is not recommended because of the necessity for inverter current protection. To control the current i_I, it is possible to use a cascaded control structure as shown in Figure 6.4 [10–13].

According to the control system presented in Figure 6.4, the filter output current i_I and filter output voltage u_s should be measured. However, using voltage sensors to measure the filter output voltage u_s in the real drive system is not recommended because it takes a long time and

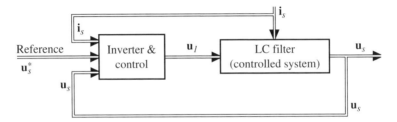

Figure 6.3 Automatic control system of the filter with disturbance compensation

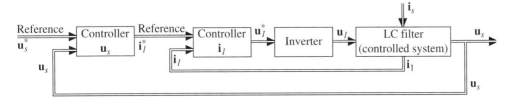

Figure 6.4 Closed-loop control, cascaded structure of the automatic filter control

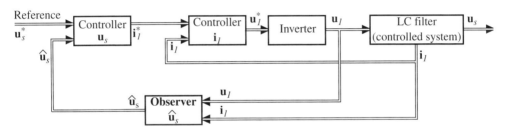

Figure 6.5 Cascaded closed-loop filter control scheme with motor voltage observer

is subject to disturbance connections between the sensors and filter, which complicates the construction and increases the cost and reliability of the whole system.

Instead of direct measurement of the signals in the real system, the voltage u_s may be estimated according to the description in Chapter 5. The filter control scheme with motor voltage observer $\hat{u}_s$ is shown in Figure 6.5.

In this way, a two-dimensional control system is established, in which the control channels in axes α and β are identical. Therefore, it could be described as homogeneous and lacking in nonlinearity. This simplified system analysis should only be used for particular control channels of the components α and β and should not be analyzed comprehensively.

The presented concept of the sinusoidal filter control was used in the drive control algorithm with a squirrel cage induction motor.

6.3 Field Oriented Control

FOC is the most popular control method of induction motors [14]. Such a control strategy ensures that the characteristics of the induction motor drive are similar to those with separately excited DC motors.

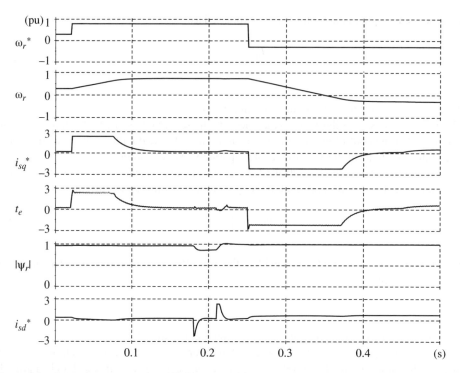

Figure 6.8 Operation of sensorless FOC system of induction motor with sinusoidal filter after step change of rotor speed, flux, and load, simulation

waveforms in Figures 6.8 and 6.9, it is possible to see worse operation of the system with the sinusoidal filter. This result is particularly evident in the torque waveform, which is associated with significant oscillation caused by the structural change in the controlled object.

Undesired effects of the filter on the FOC system could be compensated by using proper modification of the control scheme. In the system with the sinusoidal filter, additional state variables appear: capacitor voltage $u_{c\alpha}$, $u_{c\beta}$ and choke current $i_{1\alpha}$, $i_{1\beta}$. To ensure full control of the extended object, the control structure should contain additional controllers. Assuming that $u_{s\alpha} \approx u_{c\alpha}$ and $u_{s\beta} \approx u_{c\beta}$, the model of the sinusoidal filter in the rotating system dq has the following form:

$$\frac{du_{sd}}{d\tau} = \frac{i_{cd}}{C_1} \quad (6.12)$$

$$\frac{di_{1d}}{d\tau} = \frac{u_{1d} - u_{sd}}{L_1} \quad (6.13)$$

$$\frac{du_{sq}}{d\tau} = \frac{i_{cq}}{C_1} \quad (6.14)$$

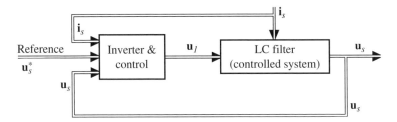

Figure 6.3 Automatic control system of the filter with disturbance compensation

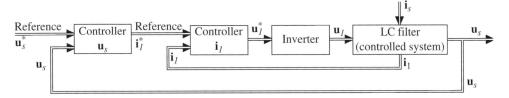

Figure 6.4 Closed-loop control, cascaded structure of the automatic filter control

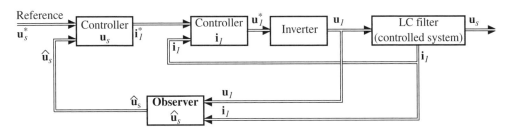

Figure 6.5 Cascaded closed-loop filter control scheme with motor voltage observer

is subject to disturbance connections between the sensors and filter, which complicates the construction and increases the cost and reliability of the whole system.

Instead of direct measurement of the signals in the real system, the voltage u_s may be estimated according to the description in Chapter 5. The filter control scheme with motor voltage observer $\hat{u}_s$ is shown in Figure 6.5.

In this way, a two-dimensional control system is established, in which the control channels in axes α and β are identical. Therefore, it could be described as homogeneous and lacking in nonlinearity. This simplified system analysis should only be used for particular control channels of the components α and β and should not be analyzed comprehensively.

The presented concept of the sinusoidal filter control was used in the drive control algorithm with a squirrel cage induction motor.

6.3 Field Oriented Control

FOC is the most popular control method of induction motors [14]. Such a control strategy ensures that the characteristics of the induction motor drive are similar to those with separately excited DC motors.

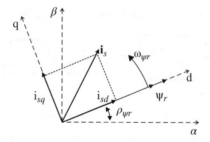

Figure 6.6 Stator current components in coordinate systems $\alpha\beta$ and dq

When using FOC, the mathematical model of the AC motor is presented in a coordinate frame rotating with a rotor flux vector (Figure 6.6).

In the FOC, vector components of the controlled variables in the rotating coordinate dq need to be identified. This requires knowing the angle $\rho_{\psi r}$, which describes the position of the rotor flux vector from the stationary system. The angle $\rho_{\psi r}$ is calculated based on the estimated rotor flux:

$$\hat{\rho}_{\psi r} = arctg \frac{\hat{\psi}_{r\alpha}}{\hat{\psi}_{r\beta}} \tag{6.1}$$

In the stationary coordinate system $\alpha\beta$, the components of the vectors in steady state are sinusoidal variables. However, the components of those vectors in the coordinate system dq at steady state are constant (DC) values.

The principle of FOC is based on the analysis of the induction motor mathematical model presented in Chapter 3. The mathematical model of the induction motor in the dq coordinate system is derived by substituting $\omega_a - \omega_{\psi r}$ as follows:

$$\frac{di_{sd}}{dt} = a_1 \cdot i_{sd} + a_2 \cdot \psi_{rd} + \omega_{\psi r} \cdot i_{sd} + a_4 \cdot u_{sd} \tag{6.2}$$

$$\frac{di_{sq}}{dt} = a_1 \cdot i_{sq} - \omega_{\psi r} \cdot i_{sd} - \omega_r \cdot a_3 \cdot \psi_{rd} + a_4 u_{sq} \tag{6.3}$$

$$\frac{d\psi_{rd}}{dt} = a_5 \cdot \psi_{rd} + a_6 \cdot i_{sd} \tag{6.4}$$

$$\frac{d\omega_r}{dt} = \frac{L_m}{L_r J} \psi_{rd} i_{sq} - \frac{1}{J} m_o \tag{6.5}$$

In the dq coordinate system rotating with the rotor flux in axis d, the q component is zero $\psi_{rq} = 0$, then:

$$|\psi_r| = \sqrt{\psi_{r\alpha}^2 + \psi_{r\beta}^2} = \psi_{rd} \tag{6.6}$$

The produced torque in the motor is then expressed as:

$$m_e = \frac{L_m}{L_r}\psi_{rd}i_{sq} = k_r|\psi_r|i_{sq} \quad (6.7)$$

where: $k_r = L_m/L_r$ is the rotor circuit coupling factor.

Equation 6.7 has a similar form to the one with the separately excited DC motor [15]. This concludes that the control of the state vector in rotating frame dq allows the induction motor to be controlled in a similar way to the DC separately excited motor.

Assuming that stator flux is controlled at a constant value, Equation 6.7 shows that the produced torque is linearly proportional to the q component of the stator current:

$$m_e \sim i_{sq}\big|_{for|\psi_r|=const.} \quad (6.8)$$

Furthermore, from Equation 6.4 it is evident that rotor flux is proportional to the d component of the stator current:

$$|\psi_r| \sim i_{sd} \quad (6.9)$$

The FOC structure of the induction motor is shown in Figure 6.7.

The block dq/αβ in the schematic presented in Figure 6.7 is an inverse Park transformation, which is transformation of variables from dq to αβ coordinates. In the block dq/αβ the command values (αβ) of the inverter output voltage are identified:

$$u_{s\alpha}^* = u_{sd}^*\cos\hat{\rho}_{\psi r} - u_{sq}^*\sin\hat{\rho}_{\psi r} \quad (6.10)$$

$$u_{s\beta}^* = u_{sd}^*\sin\hat{\rho}_{\psi r} - u_{sq}^*\cos\hat{\rho}_{\psi r} \quad (6.11)$$

Figure 6.8 presents the waveforms of the signals from the FOC system without a filter, and Figure 6.9 presents the waveforms from the system with a sinusoidal filter. In both systems, the control structure and controller parameters are identical. The effect of the observer on the system's operation was ignored by assuming access to all controlled values. By comparing the

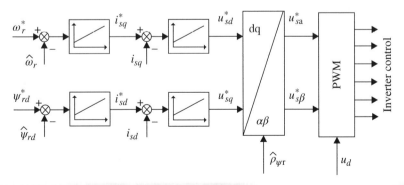

Figure 6.7 FOC structure of induction motor

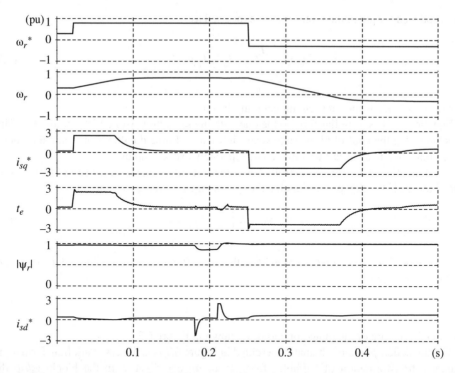

Figure 6.8 Operation of sensorless FOC system of induction motor with sinusoidal filter after step change of rotor speed, flux, and load, simulation

waveforms in Figures 6.8 and 6.9, it is possible to see worse operation of the system with the sinusoidal filter. This result is particularly evident in the torque waveform, which is associated with significant oscillation caused by the structural change in the controlled object.

Undesired effects of the filter on the FOC system could be compensated by using proper modification of the control scheme. In the system with the sinusoidal filter, additional state variables appear: capacitor voltage $u_{c\alpha}$, $u_{c\beta}$ and choke current $i_{1\alpha}$, $i_{1\beta}$. To ensure full control of the extended object, the control structure should contain additional controllers. Assuming that $u_{s\alpha} \approx u_{c\alpha}$ and $u_{s\beta} \approx u_{c\beta}$, the model of the sinusoidal filter in the rotating system dq has the following form:

$$\frac{du_{sd}}{d\tau} = \frac{i_{cd}}{C_1} \qquad (6.12)$$

$$\frac{di_{1d}}{d\tau} = \frac{u_{1d} - u_{sd}}{L_1} \qquad (6.13)$$

$$\frac{du_{sq}}{d\tau} = \frac{i_{cq}}{C_1} \qquad (6.14)$$

Control of Induction Motor Drives with LC Filters

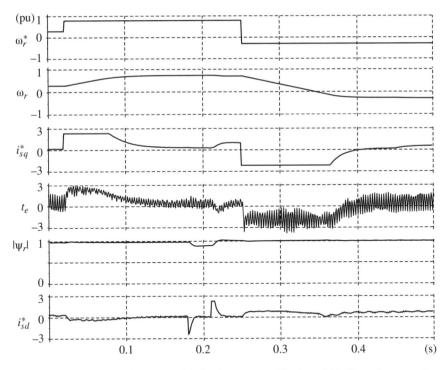

Figure 6.9 Operation of FOC system of induction motor with sinusoidal filter after step change of rotor speed, flux, and load, system without observer (simulation)

$$\frac{di_{1q}}{d\tau} = \frac{u_{1q} - u_{sq}}{L_1} \tag{6.15}$$

$$i_{cd} = i_{1d} - i_{sd} \tag{6.16}$$

$$i_{cq} = i_{1q} - i_{sq} \tag{6.17}$$

In the control system with the motor and the filter, a cascaded structure with proportional-integral (PI) controllers is used to control the components of the motor voltage u_{sd} and u_{sq} and the inverter output current i_{1d} and i_{1q} [11, 12, 16–18]. The full structure of the nonlinear vector control, taking into account the sinusoidal filter, is shown in Figure 6.10.

Figure 6.11 presents the waveforms obtained from the control structure of Figure 6.10.

In the modified FOC structure (Figure 6.11) similar results to the system without the filter (Figure 6.8) were obtained. The observed difference refers to the lower dynamics of the system with the filter which is caused by the change in the time constants of the controlled object. Nevertheless, the dynamic difference between the two schemes is not significant.

Figures 6.12 to 6.14 show examples of the waveforms for properly operating speed sensorless systems with FOC taking into account the sinusoidal filter and an observer (from Section 5.4).

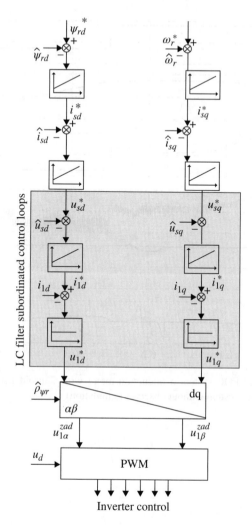

Figure 6.10 Structure of FOC system of induction motor with sinusoidal filter

6.4 Nonlinear Field Oriented Control

One of the modifications of FOC is the nonlinear vector control with decoupling between the flux and torque control channels [19]. This nonlinear field oriented control (NFOC) is a typical vector control replacing the stator current q component controller with the torque controller m_e:

$$m_e = i_{sq} \psi_{rd} \tag{6.18}$$

The electromagnetic torque m_e is noted as a variable x [19].

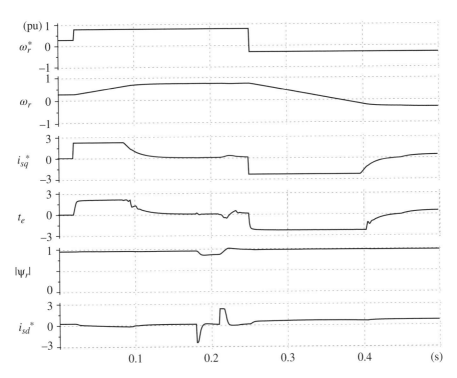

Figure 6.11 Operation of sensorless FOC system of induction motor with sinusoidal filter after step change of rotor speed, flux, and load, system without observer (simulation)

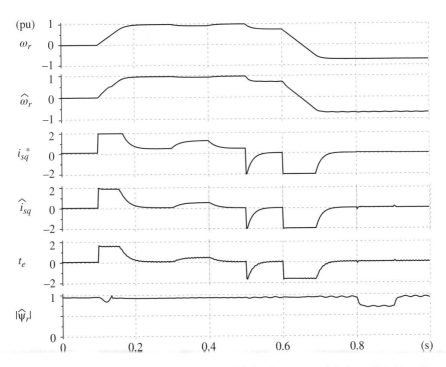

Figure 6.12 Operation of sensorless FOC system of induction motor with sinusoidal filter after step change of rotor speed, flux, and load, sensorless system with state observer from Section 5.4 (simulation)

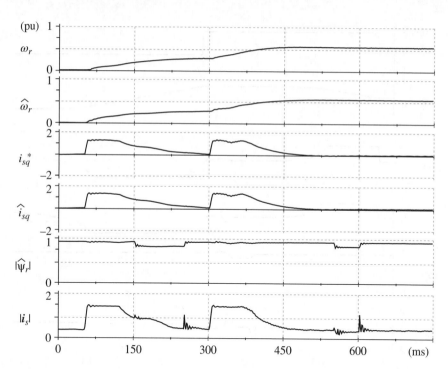

Figure 6.13 Operation of sensorless FOC system of induction motor with sinusoidal filter after step change of rotor speed, flux, and load, sensorless system with state observer from Section 5.4 (experimental)

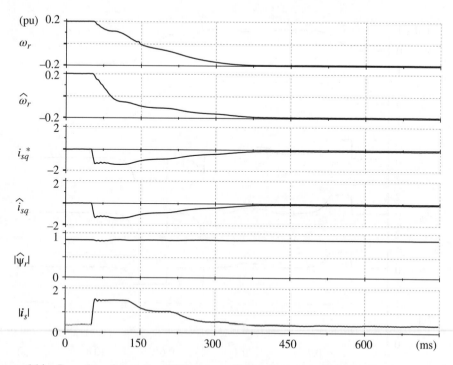

Figure 6.14 Operation of sensorless FOC system of induction motor with sinusoidal filter after speed reverse, sensorless system with state observer from Section 5.4 (experimental)

The use of torque m_e as the new state variable allows the motor model in the dq axis to be presented as:

$$\frac{di_{sd}}{d\tau} = a_1 i_{sd} + a_2 \psi_{rd} + \frac{\omega_{\psi r}}{\psi_{rd}} m_e + a_4 u_{sd} \quad (6.19)$$

$$\frac{dm_e}{d\tau} = \left(a_5 - \frac{R_s L_r}{w_\sigma}\right) m_e + a_6 \frac{i_{sd}}{\psi_{rd}} m_e - \omega_{\psi r} \psi_{rd} \left(i_{sd} + a_3 \psi_{rd}\right) + a_4 \psi_{rd} u_{sq} \quad (6.20)$$

$$\frac{d\psi_{rd}}{d\tau} = a_5 \psi_{rd} + a_6 i_{sd} \quad (6.21)$$

$$\frac{d\omega_r}{d\tau} = \frac{1}{J}\left(\frac{L_m}{L_r} m_e - m_o\right) \quad (6.22)$$

and

$$\omega_{\psi r} = a_6 \frac{i_{sq}}{\psi_{rd}} + \omega_r \quad (6.23)$$

where: i_{sd}, i_{sq}, u_{sd}, u_{sq}, and ψ_{rd} are stator current and voltage components and rotor flux components in the dq axis.

Equations 6.19 to 6.23 describe a nonlinear controlled object, in which a coupling between the flux and torque does exist. Introducing the following control signals:

$$v_1 = a_6 \frac{i_{sd}}{\psi_{rd}} m_e - \omega_{\psi r} \psi_{rd} \left(i_{sd} + a_3 \psi_{rd}\right) + a_4 \psi_{rd} u_{sq} \quad (6.24)$$

$$v_2 = \frac{\omega_{\psi r}}{\psi_{rd}} m_e + a_2 \psi_{rd} + a_4 u_{sd} \quad (6.25)$$

leads to the decomposition of Equations 6.19 to 6.23 into two linear decoupled subsystems, electromagnetic and mechanical.

The electromagnetic subsystem has the following form:

$$\frac{di_{sd}}{d\tau} = -\frac{R_s L_r^2 + R_r L_m^2}{w_\sigma L_r} i_{sd} + v_2 \quad (6.26)$$

$$\frac{d\psi_{rd}}{d\tau} = a_5 \psi_{rd} + a_6 i_{sd} \quad (6.27)$$

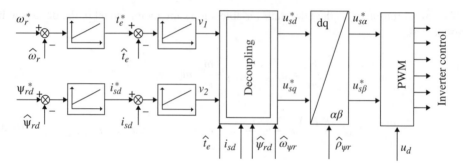

Figure 6.15 Structure of the nonlinear vector control without an inverter output filter

and the mechanical subsystem is as follows:

$$\frac{dm_e}{d\tau} = \left(a_5 - \frac{R_s L_r}{w_\sigma}\right) m_e + v_1 \tag{6.28}$$

$$\frac{d\omega_r}{d\tau} = \frac{1}{J}\left(\frac{L_m}{L_r} m_e - m_o\right) \tag{6.29}$$

The structure of the nonlinear vector control without an output filter is presented in Figure 6.15.

The structure presented in Figure 6.15 could be extended in such a way as in the classical FOC in Section 6.3. However, this method does not take into account the disturbance of the control process caused by the capacitor current of the sinusoidal filter i_c. In the control scheme of the filter state variables, this disturbance could be compensated by adding the actual capacitor current to the output signal of the motor voltage controller [20]. The full structure of the control system with disturbance compensation is presented in Figure 6.16.

With proper tuning of the controllers of both systems, the one with the basic structure (Figure 6.15), as well as the one with the extended structure (Figure 6.16) could be applied in the drive system with the sinusoidal filter. The control characteristics of the system with the filter are evidently better. This could be observed in the waveforms of both systems as seen in Figures 6.17 and 6.18.

Figures 6.17 and 6.18 present the operation of the NFOC of a 1.5-kW induction motor drive both not considering and considering the presence of the filter during the change in command speed. The system works without an observer system assuming access to all controlled variables. A clear difference is evident in the torque response. In the system without considering the presence of the filter, the real torque does not reach the maximum commanded value during the whole period. This causes a degradation of the drive dynamic. After considering the presence of the filter, the system works properly and the actual torque equals the command torque. This ensures better dynamics of the speed control compared with the system without considering the presence of the filter. In the flux response in Figure 6.18, small oscillations can be seen. This is a result of the higher system dynamic, which causes the controllers

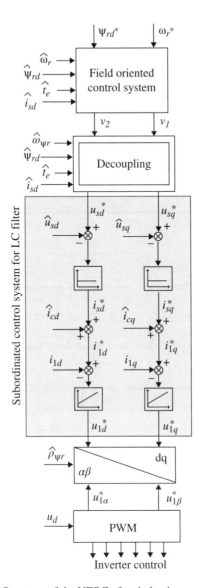

Figure 6.16 Structure of the NFOC of an induction motor with a filter

to operate in saturation mode. This disturbs the decoupling between the flux and torque control channels.

Examples of the sensorless operation of the drive with the sinusoidal filter and observer (in accordance with Section 5.3) are presented in Figures 6.19 to 6.21.

Simulation and experimental investigations were realized after step changes and the reversal of the rotor speed (Figures 6.19 to 6.21). The results show proper motor speed and torque control with stable flux maintenance.

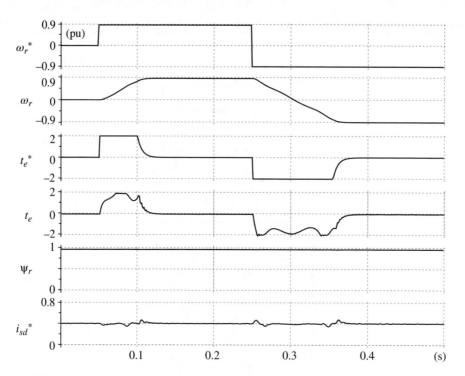

Figure 6.17 Response of the NFOC without considering the presence of the filter in the control scheme

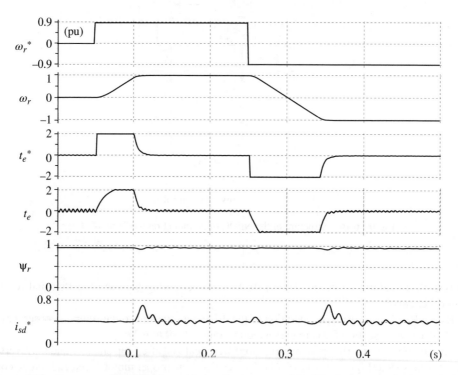

Figure 6.18 Response of the NFOC considering the presence of the filter in the control scheme during motor start-up and reversing, simulation

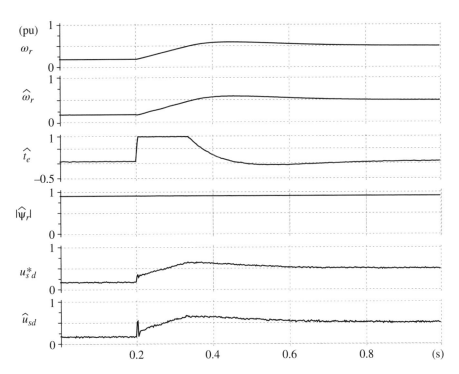

Figure 6.19 Drive system operation after step change of command speed with NFOC and considering the presence of the filter, sensorless control with state observer from Section 5.5 (simulation)

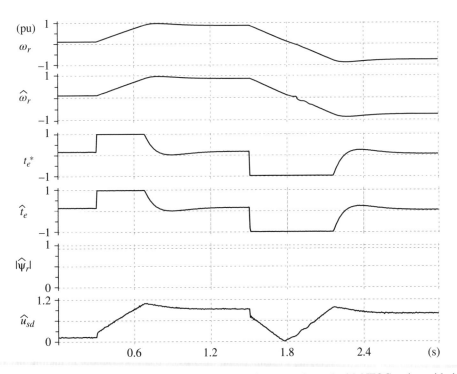

Figure 6.20 Drive system operation after step change of command speed with NFOC and considering the presence of the filter, sensorless control with state observer from Section 5.5 (simulation)

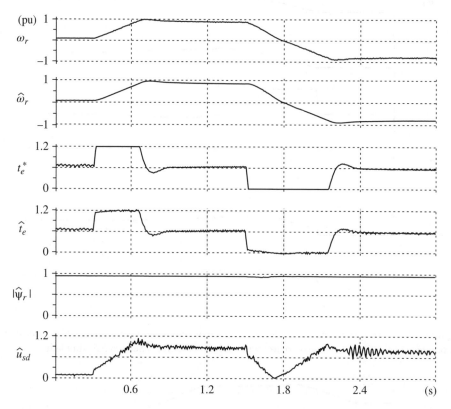

Figure 6.21 Drive system operation after step change of command speed with NFOC and considering the presence of the filter, sensorless control with state observer from Section 5.5 (experimental)

6.5 Multiscalar Control

One of the advanced control methods of an induction motor is the nonlinear control method, the so called multiscalar control [21].

This method is a modification of the FOC and direct torque control (DTC). The method provides promising results although not widespread. Detailed information about the method can be found in Krzemiński [22] and Guziński and Abu-Rub [23].

The multiscalar method is used in nonlinear systems containing couplings between the controlled variables. The basics of the multiscalar control method depends on the proper selection of the controlled variables and the use of the control function in such a way as to obtain a linear form and the decoupling of the system (Figure 6.22).

After obtaining the linear decoupling structure of the control system, it is possible to use the cascaded structure of the linear controllers.

An induction motor is a nonlinear system with decoupling between the mechanical and electromagnetic internal variables. Therefore, the use of the multiscalar control method for the induction motor drives is especially beneficial.

Similar to the FOC, considering the sinusoidal filter to be used is essential in the multiscalar control scheme. It is possible to distinguish the main control system of the motor state variables

Control of Induction Motor Drives with LC Filters

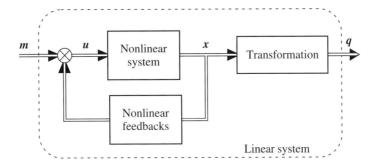

Figure 6.22 Linearization and decoupling of the nonlinear system (m, control signal of the linear system; q, state variables of the linear system; u, control signal of the nonlinear system; and x, state variables of the nonlinear system)

and the secondary (subordinate) control system of the filter state variables. These subsystems are discussed in the following sections.

6.5.1 Main Control System of the Motor State Variables

The main control system of the motor state variables is found in many references discussing the multiscalar control structure [19, 21]. In this system, the nonlinear transformation of the state variables while introducing properly selected control variables is used. The controlled new state variables are the multiscalar variables:

$$x_{11} = \omega_r \tag{6.30}$$

$$x_{12} = \psi_{r\alpha} i_{s\beta} - \psi_{r\beta} i_{s\alpha} \tag{6.31}$$

$$x_{21} = \psi_{r\alpha}^2 + \psi_{r\beta}^2 \tag{6.32}$$

$$x_{22} = \psi_{r\alpha} i_{s\alpha} + \psi_{r\beta} i_{s\beta} \tag{6.33}$$

where:

x_{11} is the rotor angular speed
x_{12} is the vector product of the stator current and rotor flux vectors (variable proportional to the motor electromagnetic torque)
x_{21} is the quadrant of the rotor linkage flux
x_{22} is the scalar product of the stator current and rotor flux vectors.

The physical meaning of multiscalar variables is clear so it is also accepted to use "natural variables" definition instead of "multiscalar variables" one [24].

The use of Equations 3.22 to 3.26 and 6.30 to 6.33 allows the motor model to be presented in the following form:

$$\frac{dx_{11}}{d\tau} = \frac{x_{12} L_m}{JL_p} - \frac{m_o}{J} \tag{6.34}$$

$$\frac{dx_{12}}{d\tau} = -\frac{1}{T_v}x_{12} - x_{11}\left(x_{22} + \frac{L_m}{w_\sigma}x_{21}\right) + \frac{L_r}{w_\sigma}u_1 \quad (6.35)$$

$$\frac{dx_{21}}{d\tau} = -2\frac{R_r}{L_r}x_{21} + 2R_r\frac{L_m}{L_r}x_{22} \quad (6.36)$$

$$\frac{dx_{22}}{d\tau} = -\frac{x_{22}}{T_v} + x_{11}x_{12} + \frac{R_rL_m}{L_rw_\sigma}x_{21} + \frac{R_rL_m}{L_r}\frac{x_{12}^2 + x_{22}^2}{x_{21}} + \frac{L_r}{w_\sigma}u_2 \quad (6.37)$$

Motor electromagnetic time constant T_v is defined as follows:

$$T_v = \frac{w_\sigma L_r}{R_r w_\sigma + R_s L_r^2 + R_r L_m^2} \quad (6.38)$$

The existing nonlinear components in Equations 6.35 and 6.37 could be compensated by using the control functions m_1 and m_2 in the form:

$$u_1 = \frac{w_\sigma}{L_r}\left[x_{11}\left(x_{22} + \frac{L_m}{w_\sigma}x_{21}\right) + m_1\right] \quad (6.39)$$

$$u_2 = \frac{w_\sigma}{L_r}\left(-x_{11}x_{12} - \frac{R_rL_m}{L_rw_\sigma}x_{21} - \frac{R_rL_m}{L_r}\frac{x_{12}^2 + x_{22}^2}{x_{21}} + m_2\right) \quad (6.40)$$

where u_1 and u_2 are the control variables of subordinated system.

After using the control signals m_1 and m_2 the commanded components of the stator voltage could be identified as:

$$u_{s\alpha} = \frac{\psi_{r\alpha}u_2 - \psi_{r\beta}u_1}{\psi_r^2} \quad (6.41)$$

$$u_{s\beta} = \frac{\psi_{r\beta}u_2 - \psi_{r\alpha}u_1}{\psi_r^2} \quad (6.42)$$

Introducing the transformation (Equations 6.30 to 6.33) and the functions (Equations 6.39 to 6.42) provides for the decomposition of the motor model to a linear form, which consists of two fully decoupled subsystems: the electromagnetic and the mechanical.

The mechanical subsystem:

$$\frac{dx_{11}}{d\tau} = \frac{L_m}{JL_r}x_{12} - \frac{1}{J}T_L \quad (6.43)$$

$$\frac{dx_{12}}{d\tau} = -\frac{1}{T_v}x_{12} + m_1 \quad (6.44)$$

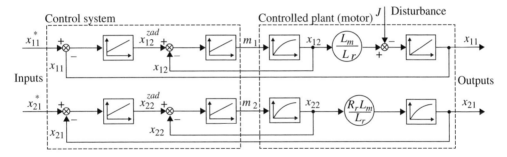

Figure 6.23 The base structure of the multiscalar control

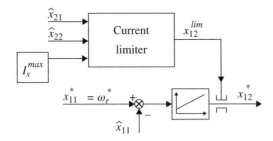

Figure 6.24 Current limiting method in the multiscalar control

The electromagnetic subsystem:

$$\frac{dx_{21}}{d\tau} = -2\frac{R_r}{L_r}x_{21} + 2\frac{R_r L_m}{L_r}x_{22} \tag{6.45}$$

$$\frac{dx_{22}}{d\tau} = -\frac{1}{T_v}x_{22} + m_2 \tag{6.46}$$

The obtained linear and decoupled form of the controlled system (Equations 6.43 to 6.46) allows the known cascaded structure of the PI controllers to be used (Figure 6.23).

In the equations of the multiscalar model (6.43 to 6.46) there is no appearance of the stator current i_s. However, this current should be controlled to protect the motor against overloading conditions (ensures the current remains below the maximum value, I_s^{max}). In the multiscalar control method, such a limit is realized by imposing the next limit on the command value of the motor torque as follows:

$$x_{12}^{lim} = \sqrt{\left(I_s^{max}\right)^2 x_{21} - x_{22}^2} \tag{6.47}$$

The signal x_{12}^{lim} is a limit of the output value of the speed controller (Figure 6.24).

The existing multiscalar variables x_{21} and x_{22} in Equation 6.47 are identified using the state observer.

6.5.2 Subordinated Control System of the Sinusoidal Filter State Variables

In the system without the filter, the control variables $u^*_{s\alpha}$ and $u^*_{s\beta}$, identified in Equations 6.41 and 6.42, are commanded directly to the pulse width modulation (PWM) block. The control algorithm ensures switchings of the transistors that enable the commanded voltage u_s at the inverter output to be obtained.

After using an inverter output filter, the voltages and currents of the inverter output and filter output (motor values) differ. This limits or even prohibits the proper operation of the drive. To solve this problem, it is possible to adopt a similar concept to the FOC of using the subordinated control system [7, 8] for the multiscalar control method, as shown in Figure 6.25.

In the subordinated control system of the filter, the motor voltage u_s and inverter output current i_1 are controlled using the cascaded structure of the PI controllers. To avoid the phase shift in each of the control channels, the filter state variables are controlled in the rotating reference frame KL that is connected to the position of the stator voltage vector u_s (Figure 6.26). The stator voltage vector position is noted as ρ_{us}. The structure of the subordinated control system of the filter is shown in Figure 6.27.

In the control structure shown in Figure 6.27, the voltage drop in the damping resistance of the filter was ignored. Considering such resistance by adding to the output of the voltage controllers from Figure 6.27, the following additional relationships are possible [7]:

$$u^*_{CK} = u^*_{SK} + i_{CK} R_C \tag{6.48}$$

$$u^*_{CL} = u^*_{SL} + i_{CL} R_C \tag{6.49}$$

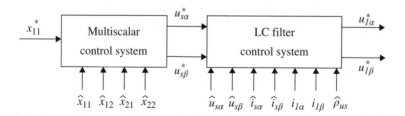

Figure 6.25 Block diagram of the multiscalar control system for the electric drive with the sinusoidal filter

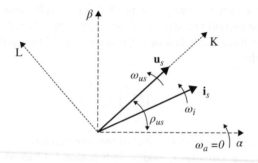

Figure 6.26 Reference coordinate systems $\alpha\beta$ and KL

Control of Induction Motor Drives with LC Filters

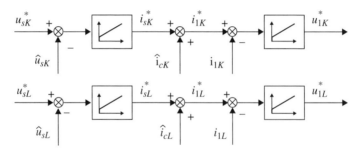

Figure 6.27 The structure of the subordinated control system of the filter state space variables

Simplification of the control structure from Figure 6.27 is possible by not taking into account the interference, which is the capacitor current i_c. Such a solution was presented in Guziński and Włas [3, 4]. This solution, however, has inferior properties compared with the solution considering the capacitor current i_c [20].

The complete multiscalar control structure of the induction motor drive system with a voltage inverter and sinusoidal filter is presented in Figure 6.28.

The waveforms obtained from the control system presented in Figure 6.28 for the drive with the sinusoidal filter and the multiscalar method are presented in Figures 6.29 to 6.33 [7,8].

It is evident from Figures 6.29 to 6.33 that the mechanical and electromagnetic variables are controlled properly and decoupled, except for the short transients in which the PI controllers reach the saturation allowed.

The variation in the variable x_{21} observed in Figure 6.30 appears in transients. This is observed during an increase as well as a decrease in the speed, Figure 6.34.

The observed reduction in the variable x_{21} is caused by the action of the control system in the observer loop. During the change in the rotor speed, errors appear in the estimation of the multiscalar variables, which cause a reaction in the control system. Those changes could be compensated by tuning the controller parameters. In the case of Figure 6.30 those parameters were not properly tuned.

The results in Figure 6.35 show the simulated system operation with different parameters of the x_{22} controller.

In Figure 6.35 the effect of the x_{21} and x_{22} controller parameters on the flux and speed waveforms is clear.

After eliminating the observer system and the estimation errors by assuming variable measurements, the mechanical and electromagnetic subsystems are decoupled, Figure 6.36.

With the same controller parameters and the same control structure with variables measurement, the flux is kept constant during speed variation. The estimation error in x_{21} is not large; however, when using the state observer in the control loop, the controllers act as observed in Figure 6.30.

In the multiscalar control system in Figure 6.28, the possible operation in the field weakening region is presented. No results are shown in this speed range because no problems were observed. The field weakening region is a comprehensive topic regardless of whether or not using a sinusoidal filter. The aim of this work is to show a method of modifying the control system and estimation algorithm to ensure proper operation of the drive with a filter. Operation of the drive system with the filter in the field weakening region is presented in Figure 6.37.

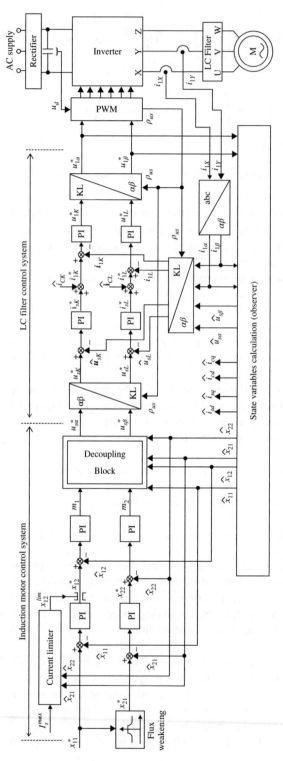

Figure 6.28 Nonlinear control structure with the controllers of filter state variables

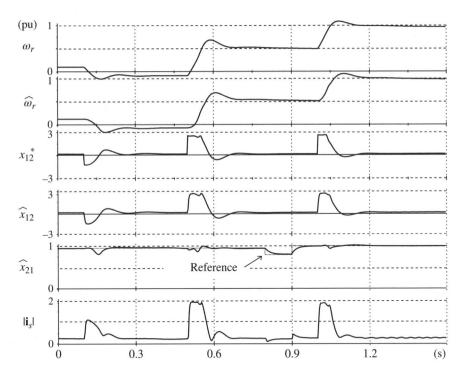

Figure 6.29 Waveforms of the multiscalar control considering the presence of the filter, sensorless operation after changing the speed and flux; simulation results

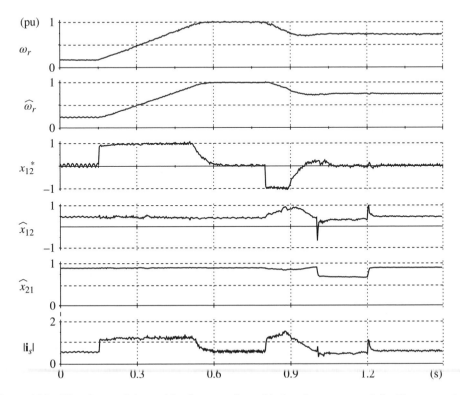

Figure 6.30 Waveforms of the multiscalar control considering the presence of the filter, sensorless operation after changing the speed and flux; experimental results

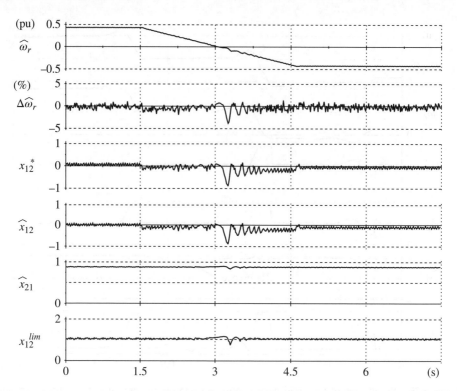

Figure 6.31 Waveforms of the multiscalar control considering the presence of the filter, sensorless operation after motor reversal; experimental results

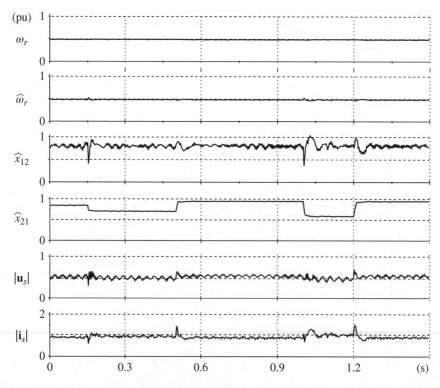

Figure 6.32 Waveforms of the multiscalar control considering the presence of the filter, sensorless operation after changing the flux; experimental results

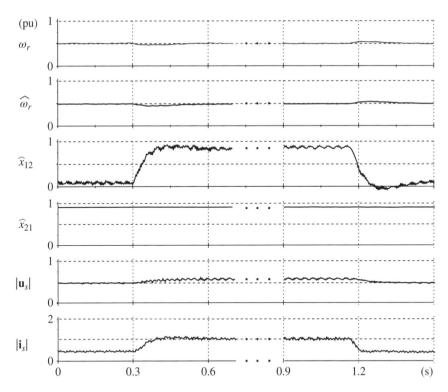

Figure 6.33 Waveforms of the multiscalar control considering the presence of the filter, sensorless operation after changing motor load; experimental results

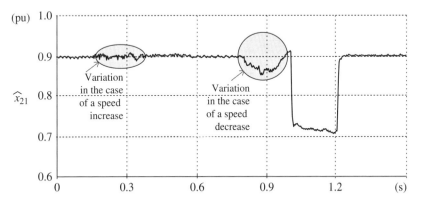

Figure 6.34 Variations of motor flux in the cases of speed changes.

The waveforms show the operation of the system after a speed increase over the base value under the no load condition. Field weakening is evident in the x_{21} waveform. Furthermore, the variable x_{21} is shown under the no load condition ($T_L = 0.01$ pu) and with a loaded motor ($T_L = 0.4$ pu)

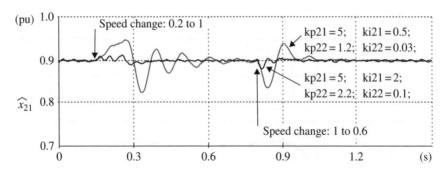

Figure 6.35 Influence of the controller gains on the flux regulation process in the case of the reference speed changes

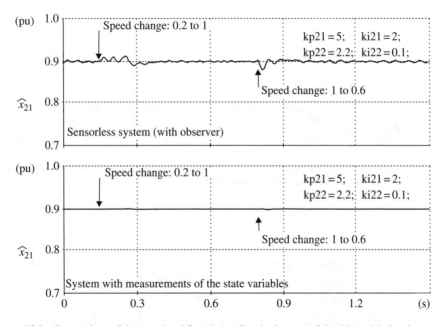

Figure 6.36 Comparison of the speed and flux decoupling in the case of the drive with the observer and with measurement of the state variables

6.6 Electric Drive with Load-Angle Control

The load angle, δ, in the induction motor is the angle between the rotor flux vector, ψ_r, and the stator current vector, i_s, Figure 6.38. The load-angle control is a well-known method for induction motor drives. An example of this type of control is one in which the DTC with a load-angle estimator was presented [25]. A similar method using the command values of the stator current in the FOC was presented in Bogalecka [26]. A method using the superordinated controllers of the multiscalar variables for a sensorless drive was presented in Abu-Rub and colleagues [27].

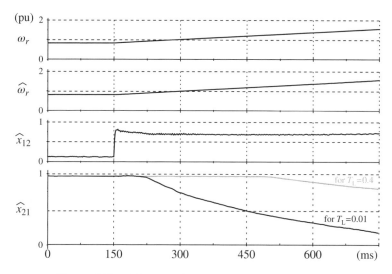

Figure 6.37 Drive operation in the field weakening region

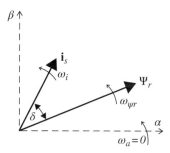

Figure 6.38 Load angle in the squirrel cage induction motor

In the solutions presented so far with load-angle control, the existence of the motor choke and its effect on the drive operation were not considered. The use of the motor choke, an externally added element, complicates the electric drive because of the difference in voltage between that generated at the inverter output and that of the motor. Additional difficulties resulting from the use of the choke relate to the action of the current controller in the control structure. This is particularly important in the case of advanced current control methods, such as predictive control [27]. Hence, the use of a motor choke forces modification of the control structure.

In the electric drive with load-angle control, the change in the vector length of the stator current and rotor flux, together with their mutual positions, allows the electromagnetic motor torque, t_e, to be controlled according to the relationship:

$$t_e = k_m \cdot \mathrm{Im}(\psi_r^* \mathbf{i}_s) = k_m |\psi_r||\mathbf{i}_s|\sin\delta \tag{6.50}$$

where k_m is the proportionality factor.

In Equation 6.50 it is possible to control the motor speed and flux using the control system presented in Figure 6.35 [26].

In the system in Figure 6.35 the superordinated controllers are the speed and rotor flux controllers. The superordinated control system identifies the command values of the current magnitude and its angular speed. In the assumed control structure, the load angle is controlled by the changing current angular speed ω_i^*:

$$\omega_i^* = \omega_2^* + \hat{\omega}_r \tag{6.51}$$

where ω_2^* is the command slip angular speed, obtained by the load-angle controller.

The command current angular speed ω_i^* together with the stator current magnitude $|i_s^*|$ are output signals for the current controller algorithm. The current controller identifies the command components of the inverter output voltage vector $\boldsymbol{u}_1$ for the PWM block. The first-order element used in the channel of the commanded current magnitude $|i_s^*|$ limits its changes because in the real system the current changes are limited by the motor time constant.

The command signals of the current magnitude $|i_s^*|$ and load angle δ^* are generated from the speed and flux controllers according to:

$$|i_s^*| = \sqrt{\left(i_{sd}^*\right)^2 + \left(i_{sq}^*\right)^2} \tag{6.52}$$

$$\delta^* = \arctg\left(\frac{i_{sq}^*}{i_{sd}^*}\right) \tag{6.53}$$

With constant values of the magnitudes of the rotor flux $|\psi_r|$ and stator current $|i_s|$ the torque could only be controlled using load angle δ variation.

The disadvantages of the system presented in Figure 6.39 are the control difficulties related to the system's nonlinearity and the coupling between the variable control channels. This is evident in the induction motor model equation represented by the amplitude-phase relationship [19]:

$$\frac{d|\mathbf{i}_s|}{d\tau} = \frac{1}{T_i}\left(|\mathbf{i}_s| - I_s^*\right) \tag{6.54}$$

$$\frac{d\delta}{d\tau} = -\frac{R_r L_m}{L_r}\frac{i_s}{|\psi_r|} + \omega_i - \omega_r \tag{6.55}$$

$$\frac{d|\psi_r|}{d\tau} = -\frac{R_r}{L_r}|\psi_r| + \frac{R_r L_m}{L_r}i_s \cos\delta \tag{6.56}$$

$$\frac{d\omega_r}{d\tau} = -\frac{1}{J}\left(\frac{L_m}{L_r}|\psi_r||i_s|\sin\delta - m_0\right) \tag{6.57}$$

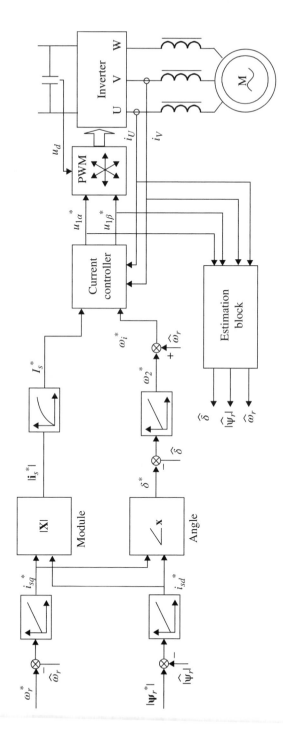

Figure 6.39 The structure of the induction motor drive with load-angle controller

where:

T_i is the time constant of the low order filter reflecting the real inertia in the channel of the stator current control channel
ω_i is the angular speed of the stator current.

It is important to pay attention to Equation 6.54, which models the physical limit of the current changes. The current generated at the inverter output through the current control strategy cannot be considered to be the control value because of its variation limits. Therefore, in the control channel an additional low order element is introduced into the commanded current loop. The current variation limit is modeled using the low-order equation (6.54). This equation is realized only in the control structure away from the motor.

Equations 6.55 and 6.56 contain nonlinearities and couplings. The analysis of these relationships makes it possible to linearize the system and to decouple the control channels. This is possible by using nonlinear control theory [21, 28]. The nonlinear control for the system with the load-angle controller was presented in Krzemiński [19] and Abu-Rub [29], where the multiscalar variables were used in the superordinated control structure.

A simpler structure of nonlinear control with load angle controller could be achieved using the current vector components in the rotating frame dq [26]. This structure will be presented next and is based on the equation analysis of the phase-amplitude model of the induction motor (Equations 6.54 to 6.57).

To guarantee the stability of the system, the function $\cos\delta$ presented in Equation 6.56 should be added. Otherwise, positive feedback will appear in the control system. To ensure a condition $\cos\delta > 0$, the load angle δ must be limited within the range $\pm\pi/2$.

In the control system, a new control variable ψ_r^* is provided, which is described as:

$$\psi_r^* = L_m |i_s| \cos\delta \tag{6.58}$$

Based on Equation 6.58 it is possible to identify the amplitude of the motor current:

$$I_s^* = \frac{\psi_r^*}{L_m \cos\delta} \tag{6.59}$$

The equation of the low-order element (Equation 6.54) may be replaced by the relationship of the rotor flux commanded reference magnitude:

$$\frac{d\psi_{ri}^*}{d\tau} = \frac{1}{T_\psi}\left(\psi_{ri}^* - \psi_r^*\right) \tag{6.60}$$

where ψ_{ri}^* is the output signal of the low order element with time constant T_ψ.

After introducing Equation 6.60, the command current magnitude I_s^* can be identified as:

$$I_s^* = \frac{\psi_{ri}^*}{L_m \cos\delta} \tag{6.61}$$

After introducing the new control signal ψ_r^*, the equation of the load-angle dynamic (Equation 6.55) will have the following form:

$$\frac{d\delta}{d\tau} = -\frac{R_r}{L_r} \frac{\psi_{ri}^*}{|\psi_r|\cos\delta} + \omega_i - \omega_r \tag{6.62}$$

Equation 6.62 is still nonlinear. Converting Equation 6.62 to linear form becomes possible if the command current angular speed ω_i^* is identified according to the following:

$$\omega_i^* = \omega_r + \frac{R_r}{L_r} \frac{\psi_{ri}^*}{|\psi_r|\cos\delta} + \frac{1}{T_\delta}\left(\delta_{ref}^* - \delta\right) \tag{6.63}$$

where:

δ_{ref}^* is the output signal of the load angle controller
$T\delta$ is the time constant adopted for the load-angle control channel.

After considering Equations 6.58 and 6.63, it is possible to obtain the equations of the load angle and rotor flux as follows:

$$\frac{d\delta}{d\tau} = \frac{1}{T_\delta}\left(\delta_{ref}^* - \delta\right) \tag{6.64}$$

$$\frac{d|\psi_r|}{d\tau} = \frac{R_r}{L_r}\left(\psi_{ri}^* - |\psi_r|\right) \tag{6.65}$$

The dynamic system (Equations 6.64 to 6.65) is linear and decoupled. The structure of the nonlinear control system with the load angle controller is presented in Figure 6.40 [23]. In this system the command load angle is as follows:

$$\delta^* = arctg\left(\frac{i_{sq}^*}{i_{sd}^*}\right) \tag{6.66}$$

The command current amplitude I_s^* and current angular frequency ω_i^* are identified in block 1 based on Equation 6.61 and in block 2 using Equation 6.63.

The current controller is presented in Chapter 7. Such a controller considers the presence of the motor choke. Variable estimation is realized in the speed observer with the disturbance model, which is presented in Chapter 5. In the observer the inductance of the motor choke is assumed.

Examples of simulation waveforms showing the operation of the nonlinear control system with load angle control are presented in Figures 6.41 to 6.44, whereas the experimental results are shown in Figures 6.45–6.48.

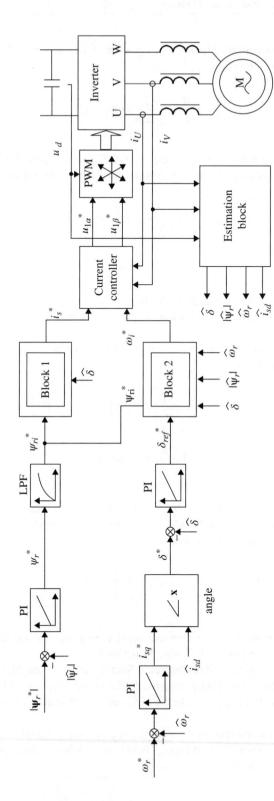

Figure 6.40 Structure of the nonlinear control system of the induction motor with load-angle control

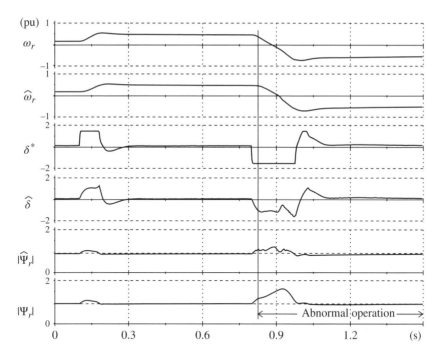

Figure 6.41 Operation of the induction motor (IM) drive system for the control structure presented in Figure 6.39 (i.e., without decoupling) in the case of the step change of command speed and reversing, simulation result

Figure 6.41 presents the operation of the complete system without using nonlinear decoupling. The structure of the control system is consistent with Figure 6.39. The system works assuming that all the control variables are accessible, whereas the parameters obtained using the observer system were not used in the control process. In steady state, the rotor speed and flux are maintained at the command values; however, during transients the coupling appears between the controlled channels. During motor reversing, observed significant flux estimation errors are observed, which prohibits practical operation of the system. In simulation, the motor flux increases significantly, which is not possible in the real motor because of the existing saturation of the magnetic field. The motor model used in the simulation is linear and assumes a linear magnetic path, therefore allowing further operation of the drive. The system operation without the decoupling scheme is possible only with significant limits on the control dynamic.

The operation results of the decoupled control system with load angle and using the controlled variables computed in the observer system are presented in Figures 6.42 to 6.44.

Figure 6.42 presents the system's operation during speed change and motor reversing. Compared with the results from Figure 6.41 a much better response is observed. The speed increase at instant 0.1 s has no effect on flux control. The estimated load angle is identical to the command value. Only during motor reversing is the flux estimated with error, which is a consequence of passing through an unstable regenerative region at low speed. This phenomenon also exists in other known sensorless drives [30]. Although there is significant instant flux estimation error, the drive changes the speed direction and continues working stably.

Figure 6.43 presents the operation of the sensorless control system after flux and torque changes. Thanks to the decoupling between the control subsystems, the flux change does not

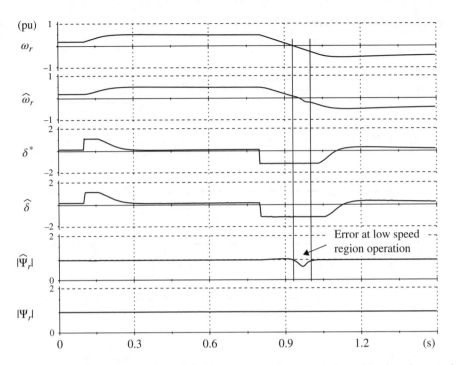

Figure 6.42 Operation of the nonlinear IM drive system with motor choke and load-angle control in accordance with Figure 6.40 after step change in command speed and reversing, simulation result

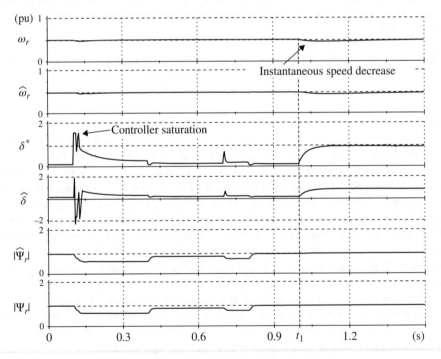

Figure 6.43 Operation of the nonlinear IM drive system with motor choke and load-angle control in accordance with Figure 6.40 after step change in command flux, simulation result

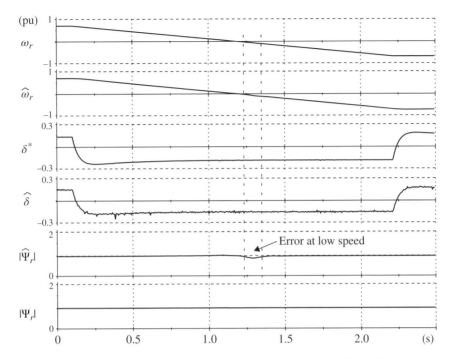

Figure 6.44 Operation of the nonlinear IM drive system with motor choke and load-angle control based on Figure 6.40 during slow reversing, simulation result

affect the speed control channel except at the instants in which the controllers reach saturation. After the step change in the motor load at instant $t_l = 1$ s, a transient speed control error appears, which is then compensated by the speed controller in 0.3 s.

Figure 6.44 presents the operation of the decoupled sensorless drive during and after slow motor reversing. Compared with the high power motor reversing presented in Figure 6.42, the flux estimation error at low speed is smaller.

Experimental verifications are shown in Figures 6.45 to 6.48. In the experiments of Figure 6.45, the command speed value and direction were changed. The speed controller was tuned in such a way so as to provide fast dynamics with low overshoot. As a result of the high-speed control, speed dynamics small flux oscillations have appeared after the fast changes in the motor torque. This is a result of the disturbance in the nonlinear control structure after controller saturation. At lower speed, the estimated speed is obtained with a small error; such a phenomenon has been observed in other works [19, 31, 32]. The drive system is resistant to the existing speed error and operates stably.

In the waveforms of Figure 6.46, small speed changes are evident after the step changes of the load torque $0 \div 0.5$ pu, which are compensated using the superordinated speed controller. Flux changes are negligibly small.

In the investigation of Figure 6.47, the rotor speed and flux were changed. The sensorless control assures consistency between the actual and command signals

Figure 6.48 presents system operation at lower speeds. Speed estimation errors exist during reversing. In this low-speed region, high nonlinear reactions of the drive system are observed.

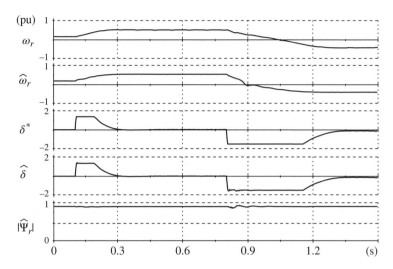

Figure 6.45 Operation of the nonlinear IM drive system with motor choke and load-angle control in accordance with Figure 6.40, experimental results

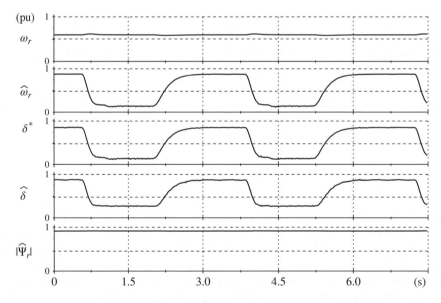

Figure 6.46 Operation of the nonlinear IM drive system with motor choke and load angle control in accordance with Figure 6.40 after changing the load torque, experimental results

These are caused by estimation errors existing at low speed during reversing. After changing the speed direction the system passes through an unstable operating region. Such an operating mode was mentioned in Kubota [33] and Suwankawin and Sangwongwanich [33]. To limit or eliminate such disadvantageous effects, additional adaptive algorithms or additional feedback are considered [30, 34].

Control of Induction Motor Drives with LC Filters

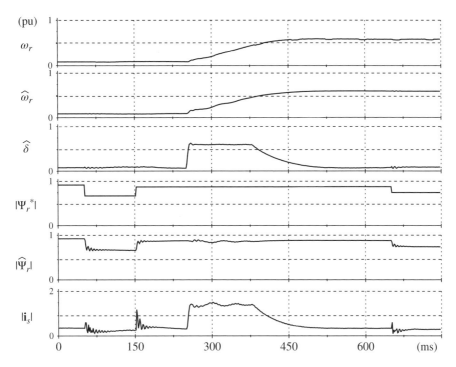

Figure 6.47 Operation of the nonlinear IM drive system with motor choke and load angle control in accordance with Figure 6.40 after speed and flux changes, experimental results

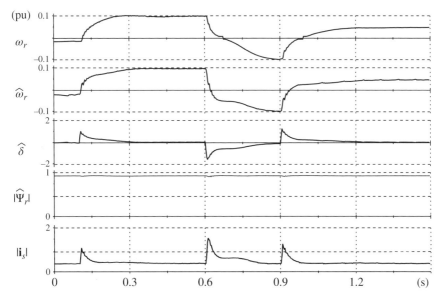

Figure 6.48 Operation of the nonlinear IM drive system with motor choke and load-angle control in accordance with Figure 6.40 after motor reversing, experimental results

The control scheme presented in Figure 6.48 did not include such a stabilizing procedure. Regardless of the existing error, the system is still resistant to the imprecise speed computation and works stably.

The simulation and experimental results prove the correct operation of the nonlinear induction motor control system with motor choke and load angle controller while using the speed and flux observer.

6.7 Direct Torque Control with Space Vector Pulse Width Modulation

In this chapter, another concept of the sensorless control method for the induction motor with voltage source inverter and LC filter is presented. Figure 6.49 shows a block diagram of the control system. The outer loops are for the motor control, whereas the inner are for the filter control. The whole control system operates in synchronous dq coordinates, oriented with stator flux position.

Motor control is based on the DTC principle which is well known from the 1980s [35, 36]. The general DTC principle is simple and interesting, especially because of the excellent torque dynamics. Because of some disadvantages, for example, starting, low-speed problems, torque ripple, and so on, the DTC is still under development [37, 38].

The main disadvantage of the classical DTC is the variable switching frequency, which is dependent on the flux and torque hysteresis controller band. To eliminate this problem, the unique DTC switching table is replaced by a pulse width modulator [39, 40]. With the PWM modulator, the torque and flux are forced by the selection of the proper voltage vector generated in the PWM algorithm instead of the particular transistor on/off combinations. Nowadays, digital control of the PWM modulator is mostly adopted because of the space vector modulation (SVM) algorithm. If SVM is applied to DTC the control method is named the direct torque control–space vector modulation (DTC-SVM) [37, 41]. With this the constant switching frequency in the inverter is obtained.

In the block scheme in Figure 6.49, two PI controllers are used for flux magnitude $|\psi_s|^*$ and torque t_e^* force. The output signals of the controllers are interpreted as commanded voltage components u_{sd}^* and u_{sq}^* where the dq coordinates are stator flux oriented [42]. That control idea is based on the use of a simplified stator model [37]:

$$u_{sd} = R_s i_{sd} + \frac{d|\psi_s|^*}{dt} \approx \frac{d|\psi_s|^*}{dt} \tag{6.67}$$

$$u_{sq} = R_s i_{sq} + \omega_{\psi s}|\psi_s|^* = k_s t_e^* + \omega_{\psi s}|\psi_s|^* \tag{6.68}$$

where $\omega_{\psi s}$ is stator flux vector angular speed and $k_s = R_s/|\psi_s|^*$.

It can be seen in Equations 6.67 and 6.68 that u_{sd} has influence on $|\psi_s|$ only. If the $\omega_{\psi s}|\psi_s|$ term in Equation 6.68 is decoupled, the voltage component u_{sq} has influence on the motor torque control [26]. To simplify the control, this decoupling term is sometimes omitted [43].

If the speed control is a control goal, the commanded torque t_e^* is set by the PI speed controller.

The DTC-SVM control structure is similar to the classical FOC control [38]. The main difference is that the DTC-SVM has fewer controllers, that is, three instead of four used in FOC.

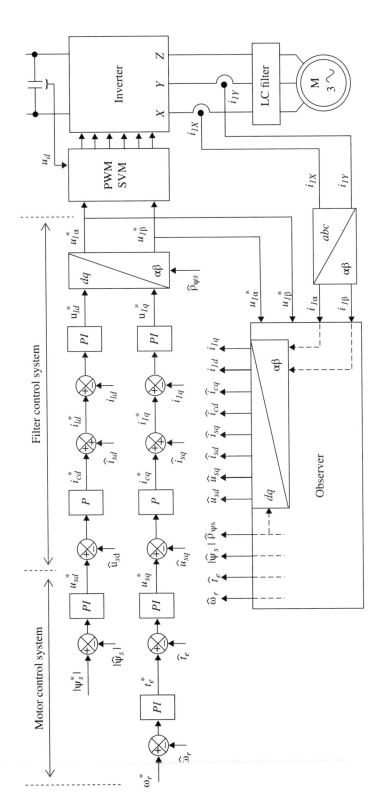

Figure 6.49 Block diagram of DTC-SVM (direct torque control–space vector modulation) control for IM drive with LC filter (dq components are stator flux oriented, $\alpha\beta$ are stationary coordinates)

Additionally, in DTC-SVM no coordinate transformation for the stator current is required. Also, the dq coordinated in the DTC-SVM are stator oriented, whereas in most FOC they are rotor oriented.

In the case of the drive with the LC filter, the components of the commanded motor voltage u_{sd}^* and u_{sq}^* would not be generated at the motor terminal because the LC filter is installed. The inverter with the PWM can apply only the filter input voltage. It can be seen that leaving the motor voltage and currents beyond the control can negatively influence drive operation. It is obvious that additional control loops have to be applied.

Filter control systems are widely used in uninterruptible power supply (UPS) where the sinusoidal output voltage is strongly required. Some UPS control solutions where multiloop control seems to be most attractive can be found in the literature [11, 44]. If this strategy is applied in the electric drive, the task for the multiloop is to achieve the required motor voltage with simultaneous control of the inverter output current. This kind of regulation is also applicable in adjustable speed drives [45].

As was presented in Loh and colleagues [44], good properties are obtained with filter multiloop control when the inverter current is used in the feedback and the motor current is used for disturbance compensation.

In the control structure in Figure 6.49, the filter control part has an outer voltage loop and inner current loop. The voltage controller is a simple P-regulator. The P controller is sufficient because the integration properties of the capacitor C_f [46]. The u_s regulators create a reference signal of the capacitor currents i_c^*. Next, the $\alpha\beta$ components of the actual i_s are added to i_c^* and finally the commanded inverter current i_1^* components are created. The goal of using i_s in the feedback is for disturbance decoupling.

Current i_1 components are controlled by PI type regulators. The regulator outputs are inverter output voltage u_1^* dq components. These components are subsequently transformed to the $\alpha\beta$ system. They are $\alpha\beta$ components of u_1^*. The reference values that are applied are provided to the PWM algorithm, which undertakes transistor control.

It is noticeable in Figure 6.49 that the obtained control structure is more complex compared to the DTC-SVM without filter—seven controllers versus three. However, additional controllers are necessary to take full control of the LC filter. The advantage of such a complex system is that the current is under control. It is worth noting that in classical DTC or DTC-SVM the current is beyond control so additional blocks are required to protect the drive. Such current protection is preserved by the multiloop inner system shown in Figure 6.49.

Examples of simulation investigations for full sensorless control systems are presented in Figures 6.50 to 6.52.

Figure 6.50 presents the steady state operation for $t_L = 0.41$ pu. It is shown that the PWM inverter output voltage has been converted into sinusoidal waveforms. The smoothed current is also observable. The motor torque remains constant with control accuracy of better than 0.02 pu. In Figure 6.51, the motor command speed was first changed, next the load torque was decreased and finally the motor flux was decreased and then increased. In Figure 6.52, motor slow reversing is presented. Constant load was used; load torque was unchanged during reversing. Based on the results, it is noticeable that the drive is working properly. The controlled variables follow the commanded variables. The estimated and actual speeds match but with small errors during the transients. The motor flux estimation error is close to 1%, even when the motor speed crosses the zero value.

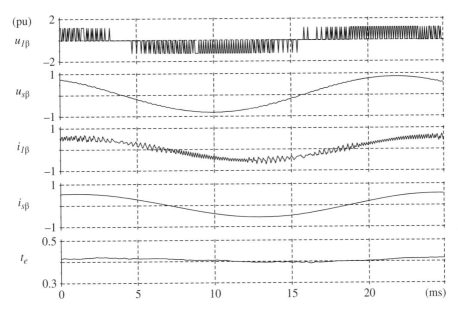

Figure 6.50 Sensorless steady state operation (load torque tl = 0.41 pu)

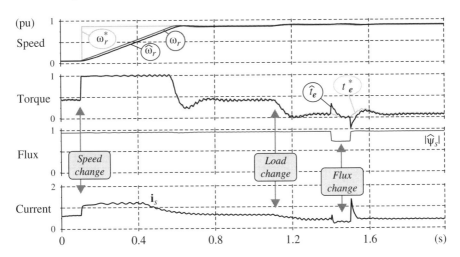

Figure 6.51 Simulation results of the sensorless drive under speed, load, and flux variation

Examples of the experimental results are presented in Figures 6.53 to 6.55. The robustness of the proposed system on the motor parameters changes and the uncertainty was tested by simulation and experimentation. The methodology of the robustness testing was that the motor parameters were changed in the observer equation while the motor was working in open loop U/f control. The robustness was tested by comparing the computed speed with the actual speed measured with an encoder. The parameters that were changed were the rotor and stator resistances and inductances as well as the mutual inductance. The range of the changes was +50% for the resistances and +20% for the inductances. Some examples are shown in Figures 6.56–6.59.

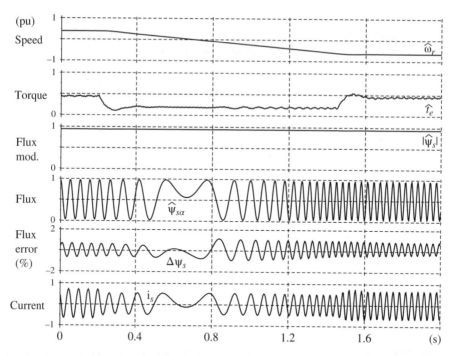

Figure 6.52 Simulation results of the sensorless drive during slow reversing with active load

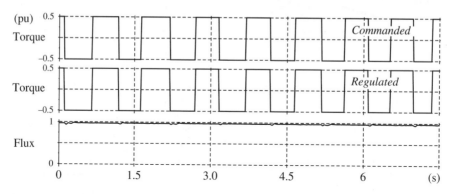

Figure 6.53 Closed loop torque control, experiments

It was observed that with the changes in the R_r and R_s parameters, the speed estimation error was less than 1%. However, inductances could have a strong influence on estimation error as observed for L_s in Figure 6.58 or for L_m in Figure 6.59.

Tests were carried out for different motor speeds and no significant differences were observed, for example, Figure 6.60.

Based on the provided experimental tests the final conclusions were collected in Table 6.1.

The motor parameter variations have no significant influence on observer robustness. The most significant parameters observed were inductances, especially L_m, which influence observer operation. However, the acceptable parameter range variation seems to be high

Control of Induction Motor Drives with LC Filters

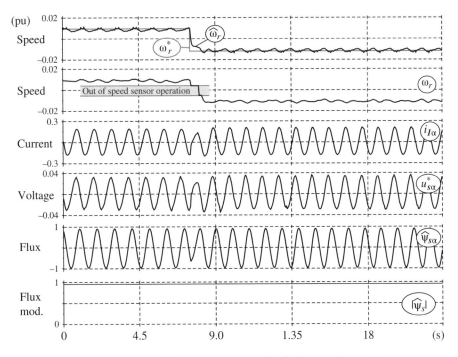

Figure 6.54 Experimental results, motor reversing in low speed operation range

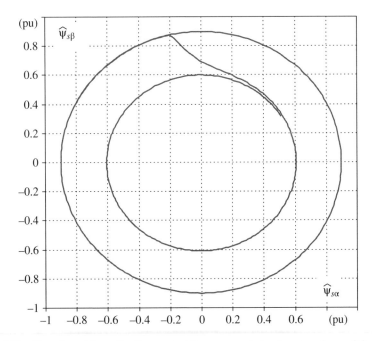

Figure 6.55 Experimental results, flux control for commanded value step change 0.9 → 0.6 pu

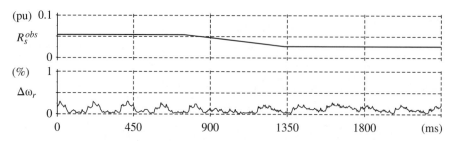

Figure 6.56 Speed estimation error for -50% R_s error in observer, experiment ($\omega_r = 0.2$ pu)

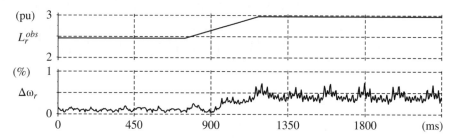

Figure 6.57 Speed estimation error for $+20\%$ L_s error in observer, experiment ($\omega_r = 0.2$ pu)

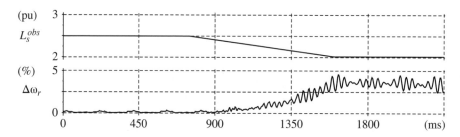

Figure 6.58 Speed estimation error for -20% L_s error in observer, experiment ($\omega_r = 0.2$ pu)

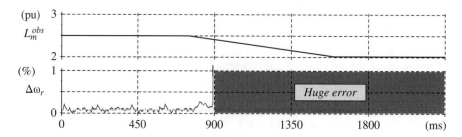

Figure 6.59 Speed estimation error for -20% L_m error in observer, experiment ($\omega_r = 0.2$ pu)

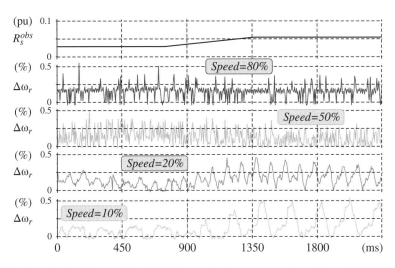

Figure 6.60 Speed estimation error for +50% R_s error in observer, experiment (ω_r = 0.1/0.2/0.5/0.8 pu)

Table 6.1 Observer robustness (the limit of 1% speed estimation error)

Parameter (pu)	Upper limit (%)	Lower limit (%)
Rs	+50	−50
Rr	+50	−50
Ls	+20	−4
Lr	+20	−12
Lm	+7	−10

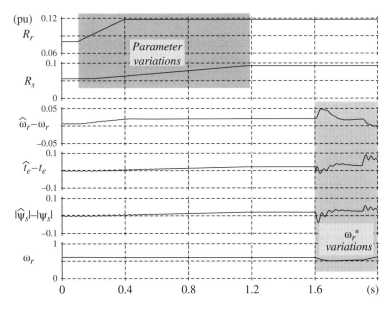

Figure 6.61 Robustness test for closed loop sensorless system, simulation (R_r and R_s grow at a constant speed 0.6 pu, then the speed regulation 0.6 → 0.5 → 0.6 pu is tested)

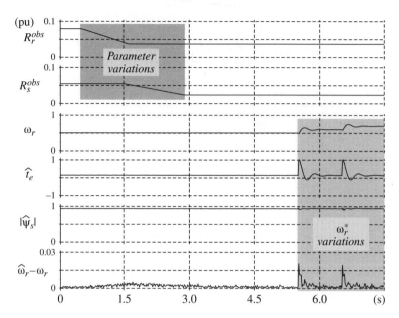

Figure 6.62 Robustness test for closed loop sensorless system, experiment (R_r and R_s decrease at a constant speed of 0.5 pu, then the speed regulation 0.5 → 0.6 → 0.7 pu is tested)

enough for correct drive operation. In all simulations and experiments, the closed loop system worked without any on-line parameter estimations.

The following tests were undertaken in simulation software for a full sensorless system. The motor parameters used in the motor model were changed. An example of the results is given in Figure 6.61.

In Figure 6.61, motor resistances R_r and R_s grow up to 150% of the nominal values. The system works at a constant speed 0.6 pu. The accuracy of the estimated speed, torque, and flux is tested. Finally, the speed regulation 0.6 → 0.5 → 0.6 pu. is tested. It is noticeable that the estimation error grew for each of the variables but did not exceed 3% at steady state.

The last step was undertaken to test the robustness of the closed loop sensorless experimental system. As in the previous experiments, the motor parameters used in the observer calculations were changed. An example of the results is given in Figure 6.62.

For the closed-loop system the acceptable parameter variations coincide with values given in Table 6.1.

6.8 Simulation Examples

6.8.1 Induction Motor Multiscalar Control with Multiloop Control of LC Filter

The model of induction motor multiscalar control with multiloop control of LC filter (*Chapter_6\Control_1\Simulator.slx*) is shown in Figure 6.63.

The model *Simulator.slx* consists of an induction motor drive with LC filter and closed-loop control scheme. For motor control the nonlinear control principle is used in addition to the

Control of Induction Motor Drives with LC Filters

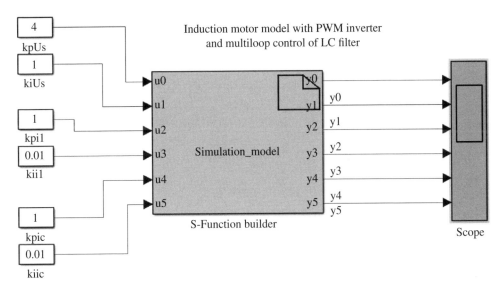

Figure 6.63 MATLAB®/Simulink model of induction motor multiscalar control with multiloop control of LC filter, *Chapter_6\Control_1\Simulator.slx*

multiloop principle for LC filter. The main simulation file is `MAIN.c`. Via Simulink parameter block it is possible to change the gains of multiloop controllers for stator voltage, inverter current, and capacitor current.

An example simulation results is presented in Figure 6.64.

In Figure 6.64, the next variables are presented (from top to bottom):

- `omegaR`, real motor speed
- `Usx`, motor supply voltage (alpha component)
- `isx`, stator current (alpha component)
- `isq`, stator current (q component)
- `i1qz`, reference inverter output current (q component)
- `i1q`, inverter output current (q component).

At the beginning the drive starts with V/f principle with ramp ($f^* = 25$ Hz), next at 0.2 s the control is switched to closed-loop mode. Next, the reference speed, reference flux, and motor load are changed in accordance with the following sequence:

```
if(timee>1100) x11z=0.8;  // speed change
if(timee>1450) x11z=-0.5; // speed change
if(timee>1700) m0=0.1;    // load torque change
if(timee>1850) m0=0.5;    // load torque chang
```

where *time* variable is the global time in milliseconds.

The initial transients of filter variables are results of the open loop control. For closed-loop control all variables are kept close to references.

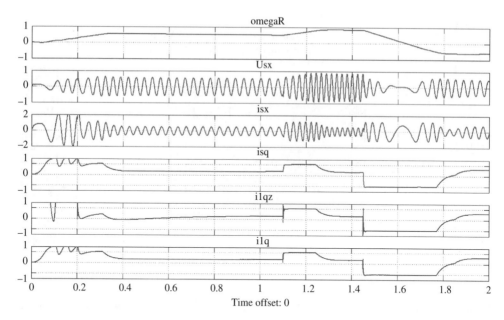

Figure 6.64 Example of the simulation results for induction drive closed-loop control (control system presented in Section 6.5)

In the example when the PWM is switched off, the inverter operates as ideal controlled voltage source. However, it is possible to switch on PWM by changing the adequate part of MAIN.C to next form:

```
// Inverter with PWM

if(cycle==1)
     {
     if (impulse<t0/2) {usx=0; usy=0;}
     else if (impulse<(t0/2+t1)) {usx=ud*ux[idu1];
       usy=ud*uy[idu1];}
     else if (impulse<(t0/2+t1+t2)) {usx=ud*ux[idu2];
       usy=ud*uy[idu2];}
     else {usx=0; usy=0;}
     }
else
     {
     if (impulse<t0/2) {usx=0; usy=0;}
     else if (impulse<(t0/2+t2)) {usx=ud*ux[idu2];
       usy=ud*uy[idu2];}
     else if (impulse<(t0/2+t2+t1)) {usx=ud*ux[idu1];
       usy=ud*uy[idu1];}
     else {usx=0; usy=0;}
     }
```

```
// Inverter without PWM
/*
usx=usxref; usy=usyref;
*/
```

The control system is working witch a sampling equals to inverter switching period:

```
/*------------------------------------------------*/
/* Control procedure sampled with Timp            */
/*------------------------------------------------*/
if (impulse>Timp)
        {
        impulse=0.0;
        // magnitude calculation
        fr=sqrt(frx*frx+fry*fry);
        is=sqrt(isx*isx+isy*isy);
        me=(frx*isy-fry*isx)*Lm/Lr;
        rofr=angle(frx,fry);
        // rotor current calculation
        irx=b1*frx-isx*b2; iry=b1*fry-isy*b2;
        // stator flux calculation
        fsx=Ls*isx+Lm*irx; fsy=Ls*isy+Lm*iry;
        fs=sqrt(fsx*fsx+fsy*fsy);
        // stator flux position
        rofs=angle(fsx,fsy);
        // multiscalar variables
        x11=omegaR;
        x12=fsx*isy-fsy*isx;
        x21=fsx*fsx+fsy*fsy;
        x22=fsx*isx+fsy*isy;
        fi=angle_pll(x22,x12);

//================================================================

        // flux and speed observer
        OBSERVER_FUNCTION(trapH);

        frrx=(Lr_o*fssx-(Ls_o*Lr_o-Lm_o*Lm_o)*isx)/Lm_o;
        frry=(Lr_o*fssy-(Ls_o*Lr_o-Lm_o*Lm_o)*isy)/Lm_o;

        fss=sqrt(fssx*fssx+fssy*fssy);
        frr=sqrt(frrx*frrx+frry*frry);

        isssx=(fssx-Kr*frrx)/ ((1-Lm_o*Lm_o/(Ls_o*Lr_o))*
          Ls_o);
        isssy=(fssy-Kr*frry)/ ((1-Lm_o*Lm_o/(Ls_o*Lr_o))*
          Ls_o);
```

```
// stator flux vector angle calculation
rofss=angle(frrx,frry);
if(rofss>(2*M_PI)) rofss-=(2*M_PI);
if(rofss<0.0)      rofss+=(2*M_PI);
// rotor flux angular frequency
if((rofss1-rofss)>(M_PI)) omegafss=(rofss+2*M_
PI-rofss1)/(trapH);
   else if((rofss-rofss1)>(M_PI)) omegafss=(rofss-2*
     M_PI-rofss1)/(trapH);
   else omegafss=(rofss-rofss1)/(trapH);
rofss1=rofss; // memorize previous value

cos_rofss=cos(rofss); // cos and sin functions in Park
   transformation
sin_rofss=sin(rofss);

// rotor slip HA-R
x12rr=(frrx*isy-frry*isx);
merr=x12rr;
x21rr=frrx*frrx+frry*frry;
omega2rr=Rr_o*Lm_o/Lr_o*x12rr/nzero(x21rr,0.01);

// motor speed
omegarr=omegafss-omega2rr;
```
//===
```
// Control system variables
if(START_REG==1) // sensorless control
    {
    x11reg=omegarrF;
    fsreg=fss;
    frreg=frr;
    rofsreg=rofrreg=rofss;
    mereg=merr;
    }
  else          // control based on measured values
    {
    x11reg=x11;
    x12reg=x12;
    x21reg=x21;
    x22reg=x22;
    fsreg=fs;
    fsxreg=fsx;
    fsyreg=fsy;
    frreg=fr;
    rofsreg=rofs;
    rofrreg=rofr;
```

```
            mereg=me;
        }
        cos_ro=cos(rofsreg); // cos and sin functions in Park
           transformation
        sin_ro=sin(rofsreg);
        // transformation xy->dq
        isd=isx*cos_ro+isy*sin_ro;
        isq=-isx*sin_ro+isy*cos_ro;

        // controllers
        TRAPEZ(Np,Nk,trapH,X);

        if(STER_SKALAR==1)
            {   // V/f control
            omegaU+=0.0005;
            if(omegaU>omegaUrozruch) omegaU=omegaUrozruch;
            US=omegaU;
            if(US<0.1) US=0.1;
            roUU+=omegaU*Timp;
            if (roUU>2*M_PI) roU-=2*M_PI;
            usxref=US*cos(roUU);
            usyref=US*sin(roUU);
            }
        else
            {   // closed loop control
            usxref=USdPWM*cos_ro-USqPWM*sin_ro;
            usyref=USdPWM*sin_ro+USqPWM*cos_ro;
            }

        PWM();
        isxm=isx; isym=isy; // measured currents
        }
```

All the controllers are calculated using function TRAPEZ(Np,Nk,trapH,X), which includes Euler method for integration. The function TRAPEZ(Np,Nk,trapH,X) is defined in the file *TRAPEZ.C*:

```
void TRAPEZ(int na, int NA, double HA, double ZA[32])
{
static double DZA[32], DXA[32], XA[32];
static int ja;

TRAPDERY(DZA,ZA);
for(ja=na;ja<=NA;ja++) XA[ja]=ZA[ja]+HA*DZA[ja];
TRAPDERY(DXA,XA);
for(ja=na;ja<=NA;ja++) { ZA[ja]=ZA[ja]+.5*(HA*DZA[ja]+
  HA*DXA[ja]);}
}
```

The subroutine TRAPDERY() includes the PI controllers of the system:

```c
/*--------------------------------------------------------------*/
/* PI controllers */
/*--------------------------------------------------------------*/
void TRAPDERY(double DV[32], double V[32])
{
// ------- Motor controllers ---------
/* x11 controller */
Ex11=x11z-x11reg;
DV[4]=Ex11*ki11;
V[4]=limit(V[4],1.0);
x12z=limit(kp11*Ex11+V[4],1.0);

/* x12 controller */
Ex12=x12z-x12reg;
DV[3]=Ex12*ki12;
V[3]=limit(V[3],2.0);
m1=limit(kp12*Ex12+V[3],2.0);

/* x21 controller */
Ex21=x21z-x21reg;
DV[2]=Ex21*ki21;
V[2]=limit(V[2],1.5);
m2=limit(kp21*Ex21+V[2],1.5);

// decoupling
U1=(m1-x11reg*(x22reg-b3_o*x21reg))/b4_o;
U2=0.5*(m2-x21reg/Tv2)+Rs_o*x22reg;

usxz=(U2*fsxreg-U1*fsyreg)/nzero(x21reg,0.001);
usyz=(U1*fsxreg+U2*fsyreg)/nzero(x21reg,0.001);

// transformation xy->dq
usdz= usxz*cos_ro+usyz*sin_ro;
usqz=-usxz*sin_ro+usyz*cos_ro;

// ------- LC FILTER MULTILOOP CONTROLLERS ---------
// Usd controller
EUsd=usdz-Usd;
V[5]=limit(V[5],2.0);
DV[5]=0.05*kiUs*EUsd;
icdz=limit(kpUs*EUsd+V[5],2.0); // PI
//icdz=limit(kpUs*EUsd,2.0); // P

// Usq controller
EUsq=usqz-Usq;
```

```
V[6]=limit(V[6],2.0);
DV[6]=kiUs*EUsq;
icqz=limit(kpUs*EUsq+V[6],2.0); // PI
//icqz=limit(kpUs*EUsq,2.0); // P

// Back-EMF decoupling for isd and isq
i1dz=icdz+isd-Cxy*omegafss*Usq;
i1qz=icqz+isq+Cxy*omegafss*Usd;

// i1d controller
Ei1d=i1dz-i1d;
V[7]=limit(V[7],1.0);
DV[7]=kii1*Ei1d;
USdPWM=limit(kpi1*Ei1d+V[7],1.0);

// i1q controller
Ei1q=i1qz-i1q;
V[8]=limit(V[8],1.0);
DV[8]=kii1*Ei1q;
USqPWM=limit(kpi1*Ei1q+V[8],1.0);

// Back-EMF decoupling for Usd and Usq
USdPWM=USdPWM+Usd-L1*omegafss*i1q;
USqPWM=USqPWM+Usq+L1*omegafss*i1d;
}
```

Each of the controllers has the same structure except Usd, Usq controller, of which it is possible to choose between PI or P controllers.

The motor model and filter model are modeled in function *F4DERY()*:

```
/*-------------------------------------------------*/
/* Load model                                      */
/*-------------------------------------------------*/
void F4DERY(double DV[32], double V[32])
{
tau=V[1];
DV[1]=1; /* Time */

/* motor model */
isx=V[2]; isy=V[3];
frx=V[4]; fry=V[5];
omegaR=V[6];

// motor directly connected to the inverter
//DV[2]=a1*isx+a2*frx+omegaR*a3*fry+a4*usx;
//DV[3]=a1*isy+a2*fry-omegaR*a3*frx+a4*usy;
```

```
// motor connected to the inverter vis LC filter
DV[2]=a1*isx+a2*frx+omegaR*a3*fry+a4*Usx;
DV[3]=a1*isy+a2*fry-omegaR*a3*frx+a4*Usy;

DV[4]=a5*frx+a6*isx-omegaR*fry;
DV[5]=a5*fry+a6*isy+omegaR*frx;
DV[6]=((frx*isy-fry*isx)*Lm/Lr-m0)/JJ;

// LC filter model
ucx=V[7];   ucy=V[8];
i1x=V[9];   i1y=V[10];

ufx=usx;
ufy=usy;

icx=i1x-isx;
icy=i1y-isy;

Usx=Rc*(i1x-isx)+ucx;
Usy=Rc*(i1y-isy)+ucy;

DV[7]=icx/Cxy;
DV[8]=icy/Cxy;
DV[9]    =(ufx-i1x*R1-icx*Rc-ucx)/(L1+LM1);
DV[10]=(ufy-i1y*R1-icy*Rc-ucy)/(L1+LM1);

// transformations xy -> dq
Usd=Usx*cos_ro+Usy*sin_ro;
Usq=-Usx*sin_ro+Usy*cos_ro;
i1d=i1x*cos_ro+i1y*sin_ro;
i1q=-i1x*sin_ro+i1y*cos_ro;
icd=icx*cos_ro+icy*sin_ro;
icq=-icx*sin_ro+icy*cos_ro;
}
```

Changing control to sensorless requires new set of gains. Doing that is left to the reader.

6.8.2 Inverter with LC Filter and LR Load with Closed-Loop Control

The model of the system with inverter, LC filter, LR load, and closed-loop control (*Chapter_6\ Reg_LCu2\Simulator.slx*) is presented in Figure 6.65.

The model *Simulator.slx* consists of an inverter, LC filter, and LR load. The filter variables (output voltage *Usd, Usq*, and input current *i1d, i1q*) are controlled. The main simulation file is MULTI.C. Via Simulink parameter block it is possible to change the gains of multiloop controllers for filter output voltage (*kpUs, kiUs*), and for inverter output current (*kpi1, kii1*).

The example simulation results are presented in Figure 6.65.

Control of Induction Motor Drives with LC Filters 195

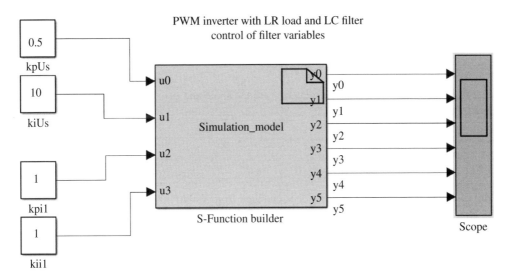

Figure 6.65 MATLAB®/Simulink model of system with inverter, LC filter, LR load, and closed loop control, *Chapter_6\Reg_LCu2\Simulator.slx*

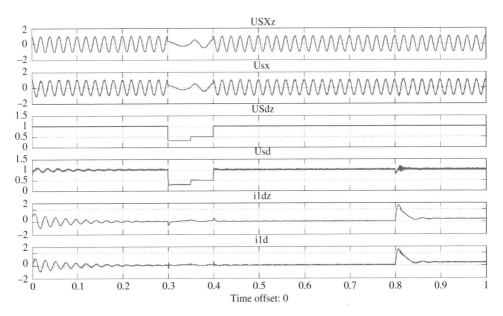

Figure 6.66 Example of the simulation results for LC filter closed loop control (subordinated multiloop control system presented in Figure 6.10)

In Figure 6.66 the next variables are presented (from top to bottom):

- USXz, reference filter output voltage (alpha component)
- Usx, real output voltage (alpha component)

- USdz, reference filter output voltage (d component)
- Usd, real output voltage (d component)
- ildz, reference inverter output current (d component)
- ild, inverter output current (d component).

The initial reference voltage and frequency are the same and equal to 1 pu. Next, the voltage and frequency are changed:

```
if(time>300) US=omegaU=0.3;
if(time>350) US=omegaU=0.5;
if(time>400) US=omegaU=1.0;
```

and finally at 0.8 s the load inductance is changed from 100 to 30%.

The model of the load and LC filter, defined in *F4DERY* function is:

```
void F4DERY(double DV[32], double V[32])
{
tau=V[1];
DV[1]=1; // time

/* RL load*/
isx=V[2]; isy=V[3];
DV[2]=(Usx-isx*Robc)/Lobc;
DV[3]=(Usy-isy*Robc)/Lobc;

/* LC filter */
ucx=V[7]; ucy=V[8];
ilx=V[9]; ily=V[10];
ufx=usx;
ufy=usy;
icx=ilx-isx;
icy=ily-isy;

DV[7]=icx/Cxy;
DV[8]=icy/Cxy;
DV[9]=(ufx-ilx*R1-icx*Rc-ucx)/(L1+LM1);
DV[10]=(ufy-ily*R1-icy*Rc-ucy)/(L1+LM1);

Usx=Rc*icx+ucx;
Usy=Rc*icy+ucy;
}
```

Parameters of the load and LC filter are (in pu):

```
// -------------- FILTER LCR ----------------------------------
// Filter parameters
double L1=0.034, Cxy=0.0347, Rc=0.0271, R1=0.001357,
  LM1=0.00597;
```

// ------------- RL load ----------------------------------
double Lobc=0.8524, Robc=0.0271;

The control procedure is included in *TRAPDERY* function:

```
void TRAPDERY(double DV[32], double V[32])
{
// angle for transformation
roUreg=roU;

// transformation xy -> dq
USdz= USXz*cos(roUreg)+USYz*sin(roUreg);
USqz=-USXz*sin(roUreg)+USYz*cos(roUreg);

// transformation xy -> dq
Usd= Usx*cos(roUreg)+Usy*sin(roUreg);
Usq=-Usx*sin(roUreg)+Usy*cos(roUreg);
ucd= ucx*cos(roUreg)+ucy*sin(roUreg);
ucq=-ucx*sin(roUreg)+ucy*cos(roUreg);

// transformation xy->dq
isd= isx*cos(roUreg)+isy*sin(roUreg);
isq=-isx*sin(roUreg)+isy*cos(roUreg);
i1d= i1x*cos(roUreg)+i1y*sin(roUreg);
i1q=-i1x*sin(roUreg)+i1y*cos(roUreg);
icd= icx*cos(roUreg)+icy*sin(roUreg);
icq=-icx*sin(roUreg)+icy*cos(roUreg);

// Usd controller
EUsd=USdz-Usd;
V[6]=limit(V[6],4.0);
DV[6]=kiUs*EUsd;
i1dz=limit(kpUs*EUsd+V[6],4.0);

// Ucq controller
EUsq=USqz-Usq;
V[7]=limit(V[7],4.0);
DV[7]=kiUs*EUsq;
i1qz=limit(kpUs*EUsq+V[7],4.0);

// i1d controller
Ei1d=i1dz-i1d;
V[8]=limit(V[8],1.5);
DV[8]=kii1*Ei1d;
USdPWM=limit(kpi1*Ei1d+V[8],1.5);
```

```
// i1q controller
Ei1q=i1qz-i1q;
V[9]=limit(V[9],1.5);
DV[9]=kii1*Ei1q;
USqPWM=limit(kpi1*Ei1q+V[9],1.5);
}
```

The filter control structure is simpler than the presented one in previous example. The influence of capacitor and load current is not taken into account.

Changing of the load from LR circuit to induction motor is left to the reader.

6.9 Summary

The aim of this chapter was to show the better operation of the drive when considering an inverter output filter. Considering the filter does not slow the speed control dynamic significantly provided that the filter parameters are properly designed, for example, according to the theory presented in Chapter 2.

When considering the filter, the complicated and time consuming task is selecting the parameters of the controllers. The complete structure with the filter requires four additional controllers, which doubles the computational complexity of the control algorithm. Nevertheless, filter is not essential since the controllers do not require significant computational power compared with the estimation algorithm.

References

[1] Guziński J. Speed observer for induction motor operating with sine-wave filter [in Polish]. Pomiary-Automatyka-Kontrola (Measurements and Automatics Control). 2007; **4**: 7–31.
[2] Adamowicz M, Guzinski J. Control of sensorless electric drive with inverter output filter. Fourth International Symposium on Automatic Control AUTSYM 2005, Wismar, Germany. September 22–23, 2005.
[3] Guziński J, Włas M. Sensorless control system of the induction motor and motor filter [in Polish]. *Sigma NOT: Przeglad Elektrotechniczny (Electrotechnical Review)*. 2006; **7–8**: 125–128).
[4] Guziński J, Włas M. Closed loop control of speed sensorless drive with motor filter. International Conference on Power Electronics, Intelligent Motions and Power Quality PCIM 2006, Nuremberg, Germany. May 30–June 2, 2006.
[5] Guziński J. Compensation of the sine-wave filter influence on control of induction motor [in Polish]. Seminarium Naukowo Techniczne (Proceedings of Scientific–Technical Seminar) Technicon 2006, Gdańsk, Poland. October 25, 2006.
[6] Guziński J. Closed loop control of induction motor with inverter output filter use [in Polish]. *Przeglad Elektrotechniczny (Electrotechnical Review)*. 2007; **3**: 11–14.
[7] Guziński J. Closed loop control of AC drive with LC filter. Thirteenth International Power Electronics and Motion Conference EPE–PEMC 2008, Poznan, Poland. September 1–3, 2008.
[8] Guziński J. Sensorless AC drive control with LC filter. The Thirteenth European Conference on Power Electronics and Applications EPE'09, Barcelona, Spain. September 8–10, 2009.
[9] Levine WS. *The control handbook*. 2nd ed. Boca Raton, FL: CRC Press; 2010.
[10] Golnaraghi F, Kuo BC. *Automatic control systems*. 9th ed. New York: John Wiley & Sons, Inc.; 2009.
[11] Seliga R Koczara W. Multiloop feedback control strategy in sine-wave voltage inverter for an adjustable speed cage induction motor drive system. European Conference on Power Electronics and Applications, EPE'2001, Graz, Austria. August 27–29, 2001.

[12] Seliga R, Koczara W. High quality sinusoidal voltage inverter for variable speed ac drive systems. Tenth International Power Electronics and Motion Control Conference, EPE–PEMC 2002, Dubrovnik, Croatia. September 9–11, 2002.

[13] Seliga R, Koczara W. Instantaneous current and voltage control strategy in low-pass filter based sine-wave voltage dc/ac converter topology for adjustable speed PWM drive system. IEEE International Symposium on Industrial Electronics, ISIE'2002, L'Aquila, Italy. June 8–11, 2002.

[14] Kaźmierkowski MP, Krishnan R, Blaabjerg F. *Control in power electronics*. San Diego, CA: Academic Press; 2002.

[15] Kaźmierkowski MP, Tunia H. *Automatic control of converter-fed drives*. Warsaw, Poland: PWN-Elsevier Science Publishers; 1994.

[16] Salomaki J, Luomi J. Vector control of an induction motor fed by a PWM inverter with output LC filter. Fourth Nordic Workshop on Power and Industrial Electronics, NORPIE'04, Trondheim, Norway. June 14–16, 2004.

[17] Salomaki J, Luomi J. Vector control of an induction motor fed by a PWM inverter with output LC filter. *EPE Journal*. 2006; **16** (1): 37–43.

[18] Grzesiak L Pawlikowski A. Vector-controlled three-phase voltage source inverter producing a sinusoidal voltage for AC motor drives. The IEEE International Conference on "Computer as a Tool," EUROCON 2007, Warsaw, Poland. September 9–12, 2007.

[19] Krzemiński Z. Differential equations of induction motor with nonlinear control synthesis with regard to saturation of main magnetic path. *Rozprawy Elektrotechniczne (Electrotechnical Theses)*. 1988; **34** (1): 117–131.

[20] Guziński J, Abu-Rub H. Asynchronous motor nonlinear control with inverter output LC filter. Second Mediterranean Conference on Intelligent Systems and Automation, Zarzis, Tunisia. March 23–25, 2009.

[21] Krzemiński Z. Nonlinear control of induction motor. Tenth World Congress on Automatic Control, IFAC'87, Munich, Germany. July 27–31, 1987.

[22] Krzemiński Z. The industrial electronics handbook. In BM Wilamowski, JD Irwin, editors. *The industrial electronics handbook*. 2nd ed., vol. **4**: Power Electronics and Motor Drives. Boca Raton, FL: CRC Press; 2011. pp. 27-1–27-18.

[23] Guziński J, Abu-Rub H. Speed sensorless induction motor drive with motor choke and predictive control. *COMPEL: The International Journal for Computation and Mathematics in Electrical and Electronic Engineering*. 2011; **30** (3): 686–705.

[24] Balogun A, Ojo O, Okafor F, Karugaba S. Determination of steady-state and dynamic control laws of doubly fed induction generator using natural and power variables IEEE *Transactions on Industry Applications*. 2013; **49** (3): 1343–1357.

[25] Dębowski A, Chudzik P, Lisowski G. State transitions in vector controlled AC tram drive. Twelfth International Power Electronics and Motion Control Conference EPE-PEMC 2006, Portoroz, Slovenia. August 30–September 1, 2006.

[26] Bogalecka E. Control system of an induction machine. EDPE 1992, High Tatras-Stara Lesna, Slovakia, 1992.

[27] Abu-Rub H, Guzinski J, Krzeminski Z, Toliyat HA. Advanced control of induction motor based on load angle estimation. *IEEE Transactions on Industrial Electronics*. 2004; **51** (1): 5–14.

[28] Isidori A. *Nonlinear control systems*. 3rd ed. London: Springer-Verlag; 1995.

[29] Abu-Rub H, Guzinski J, Toliyat HA: An advanced low-cost sensorless induction motor drive. *IEEE Transactions on Industry Applications*. 2003; **39** (6): 1757–1764.

[30] Kubota H, Sato I, Tamura Y, Matsuse K, Ohta H, Hori Y. Regenerating-mode low-speed operation of sensorless induction motor drive with adaptive observer. *IEEE Transactions on Industry Applications*. 2002; **38** (4): 1081–1086.

[31] Rajashekara K, Kawamura A, Matsuse K. *Sensorless control of AC motor drives*, New York: IEEE Press; 1996.

[32] Krzemiński Z. Sensorless control of the induction motor based on new observer. International Conference on Power Electronics, Intelligent Motions and Power Quality PCIM'00, Nuremberg, Germany. June 6–8, 2000.

[33] Suwankawin S, Sangwongwanich S. Stability analysis and design guidelines for a speed-sensorless induction motor drive. Power Conversion Conference, Nagaoka, Japan. August 3–6, 1997.

[34] Bensiali N, Etien E, Omeiri A, Champenois G. *Sensorless control of induction motor: Design and stability analysis*. World Academy of Science, Engineering and Technology; 2010.

[35] Depenbrock M. Direct self control of inverter-fed induction machines. *IEEE Transactions on Power Electronics*. 1988; **3** (4): 420–429.

[36] Takahashi I, Noguchi T. A new quick-response and high efficiency control strategy of an induction machine. *IEEE Transactions on Industrial Applications*. 1986; **22**: 820–827.
[37] Buja G, Kazmierkowski MP. Review of direct torque control of AC motors. *IEEE Transaction on Industrial Electronics*. 2004; **51** (4): 744–757.
[38] Abu-Rub H, Iqbal A, Guzinski J. *High performance control of AC drives with MATLAB/Simulink models*. New York: John Wiley & Sons, Ltd; 2012.
[39] Monmasson E, Naassani AA, Louis JP. Extension of the DTC concept. *IEEE Transactions on Industrial Electronics*. 2001; **48** (3): 715–717.
[40] Guziński J, Flisikowski Z. Practical application of induction motor direct torque control [in Polish]. Proceeding of VIII Seminarium towarzyszące VIII Targom Producentów, Kooperantów i Sprzedawców Zespołów Napędowych i Układów Sterowania (VIII Seminar of Electric Drive and Control Systems Development), Gdansk, Poland. February 12–14, 2002.
[41] Swierczynski D, Wojcik P, Kazmierkowski MP, Janaszek M. Direct torque controlled PWM inverter fed PMSM drive for public transport. Proceedings of the Tenth IEEE International Workshop on Advanced Motion Control, AMC '08, Trento, Italy. March 26–28, 2008.
[42] Xue X, Xu X, Habetler TG, Divan DM. A low cost stator flux oriented voltage source variable speed drive. Proceedings of the IEEE-IAS Annual Meeting. October 7–12, 1990.
[43] Cruz SMA Cardoso AJM. Diagnosis of rotor faults in closed-loop induction motor drives. Forty-First IEEE–IAS Annual Meeting, Industry Applications Conference, Tampa, FL. October 8–12, 2006.
[44] Loh PC, Newman MJ, Zmood DN, Holmes DG. A comparative analysis of multiloop voltage regulation strategies for single and three phase UPS systems. *IEEE Transactions on Power Electronics*. 2003; **18** (5): 1176–1185.
[45] Salomäki J, Piippo A, Hinkkanen M, Luomi J. Sensorless vector control of PMSM drives equipped with inverter output filter. Proceedings of the Thirty-Second Annual Conference of the IEEE Industrial Electronics Society IECON 2006, Paris, France November 6–10, 2006.
[46] Ryan MJ, Lorenz RD. A high performance sine wave inverter controller with capacitor current feedback and back-EMF decoupling. Proceedings of the Twenty-Sixth Annual IEEE Power Electronics Specialists Conference, PESC'95, Atlanta, GA. June 18–22, 1995.

7

Current Control of the Induction Motor

7.1 Introduction

The application of motor inductors, similar to the more complex inverter output filter systems, requires consideration of their control structure. The motor inductor, which is connected between the inverter output and motor input terminals, leads to a further voltage drop. For this reason, the motor is supplied with a lower voltage than the generated voltage of the inverter. Ignoring this drop may influence the function of many drive control systems. Especially in drive systems with a predictive current control structure, which require parameter knowledge of the controlled object, this will have a fundamental importance.

Figure 7.1 presents the general structure of the predictive current control in a drive system with an inverter-supplied squirrel cage motor.

It is possible, in predictive current control, to switch the inverter transistors with an adequate time lead. This allows the motor current waveform to be formed in a more precise way compared to other controller structures. The system response of a defined switching state can be described based on the object knowledge. Furthermore, in predictive control it is possible to reach an improved control dynamic of the motor current, compared to, for example, linear proportional-integral (PI) controllers.

Different predictive control algorithms are applied in drive systems. An overview and a comparison of predictive controller solutions are presented in other sources [1–3]. A further subdivision of the predictive control methods is described in Table 7.1 [1].

A model based predictive control (MBPC) is presented elsewhere [2], which distinguishes different control types:

- continuous control set (CS-MPC)
- finite control set (FS-MPC).

Variable Speed AC Drives with Inverter Output Filters, First Edition. Jaroslaw Guzinski, Haitham Abu-Rub and Patryk Strankowski.
© 2015 John Wiley & Sons, Ltd. Published 2015 by John Wiley & Sons, Ltd.

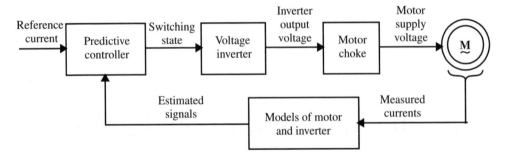

Figure 7.1 General structure of predictive current control of an inverter-fed motor

Table 7.1 Types of methods of current predictive control [1]

Predictive control		
Hysteresis-based	Trajectory-based	Model-based (MBPC)
Methods of hysteresis-based predictive control are applied in electrical drives as current control. Digital hysteresis current control can be treated as well as predictive control, whereas in digital hysteresis control, for instance, the transistor switching application is based on the foreseeable exceeding of the desired limits of the defined current in the orthogonal coordinates [4]. The most uncomplicated control version is done by relays; it is also called a *bang-bang controller*	Applying predictive control to the desired value of the trajectory current forces the course of chosen state variables according to the past calculated trajectory. These predictive control methods allow drive system control to be implemented without an additional loop of current control	In predictive control, based on the object model control dependencies, the previous states were analyzed to determine further transistor switching combinations, considering a longer time range and not only the closed algorithm step
	Among the predictive control methods, there are further systems such as direct torque control (DTC) and direct speed control (DSPC) [5]	Algorithms used for generalized predictive control (GPC) are presented elsewhere [6]. An overview of MBPC methods is given in another source [7].
The combination of hysteresis and trajectory control is known as *sliding mode control*		

In the case of CS-MPC, the control signal continuity consists of the controller output signal, which is the desired voltage of the pulse width modulation (PWM) modulator.

The following subsection will present the drive system with a motor inductor for a stator current control [2]. For this, an MBPC controller with a continuous control signal (CS-MPC) was implemented. The general structure of the MBPC controller is presented in Figure 7.2.

The main component of the MBPC controller, illustrated in Figure 7.2, is the object model, which is applied to the predictive behavior of the control object (CO). The procedure of foreseeable object behavior is divided into two parts:

Current Control of the Induction Motor

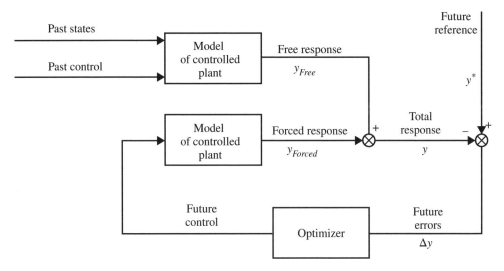

Figure 7.2 General structure of predictive current control according to object model, MBPC

- free response, which is the expected output y_{free} of the control object in a future time interval $(k + 1)$ considering the control signal for $(k + 1)$, which will be zero,
- forced response, which is an additional output component y_{forced} of the system model, corresponding to the control signal of the further time interval and noted as signal $u(k + 1)$.

The available application of the value y is the stator current, whereas the control signal is the motor supply voltage.

The sum of both values y_{free} and y_{forced} is the system response y, which is compared to the desired value y^*. Calculating the difference between y and y^* results in the control error Δy, which allows the control signal to be determined to bring the control error to zero.

The concept of two predictive current controllers: current control with prediction of the output voltage and furthermore a current controller with a simplified load model are described elsewhere [8]. Moreover, such controllers were applied in a sensorless squirrel cage-induction motor system directly supplied by a voltage feed inverter. Implementation of these controllers in a drive system with a motor inductor requires the consideration of this extra motor inductance.

7.2 Current Controller

The following subsection presents a predictive current controller that is modified for the requirements of a drive system with a motor inductor. Load model state variables in the controller are determined for the state observer, which is simultaneously used for control in a sensorless electrical drive. Thanks to that, the implementation of a further model equation is avoided. In the predictive controller of Krzemiński and Guziński [8], an induction motor was modeled as a electrical circuit with series connected inductances and electromotive force (EMFs) with, neglecting the motor resistance. Further, knowledge of the real value of the EMF is required for the right controller operation. In the system presented in Krzemiński and Guziński [8], the motor EMF was determined for simplified motor dependencies. A more precise determination of EMF is possible by applying a motor state observer. However, the application of an additional EMF controller is not necessary in the case of a current controller implementation with a sensorless motor control algorithm. This is caused by the applied observer flux control, which is used by the superior control system. It will be sufficient to expand the dependencies of the flux observer for the EMF motor calculation.

All in all, it also proved an advantage to apply one disturbed observer version, which was presented in Chapter 5 because the motor EMF is concealed in the equations. The disturbance ξ occurring in the observer could be treated as the motor EMF.

The function of the current controller is based on the assumption that the PWM of the voltage inverter operates in a mode in which the real inverter output voltage $\mathbf{u}_s$ is equal to the desired value $\mathbf{u}_s^*$.

$$\mathbf{u}_s \approx \mathbf{u}_s^* \tag{7.1}$$

By the fulfillment of the condition (Equation 7.1) and neglecting the stator resistance, the squirrel-cage motor model can be described as follows:

$$\frac{d\mathbf{i}_s}{d\tau} = \frac{\mathbf{u}_s^* - \mathbf{e}}{L_\sigma} \tag{7.2}$$

where:

e is the electromotive force vector
u_s^* is the vector of the desired inverter output voltage

In the case of a drive system with a motor inductor (dependency [Equation 7.2]) the inductance of the choke L_1 must be considered:

$$\frac{d\mathbf{i}_s}{d\tau} = \frac{\mathbf{u}_s^* - \mathbf{e}}{L_1 + L_\sigma} \tag{7.3}$$

For the notations shown in Figure 7.3 and a sufficiently short switching period T_{imp} of the transistors, Equation 7.3 becomes:

$$\frac{\mathbf{i}_s(k) - \mathbf{i}_s(k-1)}{T_{imp}} = \frac{\mathbf{u}_s^*(k-1) - \hat{\mathbf{e}}(k-1)}{L_1 + L_\sigma} \tag{7.4}$$

For the time interval $(k - 1) \ldots (k)$ in Equation 7.4, the two known variables are the desired voltage $\mathbf{u}_s^*(k-1)$ and the measured current $\mathbf{i}_s(k-1)$. The remaining variables, $\mathbf{i}_s(k)$ and $\mathbf{e}(k-1)$, are unknown and have to be determined.

In Krzemiński and Guziński [8], the value of the EMF $\mathbf{e}(k-1)$ was predicted based on the known stored values of $\mathbf{e}(k-2)$ and $\mathbf{e}(k-3)$ of the previous calculation steps. Considering the

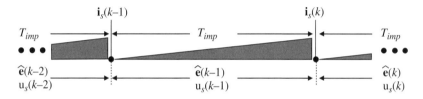

Figure 7.3 Description of signal samples for one transistor switching period T_{imp}

inductance of the motor inductor, the simplified dependencies of the EMF calculation for the intervals $(k-2)$ and $(k-3)$ are noted as follows:

$$\mathbf{e}(k-2) = \frac{\mathbf{i}_s(k-2) - \mathbf{i}_s(k-1)}{T_{imp}}(L_1 + L_\sigma) + \mathbf{u}_s(k-2) \quad (7.5)$$

$$\mathbf{e}(k-3) = \frac{\mathbf{i}_s(k-3) - \mathbf{i}_s(k-2)}{T_{imp}}(L_1 + L_\sigma) + \mathbf{u}_s(k-3) \quad (7.6)$$

A more accurate way to estimate the EMF is possible through the application of state observers of sensorless drives. Only one version of a speed observer, presented in Chapter 5, can be implemented in such drives, in which the generated disturbance is the EMF of the motor [9]:

$$\hat{\mathbf{e}} = \omega_r \hat{\mathbf{\psi}}_r = \hat{\xi} \quad (7.7)$$

Thanks to the estimated disturbance ξ, it is not necessary to determine the EMF of the simplified Equations 7.5 and 7.6.

The calculation of the electromotive motor force $\hat{\mathbf{e}}(k-2)$ in the state observer shows a significant difference, compared to those presented so far in the literature, for example, in Rodriguez and colleagues [10], where differential equations were used to calculate the EMF.

During the observer calculations, the sample EMFs $\hat{\mathbf{e}}(k-2)$ and $\hat{\mathbf{e}}(k-3)$ are stored and used in the further calculation steps of the current controller. The EMF $\hat{\mathbf{e}}(k-1)$ vector value is predicted under the assumption that the amplitude and velocity of the EMF vector position change in the calculation steps $(k-2)$ and $(k-3)$ will not change. Apart from this, the variation of the vector position EMF $\Delta\varphi_e$ is the difference from other angles noted as $\varphi_e(k-3)$ and $\varphi_e(k-2)$. Based on the relation between the arcus tangens functions, the value $\Delta\varphi_e$ can be noted as follows:

$$\Delta\varphi_e = arctg \frac{\hat{e}_\alpha(k-3)\hat{e}_\beta(k-2) - \hat{e}_\alpha(k-2)\hat{e}_\beta(k-3)}{\hat{e}_\alpha(k-2)\hat{e}_\alpha(k-3) + \hat{e}_\beta(k-2)\hat{e}_\beta(k-3)} \quad (7.8)$$

The predicted value $\hat{e}(k-1)$ is determined as follows:

$$\hat{\mathbf{e}}^{pred}(k-1) = \mathbf{C}_{EMF}\hat{\mathbf{e}}(k-2) \tag{7.9}$$

where the EMF vector transform matrix is given as:

$$\mathbf{C}_{EMF} = \begin{bmatrix} \cos(\Delta\varphi_e) & \sin(\Delta\varphi_e) \\ -\sin(\Delta\varphi_e) & \cos(\Delta\varphi_e) \end{bmatrix} \tag{7.10}$$

The predicted motor current for interval (k) is:

$$\mathbf{i}_s^{pred}(k) = \mathbf{i}_s(k-1) + \frac{\mathbf{u}_s^*(k-1) - \hat{\mathbf{e}}^{pred}(k-1)}{(L_1 + L_\sigma)} T_{imp} \tag{7.11}$$

The motor current control errors in the periods $(k-1)$ and (k) are computed as follows:

$$\Delta\mathbf{i}_s(k-1) = \mathbf{i}_s^*(k-1) - \mathbf{i}_s(k-1) \tag{7.12}$$

$$\Delta\mathbf{i}_s(k) = \mathbf{i}_s^*(k) - \mathbf{i}_s(k) \tag{7.13}$$

Moreover, the desired value of the motor current $\mathbf{i}_s^*(k)$ is described through the superior motor control. To minimize the current control errors in the interval $(k+1)$, a desired voltage vector $\mathbf{u}_s^*(k)$ is necessary:

$$\mathbf{u}_s^*(k) = \frac{\mathbf{i}_s^*(k+1) - \mathbf{i}_s^{pred}(k) + \mathbf{D}_{Is}}{T_{imp}}(L_1 + L_\sigma) + \hat{\mathbf{e}}^{pred}(k) \tag{7.14}$$

The presented output signal in Equation 7.14 of the current controller corrective behavior component $\mathbf{D}_{Is}$ is described as follows:

$$\mathbf{D}_{Is} = W_1 \mathbf{C}_{EMF} \Delta\mathbf{i}_s(k) + W_2 \mathbf{C}_{2EMF} \Delta\mathbf{i}_s(k-1) \tag{7.15}$$

and furthermore:

$$\hat{\mathbf{e}}^{pred}(k) = \mathbf{C}_{2EMF}\hat{\mathbf{e}}(k-2) \tag{7.16}$$

where:

$$\mathbf{C}_{2EMF} = \begin{bmatrix} \cos(2\Delta\varphi_e) & \sin(2\Delta\varphi_e) \\ -\sin(2\Delta\varphi_e) & \cos(2\Delta\varphi_e) \end{bmatrix} \tag{7.17}$$

Current Control of the Induction Motor

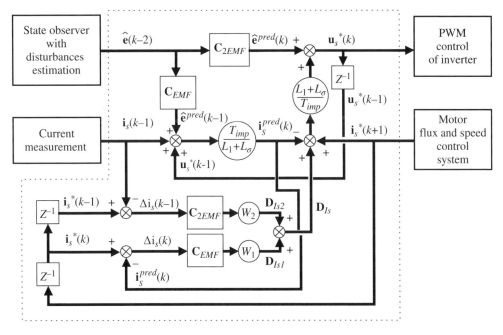

Figure 7.4 Structure of predictive current control for a drive system with motor inductor

The coefficients W_1 and W_2 noted in Equation 7.15 are fitted to minimize the current control error. A solution similar to a grid inverter operation was applied, where the coefficients W_1 and W_2 were chosen by a neuronal network [11]. However, the integral criterion was applied elsewhere [12] as:

$$I_{IAE} = \int_{t1}^{t2} |e_{rr}(t)| dt \qquad (7.18)$$

where: $e_{rr}(t)$ is the current control error in the time period $t1 \ldots t2$.

The proposed current controller structure is presented in Figure 7.4.

Corresponding to the general MBPC theory, the presented current controller includes typical MBPC blocks such as:

- predictive object model,
- costs function,
- and predictive controller that minimizes the costs function.

7.2.1 Predictive Object Model

The predicted stator current is determined for interval (k) based on Equation 7.11 in the predictive object model.

7.2.2 Costs Function

The costs function is described in Equation 7.15.

7.2.3 Predictive Controller

The predictive controller, which minimizes the costs function, is noted in Equation 7.14. The controller equation includes a costs function and the future current controller error. In practice, the CS-MPC has an infinite possibility of output controller states, which causes difficulties in the implementation of an optimal real-time control. In dependency Equation 7.14, an empirical approach was applied.

A representation of the controller with typical model predictive stator current control (MPC) blocks is shown in Figure 7.5.

For the presented current controller, the predictive time-horizon is equal to one switching period, also called the *one-step prediction horizon*, N = 1.

7.3 Investigations

In this section some example results are presented. The current waveforms presented in Figure 7.6 were recorded in open loop drive operation without a superior motor control system.

The waveforms of Figure 7.6 present a proper (0 ÷ 100 ms) and an incorrect operation (100 ÷ 250 ms) of the current controller. During correct functioning, the control error is less than 5%. Afterward, at 100 ms, the inductance value of the motor inductor L_1 is eliminated from the algorithm, which causes an immediate and significant increase of the control error that excludes further system operation. Such a reaction testifies to the need to consider the inductance L_1 in the predictive control algorithm.

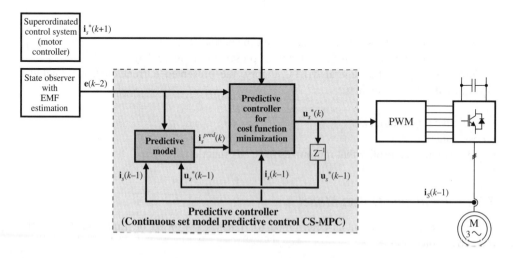

Figure 7.5 Structure of a predictive current controller for a drive system with motor inductor with differentiation of typical controller blocks of MBPC

Current Control of the Induction Motor

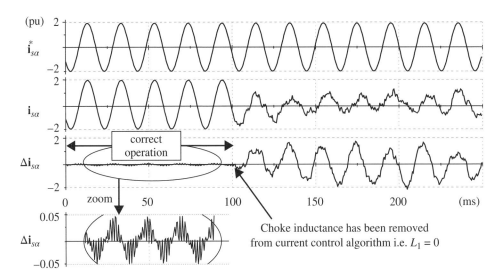

Figure 7.6 Functionality of predictive current controller. At time 100 ms, the inductor L_1 is eliminated from the controller equation, $L_1 = 0$ H (motor 1.5 kW; 1410 rpm; 400 VY; $L_\sigma = 39$ mH; inductor $L_1 = 11$ mH)

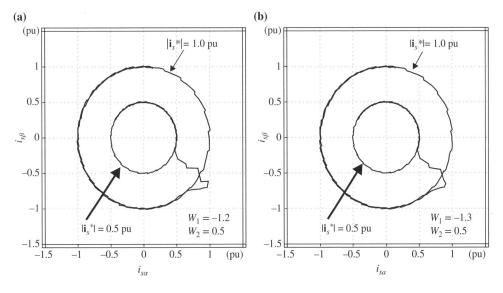

Figure 7.7 Hodograph of the motor current components during the desired amplitude changes for different values of W_1 (motor 1.5 kW; 1410 rpm; 400 V, $L_\sigma = 39$ mH; inductor $L_1 = 11$ mH)

Figure 7.7 illustrates the transient waveforms for an abrupt increase of the desired motor current amplitude with a current controller operating without a superior system control.

The step change of the desired amplitude of the motor current from 0.5 to 1 pu takes place in a few controller process steps. The current controller properties depend on the gain W_1 and the corrective term W_2. Figure 7.7 presents the exemplary influence of the gain value W_1 on the current controller properties.

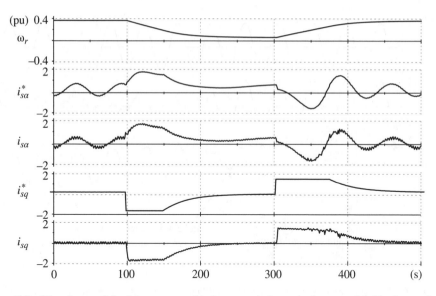

Figure 7.8 Waveforms of the drive system with the motor inductor in the case of field orientated control, connected with a predictive current controller during the variation of the motor speed (motor 1.5 kW; 1410 rpm; 400 VY; L_σ = 39 mH; inductor L_1 = 11 mH)

Figure 7.8 illustrates an exemplary representation of a field-orientated control system in connection with a current controller. During the investigations, which are presented in Figure 7.8, a change of the desired motor speed was forced. This in turn forces a modification of the desired motor currents i_{sa} and i_{sq}. A correct process of current control is discernible, the actual value follows reference signal precisely.

Figures 7.9 and 7.10 present the results of the functioning experimental system with predictive control. The controller operated with constant quantities of the desired amplitude and stator current frequency without superior control. The observer was used only for the motor EMF calculation.

Figure 7.9 demonstrates the waveforms of the desired and controlled motor currents, where the top waveform presents the current control error. In steady state this error does not exceed the 2% boundary. Figure 7.10 illustrates the current controller function in transients during a step change of the motor current amplitude. The controller operation guarantees a fast reaction to the desired value change. No overshoots or oscillations are observed.

Figure 7.11 presents the sensorless operation of the squirrel cage motor control with the application of the predictive current controller and the flux, speed, and EMF observer.

7.4 Simulation Examples of Induction Motor with Motor Choke and Predictive Control

The model of induction motor with motor choke and predictive control (*Chapter_7\Example_1\Simulator.slx*) is shown in Figure 7.12.

Current Control of the Induction Motor

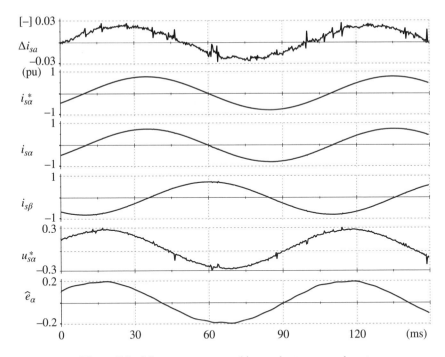

Figure 7.9 Motor current control in steady state, experiment

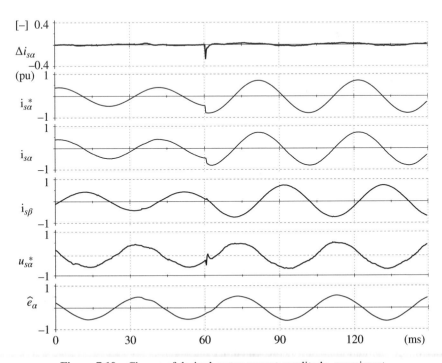

Figure 7.10 Change of desired motor current amplitude, experiment

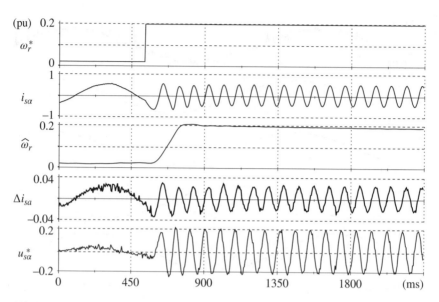

Figure 7.11 Drive system waveforms with predictive current controller and superior field-orientated control during the variation of the desired motor speed, experiment

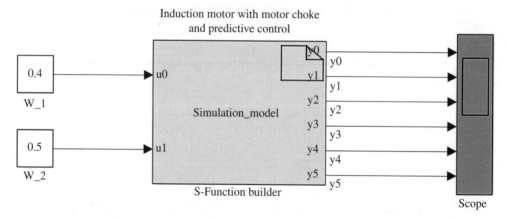

Figure 7.12 MATLAB®/Simulink model of induction motor with motor choke and predictive control, *Chapter_7\Example_1\Simulator.slx*

The model *Simulator.slx* consists of an induction motor drive with motor choke and closed-loop control. For the closed-loop control, the load angle control principle is used (see Section 6.6 and Fig. 6.39). The inner current controller is predictive one with a structure presented in Figure 7.4. The main simulation file is REG_PRED.c. Via Simulink parameter block it is possible to change the gains of current controller W_1 and W_2 (variables W_1 and W_2 in the simulation program).

An example simulation result is presented in Figure 7.13.

Current Control of the Induction Motor

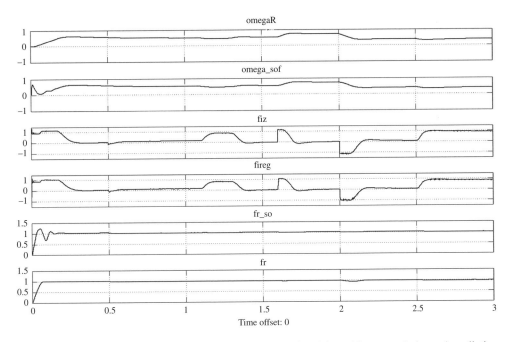

Figure 7.13 Example of the simulation results for induction drive with motor choke and predictive control (structure of predictive controller presented in Fig. 7.4)

In Figure 7.13 the next variables are presented (from top to bottom):

- omegaR, real motor speed
- omega_sof, estimated motor speed
- fiz, reference load angle
- fireg, real (controlled) load angle
- fr_so, estimated rotor flux (magnitude)
- fr, real rotor flux (magnitude)

At the beginning the closed-loop system is running based on real values of controlled signals. After 0.5 s, the drive is switched to sensorless mode. Next, the load torque and reference speed are changed.

The predictive control is performed with sample equal to inverter switching frequency. The controller input signals are αβ components of stator current:

```
//=============================================================
// reference motor current
    roI+=Timp*omegaIz;
    if( roI > 2*PI ) roI=0;
    isxref=ISz*cos(roI);
    isyref=ISz*sin(roI);
```

```
/* ------ PREDICTIVE CONTROLLER ------ */
isxm[4]=isxm[3]; isym[4]=isym[3]; // samples of measured current
isxm[3]=isxm[2]; isym[3]=isym[2];
isxm[2]=isxm[1]; isym[2]=isym[1];
isxz[2]=isxz[1]; isyz[2]=isyz[1]; // samples of reference current
isxz[1]=isxz[0]; isyz[1]=isyz[0];
isxz[0]=isxref;  isyz[0]=isyref;

Ux[4]=Ux[3]; Uy[4]=Uy[3]; // samples of reference voltage
Ux[3]=Ux[2]; Uy[3]=Uy[2];
Ux[2]=Ux[1]; Uy[2]=Uy[1];

/*
// EMF simple load model
Ex[3]=Ux[3]+LT*(isxm[3]-isxm[2]);
Ey[3]=Uy[3]+LT*(isym[3]-isym[2]);
Ex[4]=Ux[4]+LT*(isxm[4]-isxm[3]);
Ey[4]=Uy[4]+LT*(isym[4]-isym[3]);
*/

// increment of EMF vector position
K=angle(Ex[3]*Ex[4]+Ey[3]*Ey[4],-Ex[4]*Ey[3]+Ex[3]*Ey[4]);
if(K>PI) K=K-2*PI;

// EMF vector rotation
Ex[1]=Ex[3]*cos(2*K)+Ey[3]*sin(2*K);
Ey[1]=Ey[3]*cos(2*K)-Ex[3]*sin(2*K);
Ex[2]=Ex[3]*cos(K)+Ey[3]*sin(K);
Ey[2]=Ey[3]*cos(K)-Ex[3]*sin(K);

// EMF samples
Ex[4]=Ex[3];
Ey[4]=Ey[3];
//Ex[3]=EMFx; // EMF from rotor flux
//Ey[3]=EMFy;
//Ex[3]=EMFxs; // EMF from stator flux
//Ey[3]=EMFys;
Ex[3]=-1.0*zetaY; // EMF from disturbance observer
Ey[3]=zetaX;

// current prediction
isxo[1]=isxm[2]+(Ux[2]-Ex[2])/LT;
isyo[1]=isym[2]+(Uy[2]-Ey[2])/LT;

// calculation of current change
if((isxz[0]-isxz[1])*(isxz[1]-isxz[2])>0)
```

```
    {
    mode=1;
    DIx=-W_1*((isxz[1]-isxo[1])*cos(K)+(isyz[1]-isyo[1])*sin(K))
       +W_2*((isxz[2]-isxm[2])*cos(2*K)+(isyz[2]-isym[2])
       *sin(2*K));
    DIy=-W_1*((isyz[1]-isyo[1])*cos(K)-(isxz[1]-isxo[1])*sin(K))
       +W_2*((isyz[2]-isym[2])*cos(2*K)-(isxz[2]-isxm[2])
       *sin(2*K));
    }
  else
    {
    mode=-1;
    DIx=-1*((isxz[1]-isxo[1])*cos(K)+(isyz[1]-isyo[1])*sin(K));
    DIy=-1*((isyz[1]-isyo[1])*cos(K)-(isxz[1]-isxo[1])*sin(K));
    }

// reference voltage
Ux[1]=1.0*LT*(isxz[0]-isxo[1])+Ex[1]+LT*DIx;
Uy[1]=1.0*LT*(isyz[0]-isyo[1])+Ey[1]+LT*DIy;

// limit of reference voltage
usxref=limit(Ux[1],1.0); usyref=limit(Uy[1],1.0);
/* ----------- END OF CONTROLLER ------------- */
```

In the current controller procedure it is possible to change the methods of EMF calculation: from simple load model, from stator flux derivative, rotor flux derivative, or from disturbance observer.

The output signals of the predictive controller are reference voltage for PWM subroutine. The PWM subroutine is included in *REG_PRED.c* file:

```
//========== SVPWM algorithm ====================================
roU=angle(usxref,usyref);
USX=usxref; USY=usyref;
NroU=floor(roU/(PI/3));
switch(NroU)
    {
    case 0: idu1=1; idu2=2; break;
    case 1: idu1=2; idu2=3; break;
    case 2: idu1=3; idu2=4; break;
    case 3: idu1=4; idu2=5; break;
    case 4: idu1=5; idu2=6; break;
    case 5: idu1=6; idu2=1; break;
    }
wt=ux[idu1]*uy[idu2]-uy[idu1]*ux[idu2];
t1=Timp*(USX*uy[idu2]-USY*ux[idu2])/(ud*wt);
t2=Timp*(-USX*uy[idu1]+USY*ux[idu1])/(ud*wt);
```

```
t0=Timp-t1-t2;
if (t0<0) t0=0;
impulse=0;
cycle*=-1;
//================================================================
```

Based on calculated times *t1*, *t2*, and *t0* the inverter output voltage is found for actual time in subroutine of inverter model:

```
//==================== INVERTER MODEL ====================
if(cycle==1)
     {
     if (impulse<t0/2) {usx=0; usy=0;}
     else if (impulse<(t0/2+t1)) {usx=ud*ux[idu1]; usy=ud*uy[
  idu1];}
     else if (impulse<(t0/2+t1+t2)) {usx=ud*ux[idu2]; usy=ud*uy
  [idu2];}
     else {usx=0; usy=0;}
     }
  else
     {
     if (impulse<t0/2) {usx=0; usy=0;}
     else if (impulse<(t0/2+t2)) {usx=ud*ux[idu2]; usy=ud*uy[
  idu2];}
     else if (impulse<(t0/2+t2+t1)) {usx=ud*ux[idu1]; usy=ud*uy
  [idu1];}
     else {usx=0; usy=0;}
     }
//================================================================
```

The disturbance observer used for speed, flux, load torque, and EMF estimation has the structure presented in Chapter 5 (disturbance observer with complete model of the load). The structure of the observer includes the motor choke inductance which appears in observer coefficients (`Ls_so=Ls+L1;`). It is left to the reader to apply other observer structures given in Chapter 5.

7.5 Summary and Conclusions

The predictive stator current control of a squirrel-cage motor depends on knowledge of the object control parameters. However, consideration of only motor parameters in the predictive control algorithm is insufficient in the case of a drive system with a motor inductor. The motor choke inductance has an important role and must be implemented in the control procedure. Only the use of this parameter ensures an accurate control process. For a functioning predictive controller, definition of the EMF is necessary. This is possible through the introduction of equations that describe the control object model, that is, the motor and inductor. Nevertheless, in sensorless control systems, it is feasible to use an existing state observer for the EMF computations. This solution ensures a higher accuracy compared to simplified load model applications. A further advantage is given by the potential savings of the indispensable microprocessor calculations in the control system.

References

[1] Kennel R, Linder A. Predictive control of inverter supplied electrical drives. IEEE Thirty-First Annual Power Electronics Specialist Conference, PESC'03, Galway, Ireland. June 18–23, 2000.
[2] Cortés P, Kazmierkowski MP, Kennel RM, Quevedo DE, Rodríguez J. Predictive control in power electronics and drives. *IEEE Transactions on Industrial Electronics*. 2008; **55** (12): 4312–4324
[3] Kouro S, Cortes P, Vargas R, Ammann U, Rodriguez J. Model predictive control: a simple and powerful method to control power converters. *IEEE Transactions on Industrial Electronics*. 2009; **56** (6): 1826–1838.
[4] Krzemiński Z. Multiscalar model based control systems for AC machines. In BM Wilamowski, JD Irwin, editors. *The industrial electronics handbook*. 2nd ed., vol. **4**: Power Electronics and Motor Drives. Boca Raton, FL: CRC Press; 2011. pp. 27-1–27-18.
[5] Mutschler P. A new speed–control method for induction motors. Power Conversion and Intelligent Motions International Conference PCIM'98, Nuremberg, Germany. May 26–28, 1998.
[6] Clarke DW, Mohtadi C, Tuffs PS. Generalized predictive control. Part I. The basic algorithm, oraz: generalized predictive control. Part II. *Extensions and interpretations, Automatica*. 1987; **23** (2): 137–148.
[7] Clarke DW. Adaptive predictive control. *Annual Review in Automatic Programming*. 1996; **20**: 83–94.
[8] Krzemiński Z, Guziński J. Controller of output current of voltage source inverter with prediction of electromotive force. Electrical Drives and Power Electronics, Stara Leśna, Slovakia, październik. 5–7, 1999.
[9] Guziński J, Abu-Rub H. Asynchronous motor nonlinear control with inverter output LC filter. Second Mediterranean Conference on Intelligent Systems and Automation, Zarzis, Tunisia. March 23–25, 2009.
[10] Rodríguez J, Pontt J, Silva CA, Correa P, Lezana P, Cortés P, et al. Predictive current control of a voltage source inverter industrial electronics. *IEEE Transactions on Industrial Electronics*. 2007; **54** (1): 495–503.
[11] Krzemiński Z, Wojciechowski D. Control system for the PWM rectifier based on predictive current controller with neural network. International Conference on Power Conversion, Intelligent Motions and Power Quality, PCIM'2001, Nuremberg, Germany. June 19–21, 2001.
[12] Guziński J, Abu-Rub H. Speed sensorless induction motor drive with motor choke and predictive control. *COMPEL: The International Journal for Computation and Mathematics in Electrical and Electronic Engineering*. 2011; **30** (2): 686–705.

8

Diagnostics of the Motor and Mechanical Side Faults

8.1 Introduction

Assurance of the reliable operation of electric drives is an important aspect in modern industrial systems [1]. High system reliability is ensured by proper design and high-quality elements used in the design of the whole system. Various diagnostic solutions are used to increase the reliability level.

One up-to-date direction in the development of electric drives depends on the automatic detection of developing or existing faults. The principle is to ensure a diagnosis of the electric machine as well as other elements of the drives, for example, the transmission system.

This chapter deals with topics related to such motor faults, for example, rotor misalignment and eccentricity and rotor bar and bearing faults. For the diagnosis, the electromagnetic and load torques, which are computed using state observers from Chapter 5, are analyzed while considering the presence of motor filters. In this chapter, how to use the analysis of internal signals that appear in the closed-loop systems is also discussed with the drive diagnosis.

8.2 Drive Diagnosis Using Motor Torque Analysis

Bearing faults, rotor eccentricity, and drive misalignment are the most frequent faults in mechanical drives [2, 3]. Currently, bearing faults are particularly frequent compared with other types of faults. The nature of such faults was presented in Chapter 2. Many faults related to rotor eccentricity are caused by construction errors: improper balance, bearing clearance, and irregularities in the design of the bearing discs.

Motor faults can be diagnosed using many methods, such as noise analysis [4, 5] or electric current signals analysis, for example, motor current [6], instantaneous power [7], or axial flux [8]. In the diagnosis of faults in mechanical parts, electromagnetic torque analysis or load

---

Variable Speed AC Drives with Inverter Output Filters, First Edition. Jaroslaw Guzinski, Haitham Abu-Rub and Patryk Strankowski.
© 2015 John Wiley & Sons, Ltd. Published 2015 by John Wiley & Sons, Ltd.

Diagnostics of the Motor and Mechanical Side Faults

torque analysis is also used [1, 2, 6–8]. It was shown that the analysis of the torque signal in the frequency domain allows the identification of significant changes in the harmonics amplitude corresponding to the characteristic frequencies, which result from faults in the mechanical part of the drive [2]. Particular faults may cause an increase in torque harmonic magnitude. Such harmonics are directly proportional to the fault type.

The frequency that corresponds to the rotor eccentricity for a motor of p pairs of poles, fed by a voltage frequency, f_s, is as follows:

$$f_e = f_s \left[\frac{k_h}{p}(1-s) \pm 1 \right] \quad (8.1)$$

where:

s is the slip,
$k_h = 0, \pm1, \pm2, \ldots$ is the number of harmonics that correspond to the rotor eccentricity.

Information on motor torque could be obtained by measurement, for example, using a tensometer [9, 10]. However, direct measurement of the torque is rarely used in practice for economic reasons and because of the sensitivity of torque sensors to the disturbances. Inverter-fed drives are characterized by high disturbance levels. Therefore, alternative computational methods for torque identification are obviously used. Such methods may be based on the motor model in addition to the motor current and voltage [11–17].

Diagnosis of the torque transmission system becomes more difficult when using inverter output filters. The analysis method of the inverter output current cannot be adopted because the filtered current at the motor side is associated with the loss of essential information that is being used for the diagnosis.

Estimation of the motor-load torque when using a filter necessitates considering the filter similarly to the discussed speed and flux observers described in Chapter 5. It is essential to introduce additional equations that allow the load torque to be computed using the estimated speed and torque signals (Figure 8.1).

The measured stator current and the calculated rotor flux in the observer allow the actual motor torque to be computed as:

$$\hat{t}_e = \hat{\psi}_{r\alpha} \hat{i}_{s\beta} - \hat{\psi}_{r\beta} \hat{i}_{s\alpha} \quad (8.2)$$

Motor torque could be identified using the following motion equation:

$$\hat{t}_L = \hat{t}_e - J\frac{d\hat{\omega}_r}{d\tau} \quad (8.3)$$

For the identification of the motor load, it is essential to know the speed derivative, which could be computed numerically by remembering previous speed samples, for example, by using Taylor's series and the third order differential quotient:

$$\hat{\omega}'_r(k) = \frac{\hat{\omega}_r(k-2) - 4\hat{\omega}_r(k-1) + 3\hat{\omega}_r(k)}{2T_{imp}} \quad (8.4)$$

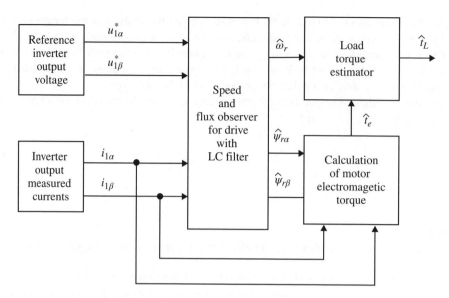

Figure 8.1 Block diagram of motor load torque estimation system

where:

$$\hat{\omega}'_r = \frac{d\hat{\omega}_r}{d\tau} \qquad (8.5)$$

With numerical differentiation, there is a risk of making high computational errors because of the subtraction of estimated speed samples with similar values. However, it is possible to obtain the speed derivative directly from the observer [18]. In the presented solution [18], the speed observer was used with the reduced disturbance model. In this book, the observer was presented in Section 5.2 while considering the filter. In this observer, the speed is computed using the following equation:

$$\hat{\omega}_r = S\sqrt{\frac{\hat{\zeta}_\alpha^2 + \hat{\zeta}_\beta^2}{\hat{\psi}_{r\alpha}^2 + \hat{\psi}_{r\beta}^2}} \qquad (8.6)$$

The speed derivative is computed analytically according to [18]:

$$\frac{d\hat{\omega}_r}{d\tau} = \frac{1}{|\hat{\psi}_r|}\left(\left(\hat{\zeta}'_\alpha \cos(\rho_\zeta) + \hat{\zeta}'_\beta \sin(\rho_\zeta)\right)S - \hat{\omega}_r\left(\hat{\psi}'_{r\alpha}\cos(\rho_{\psi r}) + \hat{\psi}'_{r\beta}\sin(\rho_{\psi r})\right)\right) \qquad (8.7)$$

where: ρ_ζ, $\rho_{\psi r}$ is the angular position of the vectors $\hat{\zeta}$ and $\hat{\psi}_r$.

Derivative values: $\hat{\zeta}'_\alpha = \frac{d\hat{\zeta}_\alpha}{d\tau}$, $\hat{\zeta}'_\beta = \frac{d\hat{\zeta}_\beta}{d\tau}$, $\hat{\psi}'_{r\alpha} = \frac{d\hat{\psi}_{r\alpha}}{d\tau}$, $\hat{\psi}'_{r\beta} = \frac{d\hat{\psi}_{r\beta}}{d\tau}$ are described using Equations 5.24 to 5.27.

Diagnostics of the Motor and Mechanical Side Faults

Similarly, it is possible to identify the speed derivative for the observers with extended and full disturbance models, in which the angular rotor speed is computed as:

$$\hat{\omega}_r = \frac{\hat{\zeta}_\alpha \hat{\psi}_{r\alpha} + \hat{\zeta}_\beta \hat{\psi}_{r\beta}}{\hat{\psi}_{r\alpha}^2 + \hat{\psi}_{r\beta}^2} \qquad (8.8)$$

Based on Equation 8.8 the derivative of estimated speed is:

$$\frac{d\hat{\omega}_r}{d\tau} = \frac{1}{\hat{\psi}_\alpha^2 + \hat{\psi}_\beta^2} \left(\hat{\psi}_\alpha \frac{d\hat{\zeta}_\alpha}{d\tau} + \hat{\psi}_\beta \frac{d\hat{\zeta}_\beta}{d\tau} - \hat{\zeta}_\alpha \frac{d\hat{\psi}_{r\alpha}}{d\tau} - \hat{\zeta}_\beta \frac{d\hat{\psi}_{r\beta}}{d\tau} \right) \qquad (8.9)$$

Equation 8.9 has a simpler form than Equation 8.7. This is more beneficial in the practical implementation of the system used for computing load torque in the microprocessor system because of the elimination of trigonometric functions from this equation.

Depending on the type of speed observer used, the motor load torque could be calculated using the drive motion equation (8.3), of which the speed derivative is computed using Equation 8.7 or 8.9.

The computation of load torque from the motion equations is a simple solution; however, it is associated with errors caused by the assumed simplifications in the mechanical system model. More precise methods could be obtained using load torque observer systems.

Motor load torque could be estimated, for example, in the state observer presented in Kadowaki and colleagues [19] and Ohishi and colleagues [20] and applied in a system used for the detection of train wheel slip. The load torque observer used is the Gopinath observer [21], designed for a system in which the disturbance (load torque) is an additional state variable. If assuming that the load torque slowly changes compared with the system dynamic, it is possible to consider its derivative as zero.

The load torque m_0 observer system used has the following form [19, 20]:

$$\frac{d}{d\tau}\begin{bmatrix} z_1 \\ z_2 \end{bmatrix} = \begin{bmatrix} 0 & -k_{1L} \\ 1 & -k_{2L} \end{bmatrix} \begin{bmatrix} z_1 \\ z_2 \end{bmatrix} + \begin{bmatrix} k_{1L} k_{2L} J \\ \left(k_{2L}^2 - k_{1L} \right) J \end{bmatrix} \hat{\omega}_r + \begin{bmatrix} k_{1L} \\ k_{2L} \end{bmatrix} \hat{m}_e \qquad (8.10)$$

$$\hat{t}_L = z_2 - k_{2L} J \hat{\omega}_r \qquad (8.11)$$

where:

k_{1L}, k_{2L} are the gain factors.
z_1, z_2 are the internal observer variables.

The variable z_1 is proportional to the motor angular speed:

$$z_1 = k_{1L} J \hat{\omega}_r \qquad (8.12)$$

whereas z_2 corresponds to the motor electromagnetic torque.

In the relationships of the torque, observer estimates of the rotor angular speed and motor torque exist, which could be identified as one of the observers presented in Chapter 5. Load torque estimation was verified experimentally using the system presented in Figure 8.2.

Experimental verification was realized using a P_n = 1.5-kW four-pole squirrel-cage induction motor. The tests were carried out first for a healthy drive, and then repeated after introducing two types of faults, being:

- angular misalignment, installation of redundant metal washers under one leg pair of the loading machine (Figure 8.3)
- imbalance of the whole drive system, weights were fixed at one end of the motor shaft; two weights of different masses were used and noted as weight no 1 and no 2 (Figure 8.4)

In the cases of the healthy and faulty drives, the load torque is computed at steady state. The computation was realized in real time by the microprocessor being the main controller of the inverter system. The computations are synchronized with the A/D converter operation and with the control system computations during a 150-µs sampling period. For the selected period of time, the computed torque values were saved in the microprocessor memory then were sent to the main computer, in which the results were analyzed *off-line*. For the computed torque,

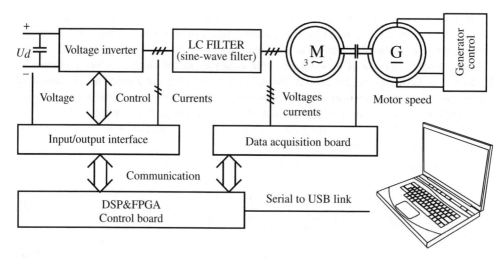

Figure 8.2 Experimental structure for the diagnosis of motor torque transmission in the sensorless electric drive with sinusoidal filter

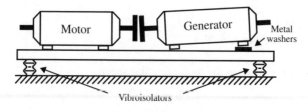

Figure 8.3 Angular misalignment, metal washers installed under one leg pair of the loading machine

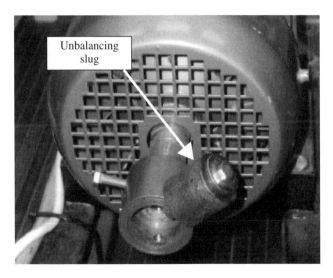

Figure 8.4 Imbalance of drive system, weight no 1 fixed to the end of the motor shaft

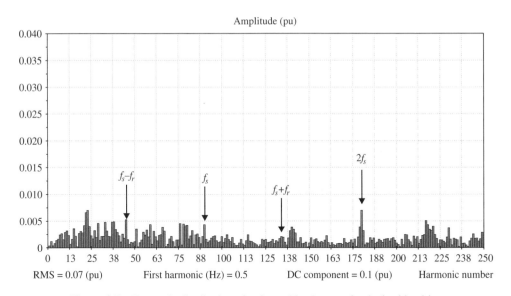

Figure 8.5 Harmonic distribution of estimated load torque for the healthy drive

the harmonic analysis was realized to search for the changes in the amplitude levels corresponding to the existing faults. Because of the artificial character of the realized faults, the analysis was limited to the frequency range up to 125 Hz. Examples of the investigation results obtained for a constant rotor speed $\omega_r = 0.9$ pu are presented in Figures 8.5 to 8.8.

The results from Figures 8.5–8.8 are compared in Figure 8.9. The symbols f_s and f_r in Figures 8.5 to 8.9 denote the supply frequency and rotor speed.

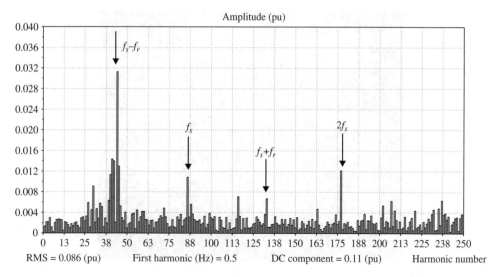

Figure 8.6 Harmonic distribution of estimated load torque for the faulty drive (angular misalignment)

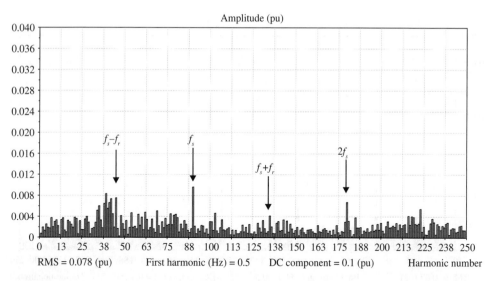

Figure 8.7 Harmonic distribution of estimated load torque for the healthy drive—drive imbalance—weight nr 1

Regardless the filter used in the drive system, the computed load torque contains information about the existing faults in the mechanical part. This is made possible by using the observers presented in Chapter 5. Those observers were modified to consider the presence of the motor filter used. In the investigated system, the rotor speed $\omega_r = 0.9$ pu corresponds to the supply voltage frequency around 45 Hz. This corresponds to 22.5-Hz rotational frequency in the drive system with a four-pole motor. During the tests, the motor was not loaded, therefore the slip could be assumed as $s \approx 0$. Therefore, the first harmonic of the signals related to the rotor

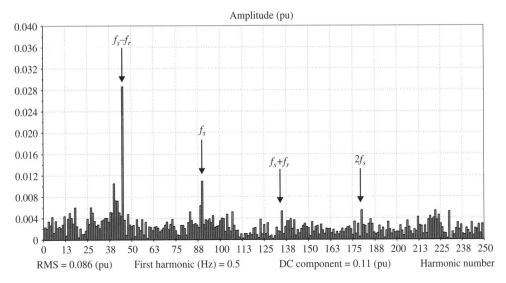

Figure 8.8 Harmonic distribution of estimated load torque for the healthy drive (drive imbalance, weight nr 2 of higher mass than nr 1)

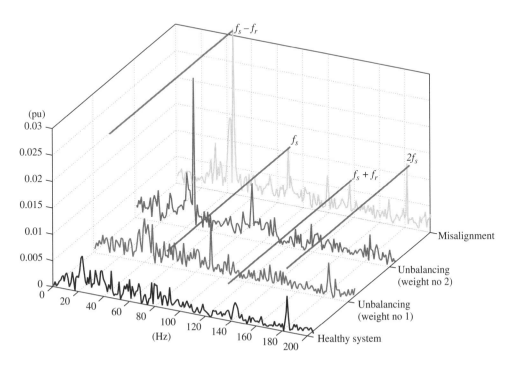

Figure 8.9 Comparison of harmonic magnitudes distribution of estimated load torque for the healthy and faulty drive

eccentricity are 22.5 and 67.5 Hz, respectively. Those frequencies are visible in the results shown in Figures 8.5 to 8.9. An increase in the torque magnitude at those frequencies with subsequent faults is shown.

For additional verification of the torque estimation, vibration measurements were taken and compared with the estimated ones under the same test conditions. The installation of vibration sensors is presented in Figure 8.10. Because of the primary upcoming frequency in the range of 1 to 130 Hz, all frequency representations were scaled to this range. The DC component of each fast Fourier transformation (FFT) was removed. For a better comparison and to find relations among different motor speeds, the figures are presented in a three-dimensional perspective.

Figures 8.11 and 8.12 present the measured vibrations of the drive system without load. The peaks of the accelerations can be found at the rotor (f_r) and stator (f_s) frequencies. Because of the number of poles, $p = 2$, the rotor frequency is half of the electrical stator frequency and had the major influence on the vibrations. Furthermore, harmonics as $2f_s$ and the addition of rotor and stator frequency $f_s + f_r$ can be found. The accelerations of the healthy drive system are relatively low with about 3 m/s² compared to the maximum peak of the fault drive system with about 60 m/s². It can be said that the behavior of the magnitude of vibrations was as expected, rising with rising speed. The disturbed drive system (Figure 8.12) shows highly increased acceleration magnitude in the vibrations. Contrary to expectation, the accelerations of the system would not increase with increased rotor speed. The peak value of the vibrations is at a rotor frequency of 22.5 Hz and could be explained by the installed vibroisolators, which come into resonance in this frequency range.

The previous evaluation of the experiment gives a reference value to the following experiment. To determine such a fault in the drive system, the industry desires a solution with no additional components. The estimated torque will be analyzed to avoid using additional sensors. Figure 8.13 shows the FFT of the calculated observer torque for the healthy drive system.

The peak of the stator frequency can be recognized in each speed state. This confirms that the drive system was free of any fault symptoms. The peaks are accurate and reproducible; however, other peaks that can be seen are not reproducible and can be declared as noise or disturbance of the observer. Figure 8.14 presents the estimated torque of the observer with artificial imbalance without load. As with the vibrations, the highest amplitude can be found at the rotor frequency f_r. These measurements verify the fault state of the drive system.

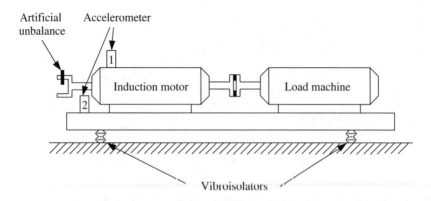

Figure 8.10 Schematic representation of the investigated test bench

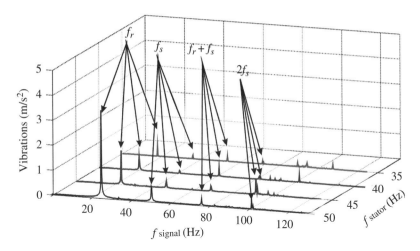

Figure 8.11 FFT analysis of the measured vibrations (sensor 1) of the healthy drive system without load

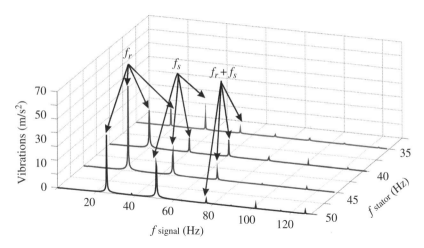

Figure 8.12 FFT analysis of the measured vibrations (sensor 1) of the drive system with artificial imbalance and without load

Figures 8.15 and 8.16 present the vibrations and the estimated torques of the drive system with artificial imbalance and loaded by the DC generator drive. The motor current I = 5A under load is slightly higher than the operating point current of the motor and it forces a stress situation. Figure 8.16 shows the operation that is close to the measurements of Figure 8.12. Nevertheless, the load affects the attenuation in the acceleration of the vibrations and decreases the rotor speed. The calculated torque of the observer shows corresponding attenuation behavior close to the measured vibrations. This comparison demonstrates that the estimated torque is a variable, which can be reliably evaluated to detect faults in an electrical machine. The peak of the stator frequency can be recognized in each speed state. This confirmed that the drive system

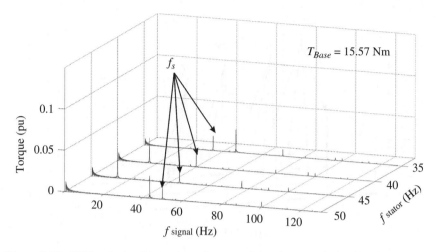

Figure 8.13 FFT analysis of the estimated torque of the observer of a healthy drive system

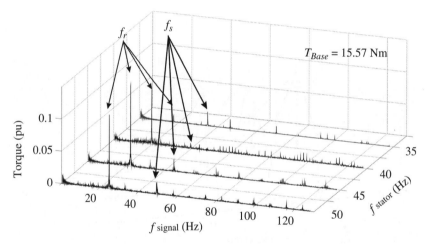

Figure 8.14 FFT analysis of the estimated torque of the observer of the drive system with artificial imbalance, faulty drive

was free of any fault symptoms. The peaks are accurate and reproducible; however, the other peaks that can be seen are not reproducible and can be declared as noise or disturbance of the observer. The illustration in Figure 8.14 presents the estimated torque of the observer with artificial imbalance without load. As with the vibrations, the highest amplitude can be found at the rotor frequency f_r. These measurements verify the fault state of the drive system.

A further way to prove the assumption of sensorless fault detection and, moreover, to investigate the impact of different observers, another observer (Observer 2) was implemented. In contrast to the first observer, the filter equations in the second observer are excluded. The experiment was undertaken by moving the measurement points (current and voltage of the inverter) directly to the induction motor.

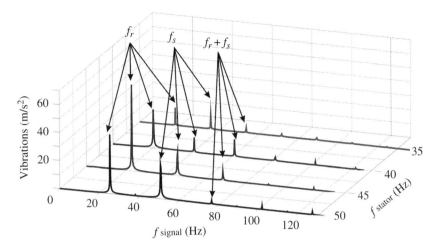

Figure 8.15 FFT analysis of vibrations (sensor 1) of the drive system with artificial imbalance and loaded with $I_s = 5A$

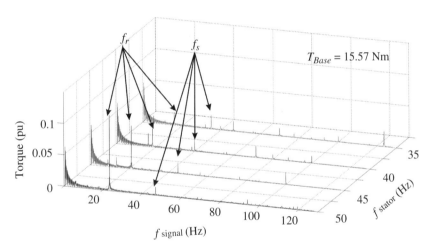

Figure 8.16 FFT analysis of the estimated torque of the drive system with artificial imbalance and loaded with $I_s = 5A$

Figures 8.17 and 8.18 present the results of the estimated torque of an observer excluding the LC filter. As shown in Figure 8.17, the behavior of the frequency analysis is comparable to the results of the first observer. The rotor frequencies, f_r, in the healthy condition are not present, but the stator frequencies, f_s, affect other amplitudes. Nevertheless, a further harmonic $4 \cdot f_r$ appears in the FFT analysis and noises are damped. In conclusion, as suggested, the change of the observer causes another analysis behavior.

Figure 8.18 shows the FFT analysis of the disturbed drive system with the second observer. It is clear to see that the rotor frequencies, f_r, appear corresponding to the previous tests. The resonant frequency in this test is at the speed of $0.8 \cdot f_s$ which could be explained by a moved

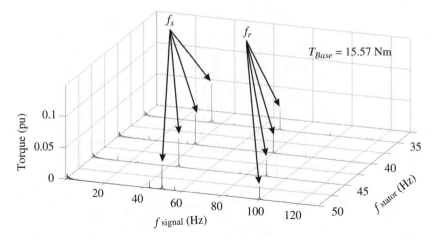

Figure 8.17 FFT analysis of the estimated torque of the healthy drive system with Observer 2

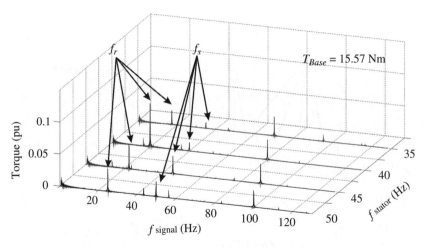

Figure 8.18 FFT analysis of the estimated torque of the drive system with artificial imbalance and Observer 2

position of the drive system. However, this demonstrates that a disturbed operation of the drive system can be recognized by analyzing the estimated torque. Moreover, this illustrates that the parameterization and creation of the observer have an influence on the behavior of the analysis, while also allowing the right interpretation of a fault in the drive system.

To identify the relevant frequencies, Equation 8.13 describes the relation of the frequencies that indicate a fault in the induction machine:

$$f_{fault} = \frac{s}{p} \cdot f_s \cdot k \tag{8.13}$$

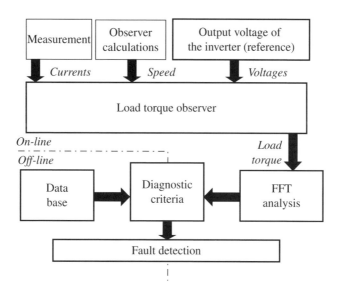

Figure 8.19 General structure of the diagnosis system of drive torque transmission

where f_{fault} is the frequency that indicates a fault; s is the slip of the induction motor; f_s is the stator frequency; and $k = 1, 2, 3, \ldots$ is the factor of the searched harmonics.

Comparing the harmonics amplitude waveforms presented on Figures 8.6 to 8.9 and 8.11 to 8.18, it is possible to see a relation between the levels of the amplitudes and the fault types. Such correlation could be used for fault diagnosis. The general concept of using the torque observer for fault diagnosis is presented in Figure 8.10.

The diagnosis system structure from Figure 8.19 is divided into two parts: *on-line* and *off-line*. The on-line system identifies the faults that are characterized by a sudden increase in selected harmonics. The off-line system is used to predict the faults while being increased. For this system, it is essential to have a database of the faults, which corresponds the particular amplitudes with related faults.

The method of diagnosis presented in this chapter uses the computed load torque and was tested for the drive without a filter for a high speed train drive [12, 22].

In the monitoring system, it is essential to use a proper diagnosis indicator, which should indicate the fault occurrence. As an example, some have proposed the use of artificial intelligence [2, 23, 24].

One of the neural diagnosis algorithms is the adaptive neuro-fuzzy inference system (ANFIS); Figure 8.20) [24].

The computations of the ANFIS system are realized based on the estimated load torque, stator current, and rotor speed. The internal structure of ANFIS was realized off-line using dedicated solutions in MATLAB/Simulink. Examples of the results showing the operation of the ANFIS diagnostic system are shown in Figure 8.21.

The results of using the ANFIS for machine diagnosis are shown in Figure 8.21. The results are taken for the healthy and faulty condition—machine misalignment. Various tests were realized for different rotor speeds. At time 0 ms, the ANFIS procedure was started to

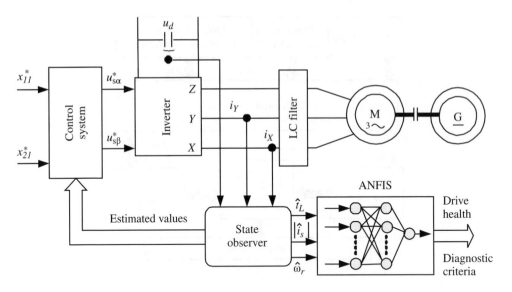

Figure 8.20 Diagnostic system based on adaptive neuro-fuzzy inference system (ANFIS)

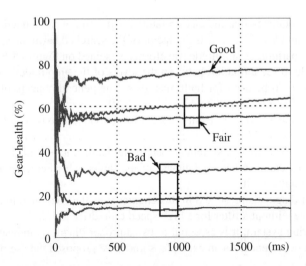

Figure 8.21 Misalignment fault detection using the adaptive neuro-fuzzy inference system (ANFIS) algorithm

compute the diagnosis scale (indicator). After 1 s, the computed diagnostic indicator reached a stable value allowing system evaluation. The percentage value of the diagnostic indicator indicates the fault level. For the healthy system this indicator does not exceed 20%, whereas for the system with large misalignment the value is around 80%.

8.3 Diagnosis of Rotor Faults in Closed-Loop Control

Faults in motors operating in closed-loop control systems could be identified by analyzing the control system's internal signals [25, 26]. The faulty rotor may be asymmetric from the electrical as well as the mechanical parameters. The existing rotor asymmetries result in the appearance of frequencies $(1 \pm 2f_2)f_s$ in the spectrum of the motor current sidebands. In the closed-loop system, motor currents and voltages are used for the estimation of some controlled state variables. Therefore, the existence of current distortion results in the appearance of related variations in the controlled variables. In such a situation, the closed-loop system tries to compensate the disturbances of the controlled variables by acting on the related controlled variables.

Because the rotor faults are permanent and characterized by specified frequencies, it is possible to distinguish the variations of the control signals caused by rotor faults from those that result from the system's reaction to the changes in the values commanded by the main controllers.

The main controlled motor variable is generally rotor angular speed, which is regulated by changing the commanded torque. The time constants of the speed controlled circuit are generally much longer than the time constants of the internal control loops and longer than those of the converters and motor. Therefore, in the speed closed-loop system, the delays in the control process, which are caused by the subordinated control system, inverter and motor, could be ignored. The closed-loop speed control with multiscalar variables has the form presented in Figure 8.22 [25].

The structure from Figure 8.22 refers to both, healthy and faulty motors. Motor faults could be treated as internal disturbances that appear in the motor block. In the closed-loop control system, those disturbances are compensated by the action of the speed controller, which are added to the disturbances that represent the correcting action of speed and flux control systems.

The detection of rotor faults working in a multiscalar control scheme was tested experimentally. The FSg 112M–8 motor type with rotor bar faults was investigated. Two adjacent bars were cut by drilling and the holes were filled with epoxy resin, to eliminate rotor imbalance. The faulty rotor is shown in Figure 8.23.

Realizing a fault in the neighboring rotor bars reflects a real fault in the motors. This could be an effect of the wrong design of the cage as a result of the appearance of air bubbles in one of the aluminum cage bars during the casting process. Motors constructed in this way heat up significantly near the bubble during normal operation because the bar resistance is higher at this location. As a result of the thermal conductivity, the neighboring bars also heat up, while excessive heat may result in interrupting their opening.

Bar faults cause an increase in rotor resistance in the machine equivalent circuit. In the case of the fault presented in Figure 8.23, an increase in the rotor resistance of around 50% was

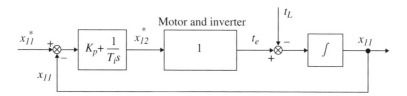

Figure 8.22 Structure of closed-loop speed control of induction motor

noticed. In Table 8.1 a comparison of the motor (FSg 112M–8) circuit parameters for the healthy and faulty (rotor bar) conditions is presented.

Figure 8.24 presents the waveforms of the command multiscalar variable x_{12}^*, registered at the output of the speed controller under the no load condition.

The time domain waveforms of x_{12}^* are different in higher harmonics as well as in the DC components. The difference in the DC component is caused by the different mechanical

Figure 8.23 Faulty rotor, motor type FSg 112M–8, 1.5 kW

Table 8.1 Equivalent circuit parameters for the faulty and healthy conditions (motor type FSg 112M–8)

Equivalent circuit parameter	Healthy motor	Motor with two faulty bars
Stator resistance Rs (Ω)	5.17	5.17
Rotor resistance Rr (Ω)	3.82	5.92
Stator inductance Ls (H)	0.273	0.283
Rotor inductance Lr (H)	0.273	0.283
Mutual inductance Lm (H)	0.253	0.256

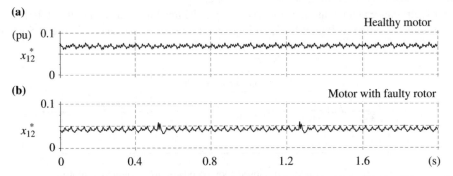

Figure 8.24 Waveforms of commanded x_{12} for the motor with healthy (a) and faulty (b) rotor bars (motor type FSg 112M–8, speed 30%, no load condition)

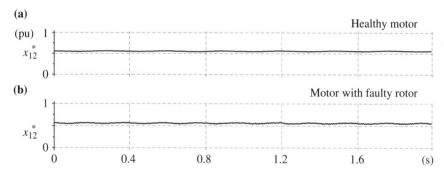

Figure 8.25 Waveforms of commanded variable x_{12}^* for the healthy (a) and faulty (b) motor rotor (motor type FSg 112M–8, speed 30%, rated load)

resistances of the motor before and after rotor replacement. However, this difference has no effect on the rotor bars fault detection, which causes an increase in the harmonics amplitudes.

In Figure 8.25 example waveforms of the command variable x_{12}^* for the motor with healthy and faulty rotor bars working at a rated load are presented. From the waveforms in Figure 8.25 it is evident that the differences in the signals of the AC component are not as clear as in Figure 8.13. This is because the AC components are negligible compared with the registered DC component.

Discovering the difference in the x_{12}^* waveforms, as a result of the rotor faults, requires harmonics analysis of the signals to be realized. In Figures 8.26 and 8.27 such analytical results for the signals from Figures 8.24 and 8.25 are presented. In Figure 8.26 the differences in the characteristic harmonics amplitudes for the motor with the healthy and faulty rotors are evident. In the case of the motor working without a load (Figure 8.26a,b), a double increase in $2f_s$ harmonic amplitudes and the existence of $4f_s$ harmonics are noted. As for the loaded motor (Figure 8.27a,b), the rotor speed frequency f_r, for which a significant increase has appeared for the faulty rotor, is shown. Additional harmonics have also appeared for the faulty motor ($4f_s$ harmonics) as well as the $6f_s$ harmonic.

In the closed-loop control system, in addition to the torque command signal, it is also possible to analyze the rotor flux or stator flux command signal. Those signals are supposed also to contain information about rotor faults.

8.4 Simulation Examples of Induction Motor with Inverter Output Filter and Load Torque Estimation

The model of induction motor with inverter output filter and load torque estimation (*Chapter_8\ Example_1\Simulator.slx*) is shown in Figure 8.28.

The model *Simulator.slx* consists of an induction motor drive with inverter output filter. The main simulation file is T_LOAD.c. The drive operates in closed-loop control with observer. Control system and observer procedure include filter model. The observer is extended with motor torque and load torque calculation. The observer is a disturbance observer with complete model of the load. Motor torque is calculated with estimated rotor flux and stator current (Equation 8.2):

```
me_so=(frx_so*isy_so-fry_so*isx_so)*Lm/Lr;
```

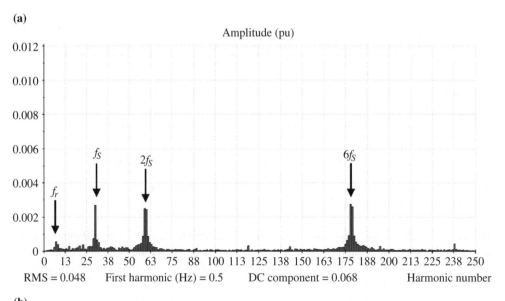

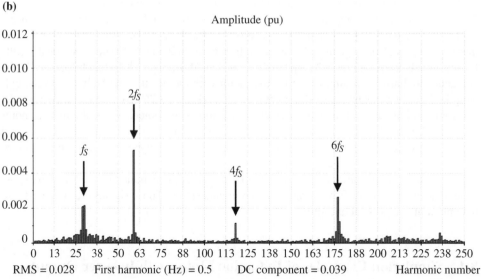

Figure 8.26 Results of the x_{12}^* harmonic analysis for motor speed 30%: healthy motor, no load (a) and faulty motor, no load (b)

The load torque is calculated using motion equation of the drive (Equation 8.3):

```
m0_so=me_so-JJ*pochR_so;
```

where *JJ* is drive inertia and *pochR_so* is the derivative of motor speed. Speed derivative is calculated by Lagrange method:

```
pochR_so=(omega_sofi2-4*omega_sofi1+3*omega_sofi)/(2*Timp);
```

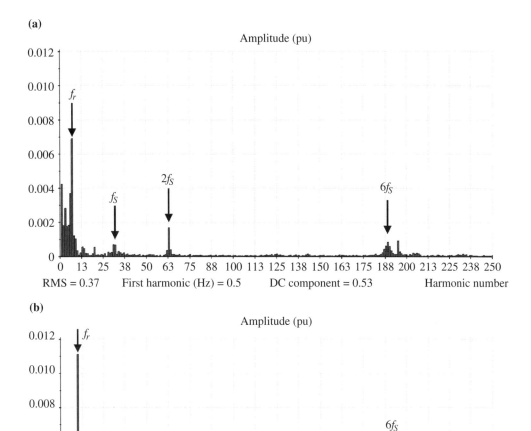

Figure 8.27 Results of the x_{12}^* harmonic analysis for motor speed 30%: (a) healthy motor, loaded motor and (b) faulty motor, loaded motor

The structure of the observer and structure of the control are as presented in Chapters 5 and 6 respectively.

Example of simulation results are presented in Figure 8.29. In Figure 8.29 the next variables are presented (from top to bottom):

- `omegaR`, real motor speed,
- `omega_sof`, estimated motor speed,

- `fr`, rotor flux (magnitude),
- `Load m0`, real load torque,
- `Torque me`, estimated motor torque,
- `Load m0_so`, estimated load torque.

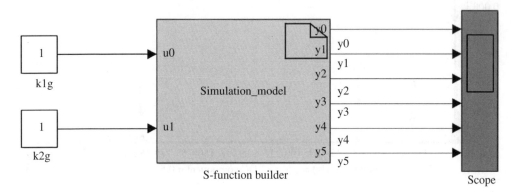

Figure 8.28 MATLAB/Simulink model of induction motor with inverter output filter and load torque estimation, *Chapter_8\Example_1\Simulator.slx*

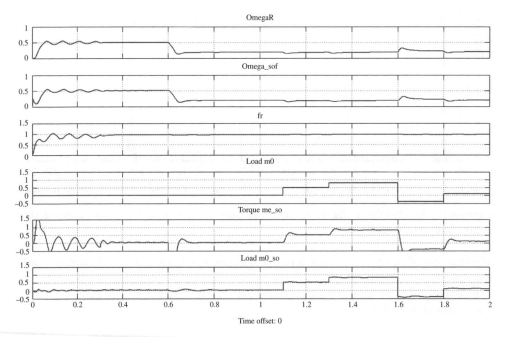

Figure 8.29 Example of simulation results for induction drive with inverter output filter and load torque estimation

After the initial transients with V/f control, the drive is switched automatically into closed-loop control (at 0.3 s). Next, the following changes of reference speed and disturbance (load torque) are imposed:

```
if(timee>600)  x11z=0.2;
if(timee>1100) m0=0.5;
if(timee>1300) m0=0.8;
if(timee>1600) m0=-0.4;
if(timee>1800) m0=0.1;
```

The implementation of the load torque Gopinath observer is left to the reader. The observer variables are declared in C file and the observer gains ($k1g$, $k2g$) could be set by block parameters in Simulink. The Gopinath observer have to be add in function void p_SPEED_OBSERVER() as successive differential equations indexed 12 and 13.

8.5 Conclusions

Fault diagnosis of electrical motors is a difficult issue; however, it is of utmost importance for modern drive technology. Currently, measured mechanical signals such as vibration signals and electrical signals such as motor phase currents are used for fault diagnosis. Because of the problems associated with signals measurement, sensorless solutions are being sought. Therefore, dedicated computational solutions are being developed, such as the computation of motor torque and load torque. Such systems operate generally in the open loop mode, that is, the estimated values are not used in the control but only for diagnosis purposes.

In the diagnosis, it is possible to use an analysis of the closed-loop system signals. In such a case it is possible to use simpler solutions without the need for additional observers.

The measured or computed values for diagnosis purposes are beneficial in the event that there are mechanisms allowing for their proper interpretation. A proper assessment should identify the symptoms and faults.

In the assessment process of the technical condition of the motor, in the main, classic methods were used in which the selected values are analyzed and assessed directly by comparing them with the defined values in the faults database. Those methods are relatively simple; however, their practical application is limited because of the interaction between different motor signals; there is no clear relationship between the fault symptoms and real faults [2]. Therefore, the use of artificial intelligence methods, such as fuzzy logic or neural networks, is currently increasing.

The application of diagnosis observers in the drive system with filters requires proper modification, as shown in Chapter 5. By including the filter equation in the modified observer, motor faults can be identified. This, however, increases the complexity of the estimation algorithms.

References

[1] Tavner PJ. Review of condition monitoring of rotating electrical machines. *IET Electric Power Applications*. 2008; **2** (4): 215–246.
[2] Wolkiewicz M, Tarchała GJ, Kowalski CT. Monitoring of stator inter-turn short circuits in the direct field oriented controlled induction motor. Sixteenth International Power Electronics and Motion Control Conference and Exposition, PEMC 2014, Antalya, Turkey, September **21–24**, 2014.

[3] International Organization for Standardization (ISO). ISO 17359: Condition monitoring and diagnostics of machines—General guidelines. Geneva, Switzerland: ISO; 2011.
[4] Gieras JF, Wang C, Lai JC. *Noise of polyphase electric motors*. Boca Raton, FL: CRC Press; 2005.
[5] Yang SJ. *Low-noise electrical motors*. Oxford, UK: Oxford University Press; 1981.
[6] Ciszewski T, Swędrowski L. Comparison of induction motor bearing diagnostic test results through vibration and stator current measurement. *Computer Applications in Electrical Engineering*. 2013; **10**: 165–170.
[7] Trzynadlowski A, Ghassemzadeh M, Legowski SF. Diagnostics of mechanical abnormalities in induction motors using instantaneous electric power. *IEEE Transactions on Energy Conversion*. 1999; **14** (4): 1417–1423.
[8] Penman J, Sedding HG, Lloyd BA, Fink WT. Detection and location of interturn shot circuits in the stator windings of operating motors. *IEEE Transactions on Energy Conversion*. 1194; **9** (4): 652–658.
[9] Prabhakar S, Sekhar AS, Mohanty AR. Crack versus coupling misalignment in a transient rotor system. *Journal of Sound and Vibration*. 2002; **256**: 773–786.
[10] Vedmar L, Andersson A. A method to determine dynamic loads on spur gear teeth and on bearings. *Journal of Sound and Vibration*. 2003; **267**: 1065–1084.
[11] Guziński J, Diguet M, Krzemiński Z, Lewicki A, Abu-Rub H. Application of speed and load torque observers in high speed train. Thirteenth International Power Electronics and Motion Conference EPE–PEMC 2008, Poznan, Poland. September 1–3, 2008.
[12] Guziński J, Abu–Rub H, Diguet M, Krzeminski Z, Lewicki A. Speed and load torque observer application in high-speed train electric drive. *IEEE Transactions on Industrial Electronics*. 2010; **57** (2): 565–574.
[13] Hace A, Jezernik K, Šabanovic A. SMC with disturbance observer for a linear belt drive. *IEEE Transactions on Industrial Electronics*. 2007; **54** (6): 3402–3412.
[14] Kia SH, Henao H, Capolino G–A. Torsional vibration monitoring using induction machine electromagnetic torque estimation. Thirty-Fourth Annual Conference of the IEEE Industrial Electronics Society IECON'08, Orlando, FL. November 10–13, 2008.
[15] Kia SH, Henao H, Capolino, G–A. Torsional vibration assessment in railway traction system mechanical transmission. The Seventh IEEE International Symposium on Diagnostics for Electric Machines, Power Electronics and Drives, SDEMPED'09, Cargèse, France. August 31–September 3, 2009.
[16] Szabat K, Orłowska–Kowalska T. Performance improvement of industrial drives with mechanical elasticity using nonlinear adaptive Kalman filter. *IEEE Transactions on Industrial Electronics*. 2008; **55** (3): 1075–1084.
[17] Trajin B, Regnier J, Faucher J. Detection of bearing faults in asynchronous motors using Luenberger speed observer. Thirty-Fourth Annual Conference of IEEE Industrial Electronics Society, IE-CON 2008, Orlando, FL. November 10–13, 2008.
[18] Lewicki A, Geniusz A. Detection of the vibration of induction motor coupled with toothed gear. Trzecia Krajowa Konferencja Postępy w Elektrotechnice Stosowanej PES-3 (Third National Conference on Progress in Applied Electrotechnics), Zakopane, Poland. June 18–22, 2001.
[19] Kadowaki S, Ohishi S, Hata T, Iida N, Takagi M, Sano T, et al. Antislip readhesion control based on speed-sensorless vector control and disturbance observer for electric commuter train—series 205–5000 of the East Japan Railway Company. *IEEE Transactions on Industrial Electronics*. 2007; **24** (4): 2001–2008.
[20] Ohishi K, Nakano K, Miyashita I, Yasukawa S. Anti–slip control of electric motor coach based on disturbance observer. Fifth International Workshop on Advanced Motion Control – AMC'98, Coimbra, Portugal. June 29–July 1, 1998.
[21] Gopinath B. On the control of linear multiple input-output systems. *The Bell Technical Journal*. 1971; **50** (3): 1063–1081.
[22] Guzinski J, Diguet M, Krzemiński Z, Lewicki A, Abu–Rub H. Application of speed and load torque observers in high–speed train drive for diagnostic purposes. *IEEE Transactions on Industrial Electronics*. 2009; **56** (1): 248–256.
[23] Kowalski CT, Wolkiewicz M. Stator fault diagnosis of the converter-fed induction motor using symmetrical components and neural network. The Thirteenth European Conference on Power Electronics and Applications, EPE'09, Barcelona, Spain. September 8–10, 2009.
[24] Guzinski J, Abu-Rub H, Iqbal A, Moin A. Shaft misalignment detection using ANFIS for speed sensorless ac drive with inverter output filter. ISIE 2011, Gdansk, Poland. June 27–30, 2011.
[25] Cruz SMA, Cardoso AJM. Diagnosis of rotor faults in closed–loop induction motor drives. Forty-First IEEE–IAS Annual Meeting, Industry Applications Conference, Tampa, FL. October 8–12, 2006.
[26] Kołodzicjck P. Non-invasive method for rotor fault diagnosis in inverter fed induction motor drive. The Eight International Conference & Exhibition on Ecological Vehicles and Renewable Energies, Monaco. March 27–30, 2013.

9

Multiphase Drive with Induction Motor and an LC Filter

9.1 Introduction

In the previous chapters, only classical three-phase drives were presented. However, here special attention is paid to the multiphase drives, where the term *multiphase* means more than three phases. Numerous research papers have been dedicated to multiphase drives. A comprehensive review of the development and application of this technology can be found in Levi, Bojoi, Profumo, Toliyat, and Williamson [1]. Most of the available literature deals with such drives without filters. Therefore, this chapter is dedicated to the multiphase drives with and without LC filters.

The attraction of multiphase drives results from numerous advantages such as high fault tolerance, lower torque pulsation and noise, lower current losses, and reduction of the rated current of power converter devices [1, 2]. A particular benefit of such drives is the higher torque density in the case of special designs of motors and appropriate control [3–6].

Multiphase drives are attractive in applications in which high robustness and a low volume-to-power ratio are required. Such applications include electric aircraft, vehicles, rail traction, and ship propulsion [1, 2, 7, 8]. Generally, multiphase drives are attractive for most high-power systems for which the reliability is essential.

A multiphase drive consists of the following components:

- multiphase motor,
- special power electronics converter,
- and appropriate control algorithm.

A special converter is required with more phases. Numerous types of power converters are being used, from two-level voltage source inverter (VSI), which is the most popular to

---

Variable Speed AC Drives with Inverter Output Filters, First Edition. Jaroslaw Guzinski, Haitham Abu-Rub and Patryk Strankowski.
© 2015 John Wiley & Sons, Ltd. Published 2015 by John Wiley & Sons, Ltd.

multilevel VSIs and matrix converters [9–12]. The most popular converter construction is a simple extension of three-phase converters with lower rated current of the switches. However, a simple extension of the pulse width modulation (PWM) algorithm from three-phase to multiphase drive is not easy to obtain because of the higher complexity of the switch control [9, 10].

The mature technology of three-phase induction motor (IM) control is applicable to multi-phase systems, for example, field-oriented control (FOC) and sensorless control systems [1–4]. However, because of the more complex structure of the multiphase motor, some modification and extensions of the classical solutions are needed [1–4, 7, 13].

Many industrial drives with vector control can operate in the sensorless mode, that is, without speed measurement. Numerous speed estimation methods exist; however, improvement and new algorithms for speed computation are still needed [13, 14].

The problems of high dV/dt in voltage source PWM inverters for three-phase drives also appear in multiphase drives. These include bearing currents, shaft voltages, insulation stress, efficiency reduction, noise, and so on [15, 16] Therefore, in many applications of multiphase drives, a passive filter is connected at the output of the PWM VSI. One such filter is the sine-wave filter that was presented in Chapter 4. Unfortunately, the presence of an LC filter at the inverter output also causes various problems for the control and estimation systems in multi-phase drives, similar to the three-phase situation [14, 17–19]

In AC electric drives with filters, the solutions for three-phase drives could be used with all control methods, for example, for FOC [2, 20], nonlinear FOC [17], multiscalar control [18], and direct torque control (DTC) [19].

In this chapter, a speed sensorless FOC drive for a five-phase induction motor with a VSI and LC filter is presented. Limiting our discussion to this type of system is justified by the fact that the five-phase drive is the most popular multiphase drive and an understanding of it gives the basis for other extensions. The general structure of the drive is shown in Figure 9.1. The FOC principle is used for motor control, whereas the multiloop LC filter control is subordinated (see Chapter 6 for three-phase drive solutions). A flux and speed observer with equations of the five-phase motor and filter models is used (see theory in Chapter 5). The only sensors used in the proposed drive system are for the DC link voltage and inverter output currents. No speed sensor is required.

In the next subsections, a dynamic model of the five-phase induction motor, the control principle, and the estimation algorithm are presented. Both control and estimation have been

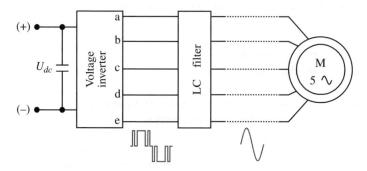

Figure 9.1 Electric drive with five-phase induction motor and inverter output LC filter

adopted in the drive with an LC filter. Furthermore, the system was investigated by simulation and experiments. The data and description of the whole system are given for the test bench used. The design and construction of the five-phase motor are also presented.

The control principle is based on the idea of using the torque produced by the first harmonic of motor currents and fluxes. No third harmonic injection is taken into account in this chapter.

9.2 Model of a Five-Phase Machine

The modeling theory of three-phase electric machines is applicable to multiphase systems; however because of the greater phase numbers, it is more complicated.

The five-phase induction machine has a spatial displacement of $\alpha = 72$ degrees between two consecutive stator phases. For the modeling, it is assumed that the number of stator and rotor phases is the same. The natural coordinates of the machine are a five-phase frame of references, $abcde$. However, preparation of the model in $abcde$ coordinates is not usually done because of calculations complexity. Another reason is that in the $abcde$ coordinates, the components of the mutual inductances matrix are dependent on the instantaneous position of the magnetic axis: the angle θ. To simplify the model calculation it is necessary to apply an appropriate coordinate transformation of the stator:

$$\mathbf{A}_s = \sqrt{\frac{2}{5}} \begin{bmatrix} 1 & \cos(-\alpha) & \cos(-2\alpha) & \cos(2\alpha) & \cos(\alpha) \\ 0 & -\sin(\alpha) & -\sin(2\alpha) & -\sin(2\alpha) & -\sin(\alpha) \\ 1 & \cos(2\alpha) & \cos(4\alpha) & \cos(4\alpha) & \cos(2\alpha) \\ 0 & \sin(2\alpha) & \sin(4\alpha) & -\sin(4\alpha) & -\sin(2\alpha) \\ \frac{1}{\sqrt{2}} & \frac{1}{\sqrt{2}} & \frac{1}{\sqrt{2}} & \frac{1}{\sqrt{2}} & \frac{1}{\sqrt{2}} \end{bmatrix} \quad (9.1)$$

and for the rotor:

$$\mathbf{A}_r = \sqrt{\frac{2}{5}} \begin{bmatrix} \cos\theta_r & \cos(\theta_r + \alpha) & \cos(\theta_r + 2\alpha) & \cos(2\alpha - \theta_r) & \cos(\alpha - \theta_r) \\ \sin\theta_r & \sin(\theta_r + \alpha) & \sin(\theta_r + 2\alpha) & \sin(\theta_r - 2\alpha) & \sin(\theta_r - \alpha) \\ 1 & \cos(2\alpha) & \cos(4\alpha) & \cos(4\alpha) & \cos(2\alpha) \\ 0 & \sin(2\alpha) & \sin(4\alpha) & -\sin(4\alpha) & -\sin(2\alpha) \\ \frac{1}{\sqrt{2}} & \frac{1}{\sqrt{2}} & \frac{1}{\sqrt{2}} & \frac{1}{\sqrt{2}} & \frac{1}{\sqrt{2}} \end{bmatrix} \quad (9.2)$$

where θ_r is the angle of the rotor position:

$$\theta_r = \int \omega_r dt \quad (9.3)$$

The stator matrix Equation 9.1 and rotor matrix Equation 9.2 are used for the power invariant transformation from the $abcde$ reference frame to the $\alpha\beta xy0$ stationary reference. Because the motor neutral point is not connected, the zero component is omitted in the next relations.

Finally, the induction motor model in the stationary frame $\alpha\beta xy$ is as follows (assuming a sinusoidal distribution of the stator windings):

$$\frac{d\psi_{s\alpha}}{dt} = u_{s\alpha} - R_s i_{s\alpha} \tag{9.4}$$

$$\frac{d\psi_{s\beta}}{dt} = u_{s\beta} - R_s i_{s\beta} \tag{9.5}$$

$$\frac{d\psi_{sx}}{dt} = u_{sx} - R_s i_{sx} \tag{9.6}$$

$$\frac{d\psi_{sy}}{dt} = u_{sy} - R_s i_{sy} \tag{9.7}$$

$$\frac{d\psi_{r\alpha}}{dt} = -R_r i_{r\alpha} - \omega_r \psi_{r\beta} \tag{9.8}$$

$$\frac{d\psi_{r\beta}}{dt} = -R_r i_{r\beta} + \omega_r \psi_{r\alpha} \tag{9.9}$$

$$\frac{d\psi_{rx}}{dt} = -R_r i_{xs} \tag{9.10}$$

$$\frac{d\psi_{ry}}{dt} = -R_r i_{sy} \tag{9.11}$$

$$i_{s\alpha} = \frac{1}{w_\sigma}\left(L_s \psi_{s\alpha} - L_m \psi_{r\alpha}\right) \tag{9.12}$$

$$i_{s\beta} = \frac{1}{w_\sigma}\left(L_s \psi_{s\beta} - L_m \psi_{r\beta}\right) \tag{9.13}$$

$$i_{sx} = \frac{\psi_{sx}}{l_{\sigma s}} \tag{9.14}$$

$$i_{sy} = \frac{\psi_{sy}}{l_{\sigma s}} \tag{9.15}$$

$$i_{r\alpha} = \frac{1}{w_\sigma}(L_s\psi_{r\alpha} - L_m\psi_{s\alpha})\qquad(9.16)$$

$$i_{r\beta} = \frac{1}{w_\sigma}(L_s\psi_{r\beta} - L_m\psi_{s\beta})\qquad(9.17)$$

$$i_{rx} = \frac{\psi_{rx}}{l_{\sigma r}}\qquad(9.18)$$

$$i_{ry} = \frac{\psi_{ry}}{l_{\sigma r}}\qquad(9.19)$$

$$\frac{d\omega_r}{dt} = \frac{L_m}{L_r J}(\psi_{r\alpha}i_{s\beta} - \psi_{r\beta}i_{s\alpha}) - \frac{1}{J}T_L\qquad(9.20)$$

where:

$$w_\sigma = L_s L_r - L_m^2\qquad(9.21)$$

and R_s, R_r, L_m, $l_{\sigma s}$, and $l_{\sigma r}$ are the motor parameters of the motor equivalent circuit (Figure 9.2). The difference between the five-phase induction motor model Equations 9.3 to 9.19, and the corresponding three-phase model is the presence of the xy component equations and circuit. In Figure 9.2, one can see that the $\alpha\beta$ circuit is decoupled from the xy stator circuit, and

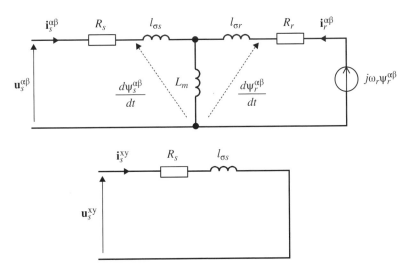

Figure 9.2 Equivalent circuit of a five-phase induction motor, assuming a sinusoidal distribution of the stator windings

the *xy* stator circuit is not coupled with the rotor. On this basis, the motor can be fully controlled using only $\alpha\beta$ components. It is also noticeable in Equation 9.21 that, assuming a sinusoidal distribution of the stator windings, the motor torque is produced only by fundamental components of the currents and fluxes.

However, it is reported that the use of *xy* component possibilities is applicable in the case of inaccuracies in the inverter operation or if the higher harmonics of the torque have to be applied. Rotor *xy* components are fully decoupled from *dq* components and from each other. Because the rotor winding is short-circuited, *xy* components cannot appear in the rotor winding. Zero sequence component equations for both stator and rotor can be omitted from further consideration as a result of the short-circuited rotor winding and star connection of the stator winding. Finally, because stator *xy* components are fully decoupled from *dq* components and from each other and vector control is applied (i.e., only *dq* axis current components are generated), the equations for *xy* components can be omitted from further consideration as well. This means that the model of the five-phase induction machine in an arbitrary reference frame becomes identical to the model of a three-phase induction machine. Hence, the same principles of control can be used as for a three-phase induction machine.

9.3 Model of a Five-Phase LC Filter

The five-phase LC filter has the structure presented in Figure 9.3 [25]. Five reactors, L_f, and five capacitors, C_f, are part of the low-pass filter. The reactors' internal resistance has been neglected.

For the purpose of description of the model, the equations of the filter for the *abcde* coordinates are written in the $\alpha\beta xy$ orthogonal frame of reference:

$$u_{1\alpha} = L_f \frac{di_{1\alpha}}{dt} + u_{s\alpha} \tag{9.22}$$

$$u_{1\beta} = L_f \frac{di_{1\beta}}{dt} + u_{s\beta} \tag{9.23}$$

$$u_{1x} = L_f \frac{di_{1x}}{dt} + u_{sx} \tag{9.24}$$

$$u_{1y} = L_f \frac{di_{1y}}{dt} + u_{sy} \tag{9.25}$$

$$C_f \frac{du_{s\alpha}}{dt} = i_{1\alpha} - i_{s\alpha} \tag{9.26}$$

$$C_f \frac{du_{s\beta}}{dt} = i_{1\beta} - i_{s\beta} \tag{9.27}$$

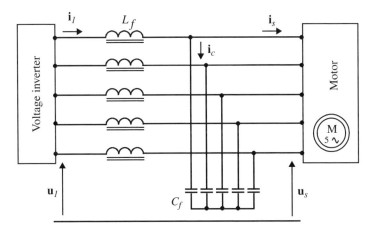

Figure 9.3 Structure of the five-phase sine-wave filter of the voltage inverter (L_f and C_f filter parameters)

$$C_f \frac{du_{sx}}{dt} = i_{1x} - i_{sx} \qquad (9.28)$$

$$C_f \frac{du_{sy}}{dt} = i_{1y} - i_{sy} \qquad (9.29)$$

$$i_{1\alpha} = i_{s\alpha} + i_{c\alpha} \qquad (9.30)$$

$$i_{1\beta} = i_{s\beta} + i_{c\beta} \qquad (9.31)$$

$$i_{1x} = i_{sx} + i_{cx} \qquad (9.32)$$

$$i_{1y} = i_{sy} + i_{cy} \qquad (9.33)$$

The voltages are related to the DC link negative terminal. The filter equivalent electrical circuits in $\alpha\beta xy$ are presented in Figure 9.4.

9.4 Five-Phase Voltage Source Inverter

The classical two-level VSI could be extended to a five-phase converter as presented in Figure 9.5.

The five-phase two-level VSI has a total of $2^5 = 32$ space vectors, 2 of which are zero vectors. The remaining 30 active vectors form three coaxial decagons in the $\alpha\beta$ and xy planes. A graphical representation of the vectors is presented in Figure 9.6.

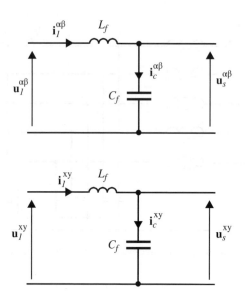

Figure 9.4 Equivalent circuits of five-phase LC filter in $\alpha\beta xy$ coordinates

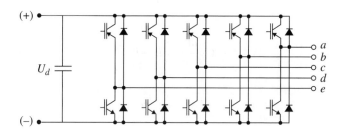

Figure 9.5 Topology of the five-phase two-level voltage source inverter

The simple extension of space vector PWM form three-phase to five-phase inverter is not applicable. The method leads to unwanted low-order harmonics in the inverter output current (Figure 9.7). The reason is lack of control in xy plane. The solution is to use space vector pulse width modulation (SVPWM) with use of four active vectors: two large vectors and two medium vectors [2]. With four active vectors, the xy plane output voltage is canceled.

For SVPWM with four active vectors, the vector switching instants can be calculated using trigonometric equations in all regions of the space vector. The components of the reference voltage vector $\mathbf{U}^*$ are [2, 20]:

$$U_x^* = \frac{|\mathbf{U}^*|\sin\left(k\cdot\dfrac{\pi}{5}-\theta\right)}{|\mathbf{U}_L|\cdot\sin\dfrac{\pi}{5}} \qquad (9.34)$$

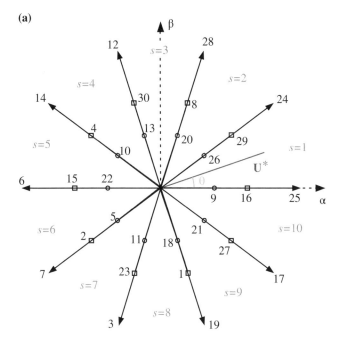

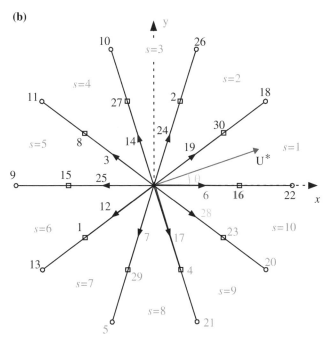

Figure 9.6 Space vector graphical representation for five-phase voltage source inverter in: **a,** $\alpha\beta$ plane and **b,** xy plane

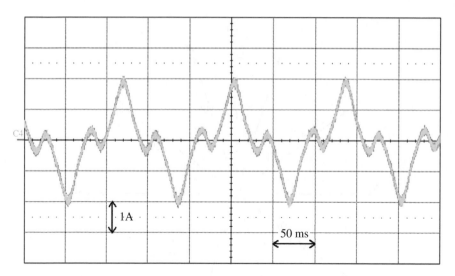

Figure 9.7 Motor current for SVPWM with two active vectors (low-order harmonics appear)

$$U_y^* = \frac{|U^*|\sin\left(\theta - (k-1)\frac{\pi}{5}\right)}{|U_L|\cdot\sin\frac{\pi}{5}} \qquad (9.35)$$

where $|U_L|$ is the magnitude of the large voltage vector; k is the number of the sector ($k = 1{-}10$); and θ is the angle of the reference vector position in relation to the α axis.

The output vector could be generated as in the PWM for a three-phase system using two adjacent longest active vectors and one zero vector. The disadvantage is that an unwanted voltage will appear in the xy plane. The forced third harmonic current for xy is limited only by R_s and $l_{\sigma s}$ elements, so its value is appreciable. It generates the additional current losses in the motor.

A more interesting issue is to use in each sequence a combination of four vectors from the available large and medium ones [2]. Proper combination can cancel the resultant voltage in the xy circuit and eliminate the related current. An example of a vector sequence for sector 1 is presented in Figure 9.8.

With regard to the ratio of vectors in $\alpha\beta$ it is possible to evaluate the relations

$$t_{x1}|U_{x1}^*| = t_{x2}|U_{x2}^*| \qquad (9.36)$$

$$t_{y1}|U_{y1}^*| = t_{y2}|U_{y1}^*| \qquad (9.37)$$

and taking into account the equations,

$$t_x = t_{x1} + t_{x2} \qquad (9.38)$$

Multiphase Drive with Induction Motor and an LC Filter

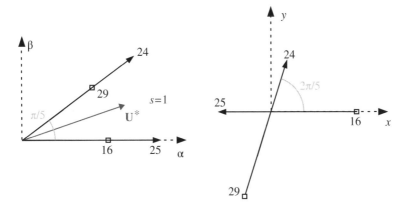

Figure 9.8 Idea of PWM with four active vectors: sector $k = 1$ case

$$t_y = t_{y1} + t_{y2} \tag{9.39}$$

it is possible to find the switching times of the particular vectors:

$$t_{x1} = 0.382 \cdot t_x \tag{9.40}$$

$$t_{x2} = 0.618 \cdot t_x \tag{9.41}$$

$$t_{y1} = 0.382 \cdot t_y \tag{9.42}$$

$$t_{y2} = 0.618 \cdot t_y \tag{9.43}$$

The zero vector switching time is:

$$t_0 = T_{imp} - t_{x1} - t_{x2} - t_{y1} - t_{y2} \tag{9.44}$$

The switching pattern used is $t_0/2 - t_{x1} - t_{x2} - t_{y1} - t_{y2} - t_0/2$ and it is reversed in the next cycle.

For PWM operation, the next simulation results are presented where the motor is directly connected to the inverter. The waveforms of the motor supply voltage for PWM with two and four active vectors are presented in Figure 9.9.

The motor currents in the case of PWM with two and four active vectors are presented in Figures 9.10a and b, respectively.

The appreciable third harmonic current in the xy plane is noticeable (Figure 9.10a), whereas its average value for the switching cycle is canceled for four active vectors (Figure 9.10b). The small transient in the xy circuit is the high harmonic component related to PWM switching.

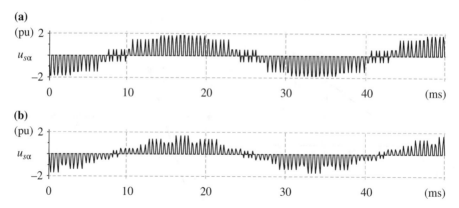

Figure 9.9 Waveform of alpha component of the inverter output voltage for PWM with two (a) active vectors and four (b) active waveforms

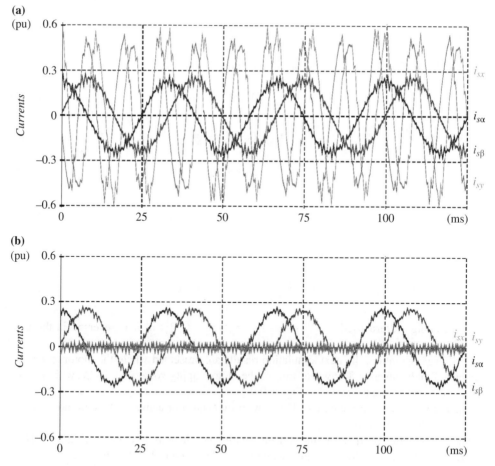

Figure 9.10 Motor currents in five-phase system for PWM with a sequence of two (a) active vectors: third harmonic currents in the xy plane; four (b) active vectors: elimination of currents in the xy plane

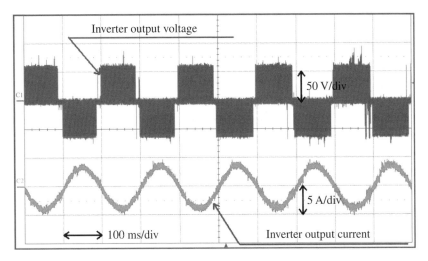

Figure 9.11 The inverter output line-to-line voltage and line currents: output frequency = 5 Hz, no load (for SVPWM with four active vectors)

An example of the inverter output voltage and currents waveforms, in case of SVPWM with four active vectors, is presented in Figure 9.11.

9.5 Control of Five-Phase Induction Motor with an LC Filter

The control structure of the system has two parts: motor control and LC filter closed-loops control. For the motor, the direct rotor flux oriented algorithm is used, whereas the voltage and current multiloop control of the LC filter are provided in the inner loops. The whole controller is shown in Figure 9.12.

For the motor control, the classical direct rotor FOC is applied. The operation in the base speed region is considered, so the rotor flux reference is kept constant. The flux and speed are controlled by i_{sd} and i_{sq} controllers. Because of the use of the LC filter, inner loops are added for the stator voltage and inverter output current control with P and proportional-integral (PI) regulators, respectively.

The whole control is done in coordinates oriented with respect to the rotor flux vector position. The control system is realized without controllers for the xy circuit. The reason is that the PWM with four vectors eliminates i_{sx} and i_{sy} currents (see Figure 9.11).

The measured signals are the inverter output current and inverter DC link voltage. The referenced values of the inverter output voltages are used as feedback signals in the observer instead of the real one. This is a result of the assumption that PWM is operating properly, and therefore the first harmonic of the inverter output is equal to the reference value. The variations of the u_d supply voltage lead to variations in the transistors' switching times to keep the reference voltage unchanged.

For the observer block, both inverter output currents and voltages are input signals. The motor feedback signals are estimated in the observer.

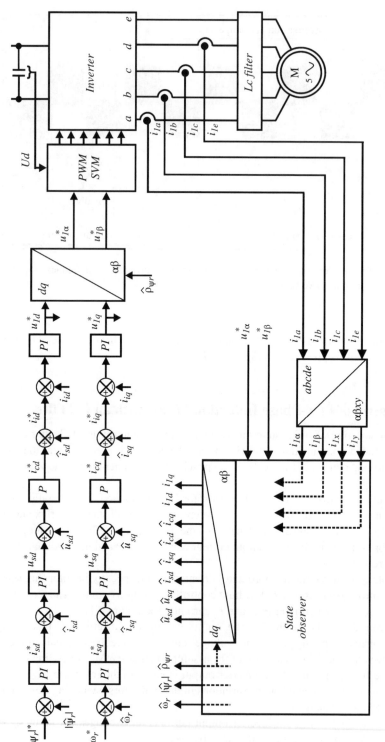

Figure 9.12 Structure of five-phase speed sensorless induction motor drive

9.6 Speed and Flux Observer

The structure of the observer has been presented by the authors in previous chapters for a three-phase motor in stand-alone operation and with LC filter operation. In this chapter an observer for a five-phase system is presented.

In addition to the estimation of $\alpha\beta$ motor variables, the xy currents are calculated. In the observer block the estimated $\alpha\beta xy$ signals are transformed to $dqDQ$ coordinates.

The observer equations including the LC filter model relations are as follows:

$$\frac{d\hat{i}_{s\alpha}}{dt} = -\frac{R_s L_r^2 + R_r L_m^2}{L_r w_\sigma}\hat{i}_{s\alpha} + \frac{R_r L_m}{L_r w_\sigma}\hat{\psi}_{r\alpha} + \frac{L_m}{w_\sigma}\hat{\xi}_\beta + \frac{L_r}{w_\sigma}\hat{u}_{s\alpha} + k_1\left(i_{1\alpha} - \hat{i}_{1\alpha}\right) \quad (9.45)$$

$$\frac{d\hat{i}_{s\beta}}{dt} = -\frac{R_s L_r^2 + R_r L_m^2}{L_r w_\sigma}\hat{i}_{s\beta} + \frac{R_r L_m}{L_r w_\sigma}\hat{\psi}_{r\beta} - \frac{L_m}{w_\sigma}\hat{\xi}_\alpha + \frac{L_r}{w_\sigma}\hat{u}_{s\beta} + k_1\left(i_{1\beta} - \hat{i}_{1\beta}\right) \quad (9.46)$$

$$\frac{d\hat{\psi}_{r\alpha}}{dt} = \frac{R_r}{L_r}\hat{\psi}_{r\alpha} + \frac{R_r L_m}{L_r}\hat{i}_{s\alpha} - \hat{\xi}_\beta - k_2 S_b \hat{\psi}_{r\alpha} + k_3 \hat{\psi}_{r\beta}\left(S_b - S_{bF}\right) \quad (9.47)$$

$$\frac{d\hat{\psi}_{r\beta}}{dt} = \frac{R_r}{L_r}\hat{\psi}_{r\beta} + \frac{R_r L_m}{L_r}\hat{i}_{s\beta} + \hat{\xi}_\alpha - k_2 S_b \hat{\psi}_{r\beta} - k_3 \hat{\psi}_{r\alpha}\left(S_b - S_{bF}\right) \quad (9.48)$$

$$\frac{d\hat{\xi}_\alpha}{dt} = \frac{R_r}{L_r}\hat{\xi}_\alpha + \frac{R_r L_m}{L_r}\hat{\omega}_r \hat{i}_{s\alpha} - \hat{\xi}_\beta - k_2 S_b \hat{\psi}_{r\alpha} + k_3 \hat{\psi}_{r\beta}\left(S_b - S_{bF}\right) \quad (9.49)$$

$$\frac{d\hat{\xi}_\alpha}{d\tau} = \frac{R_r L_m}{L_r}\hat{\omega}_r \hat{i}_{s\alpha} + \frac{R_r}{L_r}\hat{\xi}_\alpha - \hat{\omega}_r \hat{\xi}_\beta - k_4\left(i_{1\beta} - \hat{i}_{1\beta}\right) \quad (9.50)$$

$$\frac{d\hat{\xi}_\beta}{d\tau} = \frac{R_r L_m}{L_r}\hat{\omega}_r \hat{i}_{s\beta} + \frac{R_r}{L_r}\hat{\xi}_\beta + \hat{\omega}_r \hat{\xi}_\beta + k_1\left(i_{1\alpha} - \hat{i}_{1\alpha}\right) \quad (9.51)$$

$$\frac{d\hat{u}_{s\alpha}}{d\tau} = \frac{i_{1\alpha} - \hat{i}_{s\alpha}}{C_f} \quad (9.52)$$

$$\frac{d\hat{u}_{s\beta}}{d\tau} = \frac{i_{1\beta} - \hat{i}_{s\beta}}{C_f} \quad (9.53)$$

$$\frac{d\hat{i}_{1\alpha}}{d\tau} = \frac{u_{1\alpha}^* - \hat{u}_{s\alpha}}{L_f} + k_A\left(i_{1\alpha} - \hat{i}_{1\alpha}\right) - k_B\left(i_{1\beta} - \hat{i}_{1\beta}\right) \quad (9.54)$$

$$\frac{d\hat{i}_{1\beta}}{d\tau} = \frac{u_{1\beta}^* - \hat{u}_{s\beta}}{L_f} + k_A\left(i_{1\beta} - \hat{i}_{1\beta}\right) + k_B\left(i_{1\alpha} - \hat{i}_{1\alpha}\right) \quad (9.55)$$

$$\frac{dS_{bF}}{dt} = \frac{1}{T_{Sb}}\left(S_b - S_{bF}\right) \quad (9.56)$$

and $k_1, \ldots, k_4, k_A,$ and k_B are observer gains; S_b is the observer stabilizing component; S_{bF} is the S_b filtered value; and T_{Sb} is the S_b filter time constant.

Based on the estimated rotor flux and electromotive force (EMG), the motor speed is taken from:

$$\hat{\omega}_r = \frac{\hat{\xi}_\alpha \hat{\psi}_{r\alpha} + \hat{\xi}_\beta \hat{\psi}_{r\beta}}{\hat{\psi}_{r\alpha}^2 + \hat{\psi}_{r\beta}^2} \quad (9.57)$$

The second pair of dq stator current controllers requires an estimated value in the feedback loop. So for that part, additional equations of the observer are introduced:

$$\frac{d\hat{i}_{sx}}{dt} = \frac{L_r}{w_\sigma}\hat{u}_{sx} - \frac{R_s L_r}{w_\sigma}\hat{i}_{sx} + k_1\left(i_{1x} - \hat{i}_{1x}\right) \quad (9.58)$$

$$\frac{d\hat{i}_{sy}}{dt} = \frac{L_r}{w_\sigma}\hat{u}_{sy} - \frac{R_s L_r}{w_\sigma}\hat{i}_{sy} + k_1\left(i_{1y} - \hat{i}_{1y}\right) \quad (9.59)$$

$$\frac{d\hat{u}_{sx}}{dt} = \frac{i_{1x} - \hat{i}_{sx}}{C_f} \quad (9.60)$$

$$\frac{d\hat{u}_{sy}}{dt} = \frac{i_{1\beta} - \hat{i}_{sy}}{C_f} \quad (9.61)$$

$$\frac{d\hat{i}_{1x}}{d\tau} = \frac{u_{1x}^* - \hat{u}_{sx}}{L_f} + k_A\left(i_{1x} - \hat{i}_{1x}\right) - k_B\left(i_{1y} - \hat{i}_{1y}\right) \quad (9.62)$$

$$\frac{d\hat{i}_{1y}}{d\tau} = \frac{u_{1y}^* - \hat{u}_{sy}}{L_f} + k_A\left(i_{1y} - \hat{i}_{1y}\right) + k_B\left(i_{1x} - \hat{i}_{1x}\right) \quad (9.63)$$

Park and inverse Park transformation are performed with the rotor flux vector position derived based on the estimated $\alpha\beta$ flux components.

9.7 Induction Motor and an LC Filter for Five-Phase Drive

Five-phase motors are not yet commercially available in high-level production. They are rather designed according to the application requirements. The widely used way to manufacture a five-phase motor is to convert the classical three-phase motor into a five-phase one. However problems associated with the number of slots appear. In a three-phase drive the slots are a multiple of three, whereas in five-phase drive they are a multiple of five. For example, the widely used 5.5-kW four-pole motor could have 36 slots, so if the stator remains without changes some slots would have to be unused. This would lead to some asymmetry in the magnetic path.

Therefore, it is more beneficial is to design new sheets of the five-phase motor while leaving the stator core and complete rotor without changes. In the design process both the analytical calculations as well as the finite elements method are used for the optimal magnetic circuit design (Figure 9.13) [22, 23].

The induction motor can be designed with distributed or concentrated stator winding [24]. Concentrated windings are required if higher torque harmonics are to be used for improving the motor's total torque. In both cases the radial components of magnetic induction in the motor air gap have to be observed (Figure 9.14).

After the design has been completed, the stator sheets have to be cut with a CNC machine and assembled into the stator frame. Then five-phase coils have to be installed. An example of a stator sheet for a five-phase four-pole motor is presented in Figure 9.15.

When stator sheets are assembled, proper stator windings have to be put in place and the whole motor is then assembled. Examples of five-phase motor parameters are given in Tables 9.1 and 9.2 (denoted as motor 1 and motor 2).

The IM motor from Table 9.1 was designed for a low-voltage supply for use in small electric vehicles with a 48-V battery. The motor with the data given in Table 9.2 was designed to operate with a converter with a standard 400-V, 50-Hz grid voltage supply.

The LC filter for the five-phase drive has to be designed in the same way as for a three-phase motor with regard to the lower current rate. The data of the LC filter for the low-voltage motor 1 are presented in Table 9.3.

The waveforms for motor 1 and the LC filter are presented in Figure 9.16. It is noticeable that the motor voltage is smoothed.

9.8 Investigations of Five-Phase Sensorless Drive with an LC Filter

In this subsection, examples of the investigation results obtained for a drive with an induction motor (motor 1), LC filter, and voltage inverter with sensorless control are presented.

The results of the simulations are presented in Figures 9.17 to 9.19. The system was investigated for operation in closed-loop mode, where all signals estimated in the observer were used in the feedback control loop. The real speed was used only for data acquisition purposes and for the comparison with the estimated one. In Figures 9.17 and 9.18, the transients for the reference speed changes are presented. The reference speed is the same in both cases whereas the control structure is different. Figure 9.16 presents the classical FOC structure, whereas Figure 9.18 shows the proposed FOC with multiloop control. It can be observed that oscillations of the stator current torque component are eliminated when the structure with the filter

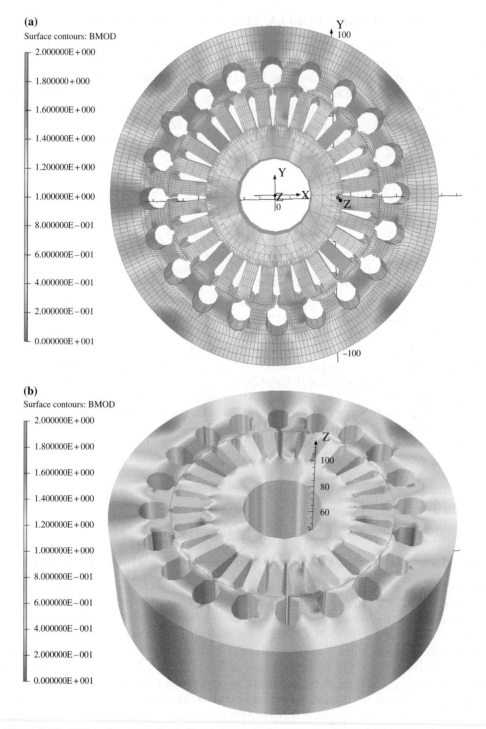

Figure 9.13 Finite element methods for the design of a five-phase induction motor: examples of magnetic field distribution in two-dimensional model (**a**) and three-dimensional model (**b**)

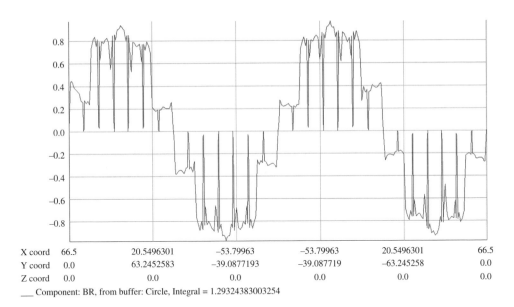

X coord	66.5	20.5496301	−53.79963	−53.79963	20.5496301	66.5
Y coord	0.0	63.2452583	−39.0877193	−39.087719	−63.245258	0.0
Z coord	0.0	0.0	0.0	0.0	0.0	0.0

___ Component: BR, from buffer: Circle, Integral = 1.29324383003254

Figure 9.14 An example of magnetic induction distribution in the air gap of a five-phase induction motor

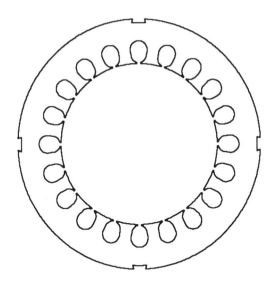

Figure 9.15 Example of stator sheet for a five-phase, four-pole induction motor (20 slots)

model is applied (see Section 9.5). Figure 9.19 proves the performance of the sensorless operation for speed reversal.

The experimental results are presented in Figures 9.20 to 9.24. The results of tests for steady state operation are presented in Figure 9.20, whereas transients for open-loop control are given in Figure 9.21 and those for closed-loop control are given in Figures 9.22 to 9.24.

Table 9.1 Parameters of five-phase induction motor for electrical vehicle (motor 1)

Parameter	Symbol	Value
Rated power	U_n	5 kW
Rated voltage (phase-to-phase)	U_n	30 V
Rated current	I_n	39.7 A
Rated frequency	f_n	50 Hz
Rated speed	n_n	2940 rpm
Poles	p	2 (one pair of poles)
Inertia	J	0.015 kgm²
Stator resistance	R_s	10 mΩ
Rotor resistance	R_r	0.731 Ω
Stator leakage inductance	$l_{\sigma s}$	47 µH
Rotor leakage inductance	$l_{\sigma r}$	47 µH
Mutual inductance	L_m	1.01 mH

Table 9.2 Parameters of five-phase induction motor for industrial application (motor 2)

Parameter	Symbol	Value
Rated power	U_n	5.5 kW
Rated voltage (phase-to-phase)	U_n	173 V
Rated current	I_n	8.8 A
Rated frequency	f_n	50 Hz
Rated speed	n_n	1425 rpm
Poles	p	4 (two pairs of poles)
Inertia	J	0.015 kgm2
Stator resistance (at 20 °C)	R_s	0.945 Ω
Rotor resistance	R_r	0.738 Ω
Stator leakage inductance	$l_{\sigma s}$	1.68 mH
Rotor leakage inductance	$l_{\sigma r}$	1.68 mH
Mutual inductance	L_m	225.3 mH

Table 9.3 Parameters of LC filter for motor 1 (Table 9.1)

Parameter	Symbol	Value
Reactor inductance	Lf	97 µH
Capacitance	Cf	30 µF
Cut-off frequency	fres	3 kHz

For open-loop control, the drive was operating under V/f control while the observer was only tested.

For steady state operation of the filter shown in Figure 9.20, it can be seen that the motor voltage is smoothed. The results presented in Figure 9.21 are for V/f control to prove the observer operation (for the observer see Section 9.6). For the steady state, the estimated speed

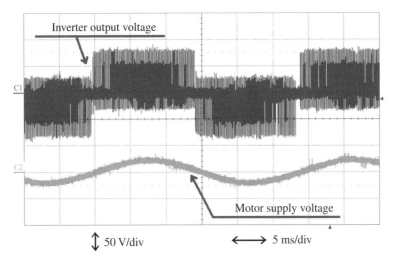

Figure 9.16 Operation of the sine-wave filter for motor 1: experimental waveforms of inverter and motor voltages

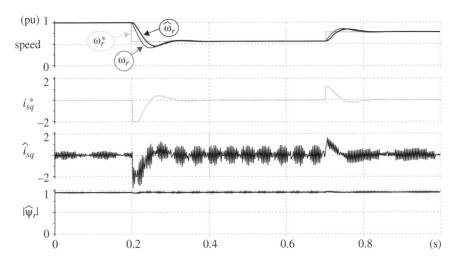

Figure 9.17 Simulation results showing a sequence with speed changes: operation for the drive with an LC filter and classical FOC structure

was compared with the measured one by a precise digital sensor. Based on this, the speed calculation error is ≤1% for the whole of the tested speed range.

The closed-loop control was initially verified with speed sensor use. Except for the speed signal, all feedback signals were taken from the observer. The results are presented in Figure 9.22. It can be seen that a steep change of motor speed gives satisfactory results with the motor torque in a limited range.

The next waveforms are for speed sensorless FOC operation of the five-phase IM drive (Figures 9.23 and 9.24). In Figure 9.23 the transients for speed control are presented, whereas

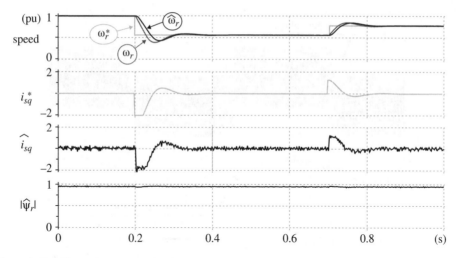

Figure 9.18 Simulation results showing a sequence with speed changes: operation for the drive with an LC filter and FOC structure with multiloop filter control

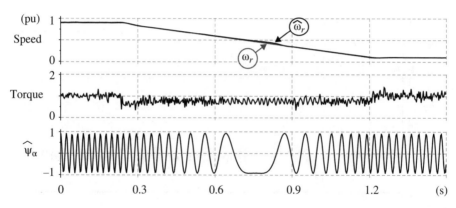

Figure 9.19 Simulation results showing a speed reversal: operation for the drive with an LC filter and proposed FOC structure with multiloop filter control

in Figure 9.24 the flux regulation is shown. In both cases a proper operation of the proposed five-phase speed sensorless drive can be noticed. It is worth underlining that in Figure 9.21 the slow change of the referenced speed is tested with acceptable results.

9.9 FOC Structure in the Case of Combination of Fundamental and Third Harmonic Currents

The multiphase motor torque can be increased in the case of a combination of the fundamental and third harmonic currents [3, 4]. The torque enhancement is about 10% depending on the design of the magnetic circuit. However, it also requires a proper extension of the control

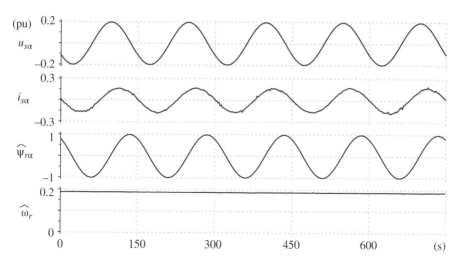

Figure 9.20 Experimental results for the open-loop control system: steady state, reference frequency = 10 Hz

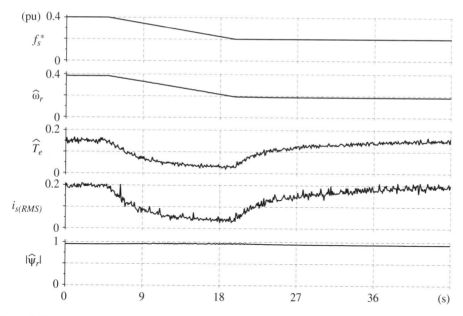

Figure 9.21 Experimental results for the open-loop control system: slow change of the motor supply frequency

system. An example of FOC with a combination of the fundamental and third harmonic currents is presented in Figure 9.25 [25].

As can be seen from Figure 9.25, two dq coordinate planes are employed in the five-phase induction motor control algorithm. By inappropriate distribution of the torque-producing and nontorque-producing harmonics of five-phase IM currents in the $d1-q1$ and

264 Variable Speed AC Drives with Inverter Output Filters

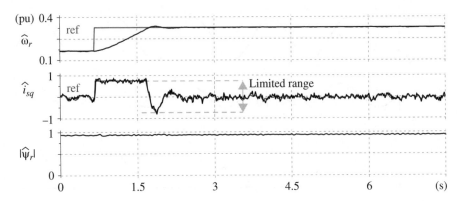

Figure 9.22 Experimental results: motor speed control for FOC with a flux observer and speed sensor

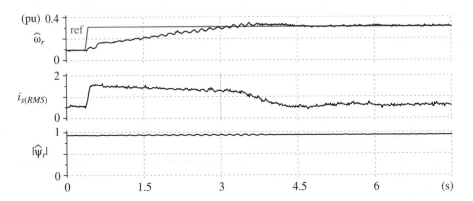

Figure 9.23 Experimental results: motor speed control for speed sensorless FOC

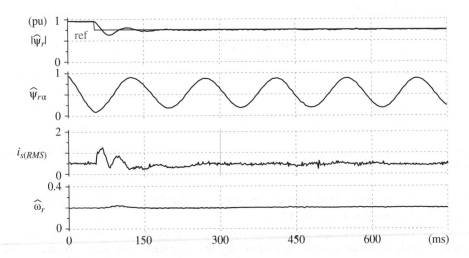

Figure 9.24 Experimental results: motor flux control for speed sensorless FOC

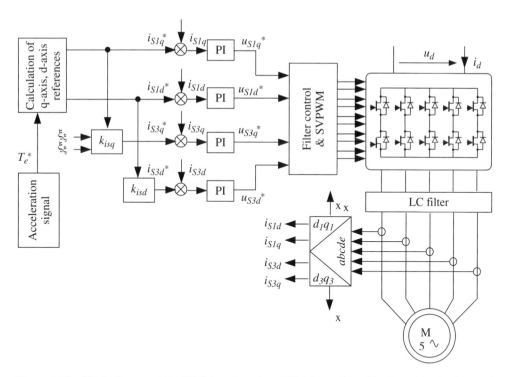

Figure 9.25 Block diagram of the FOC for a five-phase EV drive and combination of the fundamental and third harmonic currents [25]

$d3$–$q3$ reference planes, the phase current deformation may be obtained and nontorque-producing harmonics may cause additional core loss [26]. The main idea of five-phase IM control is to add a third spatial harmonic, moving synchronously with the fundamental component, to the air-gap magnetic field, which results in a decrease of its peak value and makes it possible to increase the amplitude of the fundamental component above the rated value without causing air gap flux saturation [27]. As shown in Casadei, Mengoni, Tani, Serra and Zarri [27] under the condition of synchronism, that is, $\omega_3 = 3\omega_1$, the d-axis current i_{S3d} in the $d3$–$q3$ plane should be a fraction of the i_{S1d} current, and i_{S3q} should satisfy the steady state condition:

$$i_{S3q} = \frac{3\dfrac{L_{r3}}{R_{r3}} i_{S3d}}{\dfrac{L_{r1}}{R_{r1}} i_{S1d}} i_{S1dq} \tag{9.64}$$

The properly selected proportional gains k_{isd} and k_{isq} of the $i_{S3d,ref}$ and $i_{S3q,ref}$ reference signals guarantee the fulfillment of the synchronization condition $\omega_{3h} = 3\omega_{1h}$.

9.10 Simulation Examples of Five-Phase Induction Motor with a PWM Inverter

The model of five-phase induction motor with PWM inverter (*Chapter_9\Example_1\ Simulator.slx*) is shown in Figure 9.26.

The model *Simulator.slx* consists of a five-phase induction motor, PWM inverter, and V/f control. The main simulation file is MULTI_ph.c. Because of he appropriate PWM algorithm, the third harmonic components of stator current are suppressed (see Section 9.4 for details).

Example of simulation results is presented in Figure 9.27. In Figure 9.27, the next variables are presented (from top to bottom):

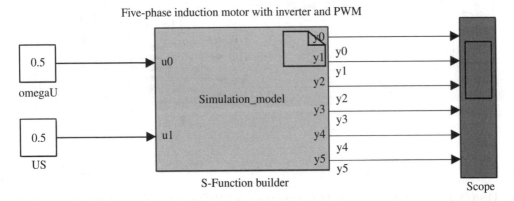

Figure 9.26 MATLAB®/Simulink model of five-phase induction motor with PWM inverter, *Chapter_9\ Example_1\Simulator.slx*

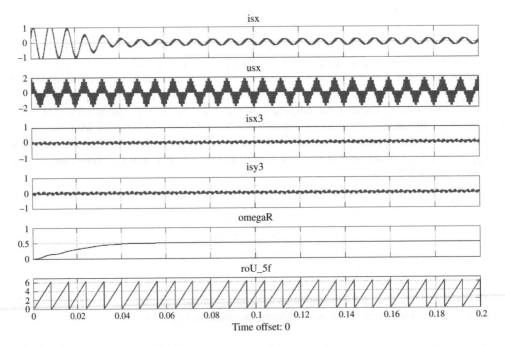

Figure 9.27 Example of simulation results for five-phase induction motor with PWM inverter

- `isx`, alpha component of stator current (fundamental),
- `usx`, alpha component of motor stator voltage,
- `isx3`, alpha component of stator current (third harmonic),
- `isy3`, beta component of stator current (third harmonic),
- `omegaR`, motor speed,
- `roU_5f`, angle position of stator voltage vector.

The PWM function is implemented in C-function *PWM5f( )*:

```
/*--------------------------------------------------------------*/
// 5-phase PWM
/*--------------------------------------------------------------*/
void PWM5f( void)
{
static double impulse=1000.0;
static int NroU;
static int idu0,idu1,idu2,idu3,idu4,idu5;
static double ta,tb,ta1,tb1,ta2,tb2;
static double t0_5f;//,t1_5f,t2_5f,t3_5f,t4_5f,t5_5f,t6_5f;

// voltage output vectors
static double ux1x[32]={0.0,0.195,-0.512,-0.316,-0.512,-0.316,-1.023,-0.828,
             0.195,0.391,-0.316,-0.121,-0.316,-0.121,-0.828,-0.632,
             0.632,0.828,0.121,0.316,0.121,0.316,-0.391,-0.195,
             0.828,1.023,0.316,0.512,0.316,0.512,-0.195,0.0};
static double uy1y[32]={0.0,-0.602,-0.372,-0.973,0.372,-0.230,0.0,-0.602,
             0.602,0.0,0.230,-0.372,0.973,0.372,0.602,0.0,0.0,
             -0.602,-0.372,-0.973,0.372,-0.230,0.0,-0.602,0.602,
             0.0,0.230,-0.372,0.973,0.372,0.602,0.0};
static double ux2x[32]={0.0,-0.512,0.195,-0.316,0.195,-0.316,0.391,-0.121,
             -0.512,-1.023,-0.316,-0.828,-0.316,-0.828,-0.121,
             -0.632,0.632,0.121,0.828,0.316,0.828,0.316,1.023,
             0.512,0.121,-0.391,0.316,-0.195,0.316,-0.195,0.512,0.0};
static double uy2y[32]={0.0,-0.372,0.602,0.230,-0.602,-0.973,0.0,-0.372,0.372,
             0.0,0.973,0.602,-0.230,-0.602,0.372,0.0,0.0,-0.372,
             0.602,0.230,-0.602,-0.973,0.0,-0.372,0.372,0.0,0.973,
             0.602,-0.230,-0.602,0.372,0.0};

impulse+=h;
if (impulse>Timp)
    {
    if(roU_5f>2*PI) roU_5f-=2*PI;
    if(roU_5f<0) roU_5f+=2*PI;
    NroU=(int)floor(roU_5f/(PI/5));

    // voltage sector selection
    switch(NroU)
        {
        case 0: idu1=16; idu2=24; idu3=25; idu4=29; sector=0.0; break;
        case 1: idu1=29; idu2=28; idu3=24; idu4= 8; sector=1.0; break;
        case 2: idu1= 8; idu2=12; idu3=28; idu4=30; sector=2.0; break;
        case 3: idu1=30; idu2=14; idu3=12; idu4= 4; sector=3.0; break;
        case 4: idu1= 4; idu2= 6; idu3=14; idu4=15; sector=4.0; break;
        case 5: idu1=15; idu2= 7; idu3= 6; idu4= 2; sector=5.0; break;
```

```
            case 6: idu1= 2; idu2= 3; idu3= 7; idu4=23; sector=6.0; break;
            case 7: idu1=23; idu2=19; idu3= 3; idu4= 1; sector=7.0; break;
            case 8: idu1= 1; idu2=17; idu3=19; idu4=27; sector=8.0; break;
            case 9: idu1=27; idu2=25; idu3=17; idu4=16; sector=9.0; break;
        }

    idu0=0;
    idu5=31;
    ta=(US_5f*sin((sector+1)*PI/5-roU_5f)/(1.35735522*ud*sin(PI/5)))*Timp;
    tb=(US_5f*sin(roU_5f-(sector)*PI/5)/(1.35735522*ud*sin
    (PI/5)))*Timp;

    ta1=0.382*ta;
    ta2=0.618*ta;
    tb1=0.618*tb;
    tb2=0.382*tb;

    t0_5f=(Timp-ta1-ta2-tb1-tb2);

    impulse=0.0;
    cycle*=-1;
    }

// actual inverter output voltage
if (cycle==1)
        {// cycle: t0/2..ta1..tb1..ta2..tb2..t0/2
        // alpha-beta      // x-y
            if(impulse<=(t0_5f/2)) {usx_5f=0; usy_5f=0; usx3_5f=0; usy3_5f=0;}
                else if (impulse<=(t0_5f/2+ta1)) usx_5f=ud*ux1x[idu1];
    usy_5f=ud*uy1y[idu1];
                    usx3_5f=ud*ux2x[idu1]; usy3_5f=ud*uy2y[idu1];}
                else if (impulse<=(t0_5f/2+ta1+tb1)) {usx_5f=ud*ux1x[idu2];
                    usy_5f=ud*uy1y[idu2]; usx3_5f=ud*ux2x[idu2];
    usy3_5f=ud*uy2y[idu2];}
                else if (impulse<=(t0_5f/2+ta1+tb1+ta2)) {usx_5f=ud*ux1x[idu3];
                    usy_5f=ud*uy1y[idu3]; usx3_5f=ud*ux2x[idu3];
    usy3_5f=ud*uy2y[idu3];}
                else if (impulse<=(t0_5f/2+ta1+tb1+ta2+tb2)) {usx_5f=ud*ux1x[idu4];
                    usy_5f=ud*uy1y[idu4]; usx3_5f=ud*ux2x[idu4];
    usy3_5f=ud*uy2y[idu4];}
                else {usx_5f=0; usy_5f=0; usx3_5f=0; usy3_5f=0;}
        }
    else
        {// cycle: t0/2..tb2..ta2..tb1..ta1..t0/2
        // alpha-beta      // x-y
        if (impulse<=(t0_5f/2)) {usx_5f=0; usy_5f=0; usx3_5f=0; usy3_5f=0;}
            else if (impulse<=(t0_5f/2+tb2)) {usx_5f=ud*ux1x[idu4];
                usy_5f=ud*uy1y[idu4]; usx3_5f=ud*ux2x[idu4];
    usy3_5f=ud*uy2y[idu4];}
            else if (impulse<=(t0_5f/2+tb2+ta2)) {usx_5f=ud*ux1x[idu3];
                usy_5f=ud*uy1y[idu3]; usx3_5f=ud*ux2x[idu3];
    usy3_5f=ud*uy2y[idu3];}
            else if (impulse<=(t0_5f/2+tb2+ta2+tb1)) {usx_5f=ud*ux1x[idu2];
                usy_5f=ud*uy1y[idu2]; usx3_5f=ud*ux2x[idu2];
    usy3_5f=ud*uy2y[idu2];}
```

```
            else if (impulse<=(t0_5f/2+tb2+ta2+tb1+ta1)) {usx_5f=ud*ux1x[idu1];
                usy_5f=ud*uy1y[idu1]; usx3_5f=ud*ux2x[idu1];
    usy3_5f=ud*uy2y[idu1];}
            else {usx_5f=0; usy_5f=0; usx3_5f=0; usy3_5f=0;}
        }
}
```

The model of five-phase induction motor is available in function *F4DERY()*c. It is modeled with a structure as shown in Figure 9.2. The circuit of third harmonic is not influencing the total motor torque.

References

[1] Levi E, Bojoi R, Profumo F, Toliyat HA, Williamson S. Multiphase induction motor drives—a technology status review. IET Electric Power Applications. 2007; 1 (4): 489–516.

[2] Abu-Rub H, Iqbal A, Guzinski J. High performance control of AC drives with MATLAB/Simulink models. Chichester, UK: John Wiley & Sons, Ltd; 2012.

[3] Xu H, Toliyat HA, Petersen LJ. Five-phase induction motor drives with DSP-based control system. IEEE Transactions on Power Electronics. 2002; 17 (4): 524–533.

[4] Xu H, Toliyat HA, Petersen LJ. Rotor field oriented control of five-phase induction motor with the combined fundamental and third harmonic currents. Sixteenth Annual IEEE Applied Power Electronics Conference and Exposition, APEC 2001, Anaheim, CA. March 4–8, 2001.

[5] Jiang D, Qian W. Study of five-phase space vector PWM considering third order harmonics. (2013). IEEE Energy Conversion Congress and Exposition, ECCE 2013, Denver, CO. September 15–19, 2013.

[6] Kim M-H, Kim N-H, Baik W-S. A five-phase IM vector control system including 3rd current harmonics component. IEEE 8th International Conference on Power Electronics and ECCE Asia (ICPE & ECCE), The Shilla Jeju, Korea. May 30–June 3, 2011.

[7] Dujic D, Jones M, Levi E. Space vector PWM for nine-phase VSI with sinusoidal output voltage generation: analysis and implementation. Thirty-third Annual Conference of the IEEE Industrial Electronics Society, IECON, Taipei, Taiwan. November 5–8, 2007.

[8] Temen F, Siala S, Noy P. Multiphase induction motor sensorless control for electric ship propulsion. Conference on Power Electronics, Machines and Drives, PEMD 2004, Edinburgh, UK. March 31–April 2, 2004.

[9] Iqbal A, Levi E, Jones M, Vukosavic SN. A PWM scheme for a five-phase VSI supplying a five-phase two-motor drive. Thirty-Second Annual Conference of the IEEE Industrial Electronics Society, IECON 2006, Paris, France. November 7–10, 2006.

[10] de Silva PSN, Fletcher JE, Williams BW. Development of space vector modulation strategies for five phase voltage source inverters. Second International Conference on Power Electronics, Machines and Drives, PEMD 2004, Edinburgh, UK. March 31–April 2, 2004.

[11] Jones M, Dujic D, Levi E, Vukosavic SN. Dead-time effects in voltage source inverter fed multi-phase AC motor drives and their compensation. Thirteenth European Conference on Power Electronics and Applications, EPE, Barcelona, Spain. September 8–10, 2009.

[12] Saleh M, Iqbal A, Moin SK, Kalam A. Matrix converter based five-phase series connected induction motor drive. Australian Universities Power Engineering Conference (AUPEC), Bali, Indonesia. September 26–29, 2012.

[13] Mengoni M, Zarri L, Tani A, Serra G, Casadei D. Sensorless multiphase induction motor drive based on a speed observer operating with third-order field harmonics. IEEE Energy Conversion Congress and Exposition (ECCE), Phoenix, AZ. September 17–22, 2011.

[14] Guzinski J, Abu-Rub H. Speed sensorless control of induction motors with inverter output filter. *International Review of Electrical Engineering*. 2008; **3** (2): 337–343.

[15] Busse DF, Erdman J, Kerkman R, Schlegel D, Skibinski G. Bearing currents and their relationship to PWM drives. IEEE Transactions on Power Electronics. 1997; 12 (2): 243–252.

[16] Muetze A, Binder A. High frequency stator ground currents of inverter-fed squirrel-cage induction motors up to 500 kW. Tenth European Conference on Power Electronics and Applications, EPE'03, Toulouse, France. September 2–4, 2003.

[17] Guzinski J. Sensorless AC drive control with LC filter. Thirteenth European Conference on Power Electronics and Applications, EPE 2009, Barcelona, Spain. September 8–10, 2009.
[18] Guzinski J, Abu-Rub H, Zobaa FA. Electric motor drive with inverter output filter. Workshop on Power Electronics for Industrial Applications and Renewable Energy Conversion, PEIA 2011, Doha, Qatar, November 3–4, 2011.
[19] Guzinski J. Sensorless direct torque control of induction motor drive with LC filter. Fifteenth International Power Electronics and Motion Conference and Exposition, EPE-PEMC 2012 ECCE Europe, Novi Sad, Serbia. September 4–6, 2012.
[20] Abu-Rub H, Rizwan Khan M, Iqbal A, Moin Ahmed SK. MRAS-based sensorless control of a five-phase induction motor drive with a predictive adaptive model. IEEE International Symposium on Industrial Electronics ISIE 2010, Bari, Italy. July 4–7, 2010.
[21] Stec P, Guzinski J, Strankowski P, Iqbal A, Abduallah AA, Abu-Rub H. Five-phase induction motor drive with sine-wave filter. IEEE International Symposium on Industrial Electronics, ISIE 2014, Istanbul, Turkey. June 1–4, 2014.
[22] Williamson S, Lim LH, Robinson MJ. Finite-element models for cage induction motor analysis. IEEE Industry Applications Society Annual Meeting, San Diego, CA. October 1–5, 1989.
[23] Rachek M, Merzouki T. Finite element method applied to the modelling and analysis of induction motors. In: P Miidla, editor. Numerical Modelling. Rijeka, Croatia: Intech; 2013. pp. 203–226.
[24] Toliyat HA, Lipo TA. Analysis of concentrated winding induction machines for adjustable speed drive applications—Part 1 (motor analysis). IEEE Transactions on Energy Conversion. 1994; 6 (4): 679–683.
[25] Adamowicz M, Guzinski J, Stec P. Five-phase EV drive with switched-autotransformer (LCCAt) inverter. Vehicle Power and Propulsion Conference, VPPC 2014, Coimbra, Portugal. October 27–30, 2014.
[26] Lu S, Corzine K. Multilevel multi-phase propulsion drives. IEEE Electric Ship Technologies Symposium, Philadelphia, PA. July 25–27, 2005.
[27] Casadei D, Mengoni M, Tani A, Serra G, Zarri L. High torque-density seven-phase induction motor drives for electric vehicle applications. IEEE Vehicle Power and Propulsion Conference, VPPC 2010, Lille, France. September 1–3, 2010.

10

General Summary, Remarks, and Conclusion

This book presents various issues related to the problems of electric drives fed by power electronic inverters with output passive filters.

This is a broad topic, covering many complex issues. This book contains selected issues related to the AC drives with LC filters. The main emphasis is on the issue of drives with sinusoidal filters. Sinusoidal filters have a significant impact on the control operation of induction motor drive systems. Therefore, the major part of this work concerns the control and estimation problems in drives with such filters. The book also raises the issues of electric drives fault diagnostics when using filters.

The presented solutions are dedicated mainly for sensorless drives with squirrel cage induction motors. Operation of sensorless drives is based only on the use of readily measured signals available in the inverter, which are the inverter output currents and input voltage. By using these signals, it is possible to have closed-loop regulation but only using appropriate methods of estimation of unmeasured variables. With the introduction of a passive filter, the motor voltages and currents differ from that of the inverter output (filter input). Therefore, the use of such filters negatively affects the available sensorless control methods. The solution to these problems is to take into account the presence of the filter in both the control and estimation of the drive variables.

In this book, the properties of selected control structures of squirrel cage induction motors are tested while proposing new solutions for the control and estimation. Various systems were investigated with close-loop sensorless mode, such as nonlinear control structures and field-oriented control.

Several control schemes with sinusoidal filter are presented in this book. Also, estimators containing filters are presented, which allow calculating the required variables by using the original inverter sensors.

Variable Speed AC Drives with Inverter Output Filters, First Edition. Jaroslaw Guzinski, Haitham Abu-Rub and Patryk Strankowski.
© 2015 John Wiley & Sons, Ltd. Published 2015 by John Wiley & Sons, Ltd.

An important element of the work is the description of nonlinear control method of the drive inverter with LC filter and using load angle control. It also proposes a predictive controller for the stator current of the motor with choke. As an addition to other available solutions, a method of improving the properties of the controller is introduced, which is based on the use of the same state observer that is simultaneously used to estimate the state variables in the control system.

The developed sensorless motor control system does not require any additional sensors, in addition to sensors being used in the conventional systems with voltage inverters. Also, the use of filter does not require the use of additional current and voltage sensors for the drive application.

Fault diagnostic issues in the drive system of LC filter are also presented. Observer structures can be modified to act as signals computing observers, which could be used to drive diagnostics. State observers are developed to calculate the electromagnetic torque and load torque of the motor. Analysis of these torques enables the user to diagnose the mechanical system of the drive. For example, it is possible to use such analysis to detect machine misalignment and drive system unbalance.

For drive systems with a filter, it is also possible to diagnose the fault using analysis of command signals in the closed-loop control system. Results are shown to indicate the suitability of this solution for detecting the damage to the rotor cage.

The book demonstrates that the introduction of the filter between the inverter and the motor does not cause malfunction of the drive, if the structure of the control and estimation take into account the model equations of the used filter. This extends the scope of used algorithms for sensorless control of induction motor drives with passive filters.

The book presents various estimation and control structures, illustrating their properties both in simulation and experimentally. The study was conducted using advanced and the latest microprocessor systems technology.

Appendix A: Synchronous Sampling of Inverter Output Current

During the implementation of control methods in electrical drives with an inverter supply, the information about the first harmonic of the inverter output current is required. Because of the pulsating character of the motor supply, the output inverter current contains higher harmonics, which are related to the transistor switching frequency. One of the most used methods to eliminate those harmonics is the synchronous sampling method with pulse width modulator function. This comes from the high accuracy and simplicity of the method. Synchronous sampling requires a load time constant that is much greater than the inverter pulse period. This condition is fulfilled for most low- and medium-power drive systems.

If the current is synchronically sampled with the pulse width modulation (PWM) operation and the measurement moments are described through extremes of the carrier function, the recorded current samples are equal to the first current harmonic. The carrier wave emerges in an explicit way, during the sinusoidal modulation. For space vector modulation (SVM), the half-time moments of zero vectors duration are corresponding to the moment extreme of the modulation function [1, 2].

First, the case of a single-phase inverter (Figure A.1) was considered with the constant desired output voltage value, in other words, a constant pulse width factor. The bridge switches are turned on in pairs: S_1-S_4 and S_2-S_3. The voltage u_s takes the value $\pm U_d$, whereby the current $i(t)$ is described as follows:

$$i(t) = \frac{1}{R} u_s(t) \cdot \left(1 - e^{(-t/\tau)}\right) + i_0 \cdot e^{(-t/\tau)} - \frac{E}{R} \qquad (A.1)$$

where i_0 is the current initial value, that is, in time t = 0, whereas τ is the receiver time constant $\tau = L/R$.

Variable Speed AC Drives with Inverter Output Filters, First Edition. Jaroslaw Guzinski, Haitham Abu-Rub and Patryk Strankowski.
© 2015 John Wiley & Sons, Ltd. Published 2015 by John Wiley & Sons, Ltd.

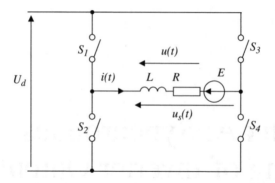

Figure A.1 Equivalent circuit of single-phase inverter (H bridge)

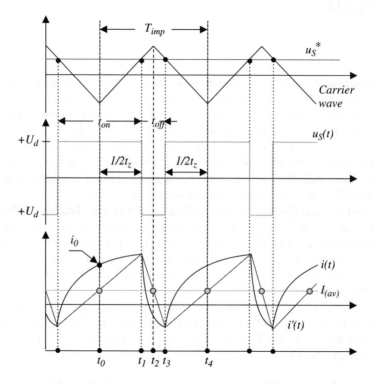

Figure A.2 Voltage and current waveforms in single-phase inverter

The corresponding waveforms are shown in Figure A.2. The current, i_S, and the forced voltage, u_S, have DC and AC components. From Equation A1 follows that the AC component of i_S is not dependent on E, whereby E does not have an influence on sampling error. The synchronous sampling method is based on the assumption, that $\tau \gg T_{imp}$. In such a case, the load current is described by Equation A2:

$$i'(t) \approx \frac{1}{L}\int u(t)dt \qquad (A.2)$$

The current, $i'(t)$, changes linearly in accordance with the dependencies:

- by rising current

$$i'(t) = \frac{U_d - E}{L} t \qquad (A.3)$$

- by falling current

$$i'(t) = \frac{-U_d - E}{L} t \qquad (A.4)$$

The current samples, $i'(t)$ for t_0, t_2 i t_4, ... , cover the mean current value, $I_{(AV)}$.

In case of an AC output voltage generation, in other words by changing width factor, the current samples are measured in the half-period of zero vector duration and not fulfilling the mean current value, where (other than in Figure A.2) the system does not operate in continuous state. To ensure an accurate current mean measurement, the sampling times should be changed for further sampling periods. Nevertheless, considering the practical aspect, this would not be realized. It will be assumed, that a change of the pulse width factor for further transistor switching periods changes insignificantly. In case of high-pulse frequency, the current sampling error can be practically neglected.

To illustrate the aforementioned issues, two identical induction motors with simultaneous supply were modeled in the simulation program (Figure A.3):

- sinusoidal voltage, gave for three-phase voltage inverter (index SIN),
- and rectangular pulse waveform of inverter output voltage (index PWM). The modeled system structure is presented as follows:

Exemplary voltage and current waveforms are shown in Figure A.4.

The comparison of the currents, $i_{s\alpha(SIN)}$ and $i_{s\alpha(PWM)}$, for one time period in Figure A.4 is presented in Figure A.5.

It can be seen in Figure A.5, that the current waveforms, $i_{s\alpha(SIN)}$ and $i_{s\alpha(PWM)}$, at the half time of the further passive vectors are the same. This result is the same any time the sequence of the simulation waveforms is repeated.

Synchronous sampling errors of the current become important for high-power drive systems, which have a low switching frequency (e.g., in the range of 100 Hz) [3].

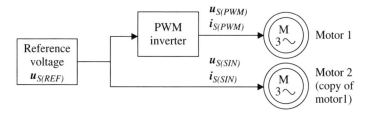

Figure A.3 Simulation model for simulation verification of the synchronous sampling principle

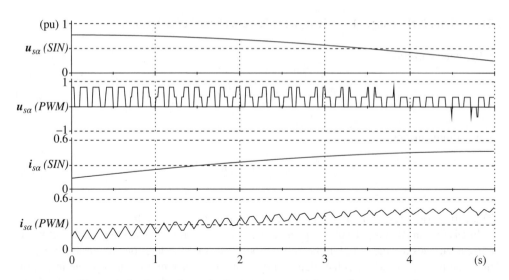

Figure A.4 Comparison of voltages and currents in synchronous sampling

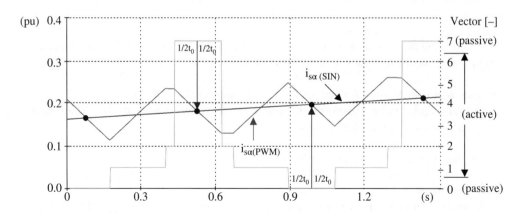

Figure A.5 Detailed comparison of currents in synchronous sampling

References

[1] Blasko V, Kaura V, Niewiadomski W. Sampling of discontinuous voltage and current signals in electrical drives—a system approach. *IEEE Transactions on Industry Applications*. 1998; **34** (5):1123–1130.
[2] Briz F, Díaz-Reigosa D, Degner MW, Garcia P, Guerrero JM. Current sampling and measurement in PWM operated ac drives and power converters. The 2010 International Power Electronics Conference IPEC, Sapporo, Japan. June 21–24, 2010.
[3] Holtz J, Oikonomou N. Estimation of the fundamental current in low-switching-frequency high dynamic medium-voltage drives, *IEEE Transactions on Industrial Applications*. 2008; **44** (5):1597–1605.

Appendix B: Examples of LC Filter Design

B.1 Introduction

The design process of the LC filter was presented in Chapter 4. In this appendix, some examples of the filter design are given. The structure of the LC filter that will be designed in this appendix is given in Figure B.1.

Figure B.1 presents the differential mode filter with L_1, C_1, and R_C elements. The L_1 and C_1 will be selected to create low-pass filter with cut-off frequency below the inverter switching frequency. The resistor (energy dispenser) R_C is added for damping filter resonances.

Example B.1 LC Filter for 1.5-kW Induction Motor

The example presents the design process for standard, low power, three-phase induction motor. The motor data are: $P_n = 1.5$ kW, $U_n = 400$ V (Y connected), $I_n = 3.5$ A, $f_n = 50$ Hz. The inverter DC link voltage is 560 V and the switching frequency is $f_{imp} = 5$ kHz. It is assumed that the fundamental of inverter output voltage will not exceed $f_{out\ 1h} = 100$ Hz. The selected LC filter have to assure that under nominal load the motor current ripple will be within $\Delta I_s = 20\%$.

Solution:
The first calculated element is inductor L_1. The motor ripple current is:

$$\Delta I_s = 0.2 \cdot I_n = 0.2 \cdot 3.5 = 0.7 \text{ A} \tag{B.1}$$

According to Equation 4.17, the L_1 inductance is:

$$L_1 > 2 \frac{U_d}{\sqrt{2} \cdot 2 \cdot \pi \cdot f_{imp} \cdot 3 \cdot \Delta I_s} = 2 \frac{560}{\sqrt{2} \cdot 2 \cdot \pi \cdot 5000 \cdot 3 \cdot 0.7} = 12 \text{ mH} \tag{B.2}$$

Variable Speed AC Drives with Inverter Output Filters, First Edition. Jaroslaw Guzinski, Haitham Abu-Rub and Patryk Strankowski.
© 2015 John Wiley & Sons, Ltd. Published 2015 by John Wiley & Sons, Ltd.

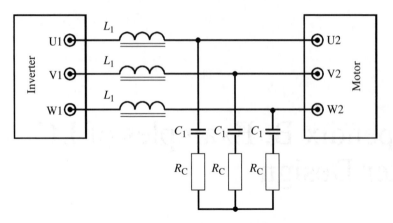

Figure B.1 Sinusoidal filter with damping resistances

For motor nominal voltage frequency (f_n) and under nominal load (I_n) inductor (L_1) will produce voltage drop:

$$\Delta U_1 = I_n 2\pi f_n L_1 = 3.5 \cdot 2\pi \cdot 50 \cdot 0.012 = 13.2 \text{ V} \tag{B.3}$$

Calculated voltage drop $\Delta U_1 = 13.2$ V, it is 5.7% of motor nominal voltage (phase voltage is 230 V). The calculated value of $L_1 = 12$ mH satisfies the limiting current ripple and acceptable voltage drop (below 10%). The inductor L_1 nominal current is equal to the motor nominal $I_{L1n} = I_n = 3.5$ A.

The next selected element will be capacitor (C_1). The C_1 value is calculated assuming the acceptable resonant frequency (f_{res}) of the filter. So the value of the f_{out1h} should satisfy:

$$10 \cdot f_{out\,1h} < f_{res} < \frac{1}{2} \cdot f_{imp} \tag{B.4}$$

$$10 \cdot 100 < f_{res} < \frac{1}{2} \cdot 5000 \tag{B.5}$$

$$1000 < f_{res} < 2500 \tag{B.6}$$

The middle value of Equation B6 is taken:

$$f_{res} = 2000 \tag{B.7}$$

hence, the capacitor (C_1) have to be:

$$C_1 = \frac{1}{4\pi^2 f_{res}^2 L_1} = \frac{1}{4 \cdot \pi^2 \cdot 2000^2 \cdot 0.012} = 0.53 \text{ μF} \tag{B.8}$$

The nearest standard value of the capacitance $C_1 = 0.5$ μF is selected.

Appendix B: Examples of LC Filter Design

For the selected L_1 and C_1, the real resonant frequency will be:

$$f_{res} = \frac{1}{2\pi\sqrt{L_1 C_1}} = \frac{1}{2\pi\sqrt{0.012 \cdot 0.0000005}} = 2056 \text{ Hz} \quad (B.9)$$

Resonant frequency calculated in Equation B9 satisfies Equation B4 condition.

The capacitor voltage will have nearly sinusoidal shape not greater than motor root mean square (RMS) voltage. So the nominal voltage of the capacitor (C_1) should be as follows:

$$U_{C1} > U_n = 400 \text{ V} \quad (B.10)$$

The capacitor current will have two components: fundamental (equal to motor voltage fundamental frequency) and high frequency $f_{imp} = 2$ kHz. To keep the capacitor temperature in the safe range, the used capacitor should have proper RMS current. It is advised to use special type filter capacitors.

The values of L_1 and C_1 calculated in the preceding way ensure that total harmonic distortion (THD) of the filter output voltage is below 5%.

The last element to select is damping resistor R_C. The resistor is calculated taking into account that filter quality factor Q have to be in the range of $Q = 5$–8. The restricted Q gives a good filter quality and satisfies the damping properties.

For the selected L_1 and C_1, the filter characteristic impedance is:

$$Z_0 = \sqrt{\frac{L_1}{C_1}} = \sqrt{\frac{0.012}{0.0000005}} = 155 \text{ }\Omega \quad (B.11)$$

For the highest quality factor, $Q = 8$, the damping resistance is:

$$R_C = \frac{Z_0}{Q} = \frac{155}{8} = 19.3 \text{ }\Omega \quad (B.12)$$

The nearest standard value of the resistor is $R_C = 20$ Ω so the real quality factor will be:

$$Q = \frac{Z_0}{R_C} = \frac{155}{20} = 7.75 \text{ }\Omega \quad (B.13)$$

The nominal voltage of R_C is the same as for C_1:

$$U_{RC} > U_n = 400 \text{ V} \quad (B.14)$$

The total power of R_C is the sum of two components related to $f_{out1har}$ and f_{imp}. For both frequencies the impedances of C_1 and R_1 connected in series are:

$$Z_{RC\ 1har} \approx X_{C1\ 1har} = \frac{1}{2\pi f_{out\ 1har} C_1} = \frac{1}{2\pi \cdot 100 \cdot 0.5e-6} = 3184 \text{ }\Omega \quad (B.15)$$

$$Z_{RC\ imp} \approx X_{C1\ imp} = \frac{1}{2\pi f_{imp} C_1} = \frac{1}{2\pi \cdot 5000 \cdot 0.5e-6} = 63.7\ \Omega \quad \text{(B.16)}$$

It is noticeable that the most important part of R_C power is related to the ripple current. So finally the power of R_C has to be:

$$P_{RC} > R_C \left[\left(\frac{U_n}{\sqrt{3} Z_{RC\ 1har}} \right)^2 + \left(\frac{U_{THD}}{Z_{RC\ imp}} \right)^2 \right] = 20 \left[\left(\frac{400}{\sqrt{3} \cdot 3184} \right)^2 + \left(\frac{0.05 * 560}{63.7} \right)^2 \right] = 1.8\ W \quad \text{(B.17)}$$

and, for example, $P_{R1} = 2$ W should satisfy the power dissipation. Probably the 2-W resistor will have too low voltage level so the higher-power resistor will be used to satisfy the Equation B14 condition.

Example B.2 LC Filter for 5.5-kW Induction Motor

In this example, the differential filter (normal mode filter) with L_1, C_1, and R_C elements will be calculated. The filter is designed for a drive with three-phase induction motor ($P_n = 5.5$ kW, $U_n = 400$ V [Y connected], $I_n = 11$ A, $f_n = 50$ Hz) and back-to-back converter. It is worth noting that back-to-back DC link voltage is higher than in a classical frequency converter. The common dc link voltage is 650 V. The converter switching frequency is $f_{imp} = 4$ kHz and the upper frequency of the output voltage is $f_{out\ 1h} = 100$ Hz. It is desired to have switching ripple current $\Delta I_s = 20\%$.

Solution:
The design process is similar as in Example B.1.
 The motor ripple current:

$$\Delta I_s = 0.2 \cdot I_n = 0.2 \cdot 11 = 2.2\ A \quad \text{(B.18)}$$

The L_1 inductance is:

$$L_1 > 2 \frac{U_d}{\sqrt{2} \cdot 2 \cdot \pi \cdot f_{imp} \cdot 3 \cdot \Delta I_s} = 2 \frac{650}{\sqrt{2} \cdot 2 \cdot \pi \cdot 4000 \cdot 3 \cdot 2.2} = 5.5\ mH \quad \text{(B.19)}$$

For the motor nominal frequency, the voltage drop on L_1 is:

$$\Delta U_1 = I_n 2\pi f_n L_1 = 11 \cdot 2\pi \cdot 50 \cdot 0.0055 = 19\ V \quad \text{(B.20)}$$

which is 8% of motor nominal voltage (phase voltage 230 V).
 The resonant frequency f_{res} should fulfill condition:

$$1000 < f_{res} < 2000 \quad \text{(B.21)}$$

so f_{res} was selected:

$$f_{res} = 1500\ Hz \quad \text{(B.22)}$$

Appendix B: Examples of LC Filter Design

Capacitance (C_1) is:

$$C_1 = \frac{1}{4\pi^2 f_{res}^2 L_1} = \frac{1}{4\cdot\pi^2 \cdot 1500^2 \cdot 0.0055} = 2 \text{ μF} \quad \text{(B.23)}$$

The characteristic impedance is:

$$Z_0 = \sqrt{\frac{L_1}{C_1}} = \sqrt{\frac{5.5e-3}{2e-6}} = 52 \text{ Ω} \quad \text{(B.24)}$$

For the quality factor $Q = 8$, the R_C is:

$$R_C = \frac{Z_0}{Q} = \frac{52}{8} = 6.5 \text{ Ω} \quad \text{(B.25)}$$

The nearest standard value of the resistance is $R_C = 6.8$ Ω.
The impedances Z_{RC1har} and $Z_{RC\,imp}$ are:

$$Z_{RC\,1har} \approx X_{C1\,1har} = \frac{1}{2\pi f_{out\,1har} C_1} = \frac{1}{2\pi \cdot 100 \cdot 2e-6} = 796 \text{ Ω} \quad \text{(B.26)}$$

$$Z_{RC\,imp} \approx X_{C1\,imp} = \frac{1}{2\pi f_{imp} C_1} = \frac{1}{2\pi \cdot 5000 \cdot 2e-6} = 15.9 \text{ Ω} \quad \text{(B.27)}$$

The power of R_1 should be no less than:

$$P_{RC} > R_C \left[\left(\frac{U_n}{\sqrt{3} Z_{RC\,1har}} \right)^2 + \left(\frac{U_{THD}}{Z_{RC\,imp}} \right)^2 \right] = 6.8 \left[\left(\frac{400}{\sqrt{3} \cdot 796} \right)^2 + \left(\frac{0.05 * 650}{15.9} \right)^2 \right] = 58.6 \text{ W} \quad \text{(B.28)}$$

Power $P_{R1} = 100$ W was selected.

Appendix C: Equations of Transformation

The natural frame of coordinates for a three-phase system is a three-axis coordinate with a 120-degree shift, but for a better analysis of the system, the use of the orthogonal coordinates is more useful and more understandable. Therefore, for the next analysis, proper coordinate transformations are presented. Two kinds of transformations are used: one for constant vectors magnitudes and another for constant power of the three-phase and two-phase frames of reference.

The transformation matrix for the conversion from a three-phase ABC coordinate to a two-phase $\alpha\beta 0$ for constant vector magnitude is described as follows [1]:

$$\mathbf{A}_W = \begin{bmatrix} \dfrac{1}{3} & \dfrac{1}{3} & \dfrac{1}{3} \\ \dfrac{2}{3} & -\dfrac{1}{3} & -\dfrac{1}{3} \\ 0 & \dfrac{1}{\sqrt{3}} & -\dfrac{1}{\sqrt{3}} \end{bmatrix} \qquad (C.1)$$

and the transformation matrix for constant power of the system is [1, 2]:

$$\mathbf{A}_P = \begin{bmatrix} \dfrac{1}{\sqrt{3}} & \dfrac{1}{\sqrt{3}} & \dfrac{1}{\sqrt{3}} \\ \dfrac{\sqrt{2}}{\sqrt{3}} & -\dfrac{1}{\sqrt{6}} & -\dfrac{1}{\sqrt{6}} \\ 0 & \dfrac{1}{\sqrt{2}} & -\dfrac{1}{\sqrt{2}} \end{bmatrix} \qquad (C.2)$$

Variable Speed AC Drives with Inverter Output Filters, First Edition. Jaroslaw Guzinski, Haitham Abu-Rub and Patryk Strankowski.
© 2015 John Wiley & Sons, Ltd. Published 2015 by John Wiley & Sons, Ltd.

Appendix C: Equations of Transformation

Using Equation C1 or C2, each of the model variables x could be transformed from ABC to two-phase $\alpha\beta 0$ coordinates and vice-versa according to the relations:

$$\begin{bmatrix} x_0 \\ x_\alpha \\ x_\beta \end{bmatrix} = \mathbf{A}_W \begin{bmatrix} x_A \\ x_B \\ x_C \end{bmatrix} \tag{C.3}$$

$$\begin{bmatrix} x_A \\ x_B \\ x_C \end{bmatrix} = \mathbf{A}_W^{-1} \begin{bmatrix} x_0 \\ x_\alpha \\ x_\beta \end{bmatrix} \tag{C.4}$$

where the variables keep the same magnitude in each coordinate.

The next relations are used when constant power of both systems is fulfilled:

$$\begin{bmatrix} x_0 \\ x_\alpha \\ x_\beta \end{bmatrix} = \mathbf{A}_P \begin{bmatrix} x_A \\ x_B \\ x_C \end{bmatrix} \tag{C.5}$$

$$\begin{bmatrix} x_A \\ x_B \\ x_C \end{bmatrix} = \mathbf{A}_P^{-1} \begin{bmatrix} x_0 \\ x_\alpha \\ x_\beta \end{bmatrix} \tag{C.6}$$

where the power of the system is kept constant in each coordinate system.

The appropriate inverse transformation matrix is:

$$\mathbf{A}_W^{-1} = \begin{bmatrix} 1 & 1 & 0 \\ 1 & -\dfrac{1}{2} & \dfrac{\sqrt{3}}{2} \\ 1 & -\dfrac{1}{2} & -\dfrac{\sqrt{3}}{2} \end{bmatrix} \tag{C.7}$$

or:

$$\mathbf{A}_P^{-1} = \begin{bmatrix} \dfrac{1}{\sqrt{3}} & \dfrac{\sqrt{2}}{\sqrt{3}} & 0 \\ \dfrac{1}{\sqrt{3}} & -\dfrac{1}{\sqrt{6}} & \dfrac{\sqrt{2}}{2} \\ \dfrac{1}{\sqrt{3}} & -\dfrac{1}{\sqrt{6}} & -\dfrac{\sqrt{2}}{2} \end{bmatrix} \tag{C.8}$$

The use of numerical methods for differential equations solution requires mathematical notations in which all derivatives are left side of the equations. For that notation, the equations for three-phase system are as follows:

$$\frac{d\mathbf{x}_{ABC}}{dt} = \mathbf{W}_{ABC}\mathbf{x}_{ABC} + \mathbf{V}_{ABC}\mathbf{u}_{ABC} \qquad (C.9)$$

where $\mathbf{x}_{ABC}$ is state vector, $\mathbf{u}_{ABC}$ is input vector, and $\mathbf{W}_{ABC}$ and $\mathbf{V}_{ABC}$ are coefficient matrices.

Double-sided multiplication of Equation C9 with matrix **A** enable the model transformation from three-phase coordinates to orthogonal coordinates:

$$\frac{d\mathbf{A}\mathbf{x}_{ABC}}{dt} = \mathbf{A}\mathbf{W}_{ABC}\mathbf{x}_{ABC} + \mathbf{A}\mathbf{V}_{ABC}\mathbf{u}_{ABC} \qquad (C.10)$$

Based on Equations C3 to C6, it is possible to get:

$$\frac{d\mathbf{x}_{\alpha\beta 0}}{dt} = \mathbf{W}_{\alpha\beta 0}\mathbf{x}_{\alpha\beta 0} + \mathbf{V}_{\alpha\beta 0}\mathbf{u}_{\alpha\beta 0} \qquad (C.11)$$

The coefficient matrices that appear in Equation C11 are as follows:

$$\mathbf{W}_{\alpha\beta 0} = \mathbf{A}\mathbf{W}_{ABC}\mathbf{A}^{-1} \qquad (C.12)$$

$$\mathbf{V}_{\alpha\beta 0} = \mathbf{A}\mathbf{V}_{ABC}\mathbf{A}^{-1} \qquad (C.13)$$

In case of filter parameter transformations, the resistances matrix is:

$$\mathbf{R}^{W}_{ABC} = \mathbf{R}^{P}_{ABC} = \begin{bmatrix} R & 0 & 0 \\ 0 & R & 0 \\ 0 & 0 & R \end{bmatrix} \qquad (C.14)$$

The three-phase choke with symmetrical coils on the toroidal core is:

$$\mathbf{M}^{W}_{ABC} = \mathbf{M}^{P}_{ABC} = \begin{bmatrix} 3M & 0 & 0 \\ 0 & 0 & 0 \\ 0 & 0 & 0 \end{bmatrix} \qquad (C.15)$$

and for a three-phase choke with an E-shaped core is:

$$\mathbf{M}^{W}_{ABC} = \mathbf{M}^{P}_{ABC} = \begin{bmatrix} 0 & 0 & 0 \\ 0 & \frac{3}{2}M & 0 \\ 0 & 0 & \frac{3}{2}M \end{bmatrix} \qquad (C.16)$$

In Equation C15, it is evident that the toroidal three-phase symmetrical choke has parameters only for the common mode, whereas in Equation C16, the E core choke has parameters only for the differential mode. That conclusion is essential for choosing the choke core in the filters design process.

References

[1] Krause P, Wasynczuk O, Sudhoff S, Pekarek S. *Analysis of Electric Machinery and Drive Systems*. 3rd ed. Hoboken, NJ: Wiley-IEEE Press; 2013.
[2] Fortescue CL. Method of symmetrical co-ordinates applied to the solution of polyphase networks. *AIEE Transactions*. 1918; **37** (Part II):1027–1140.

Appendix D: Data of the Motors Used in Simulations and Experiments

Examples of motors data used in simulation and experimental investigations are given in Tables D.1 to D.6.

Table D.1 Parameters of induction motor type Sg 90L–4

Parameter	Value
Power, P_n	1.5 kW
Voltage, U_n	380 V, star connected
Frequency, f_n	50 Hz
Current, I_n	3.7 A
Power factor, φ_n	0.8
Efficiency, η_n	77%
Mechanical speed, n_n	1420 rpm
Stator resistance, R_s	3.3 Ω
Rotor resistance, R_r	3.4 Ω
Mutual inductance, L_m	409.3 mH
Stator inductance, L_s	423.5 mH
Rotor inductance, L_r	423.5 mH

Variable Speed AC Drives with Inverter Output Filters, First Edition. Jaroslaw Guzinski, Haitham Abu-Rub and Patryk Strankowski.
© 2015 John Wiley & Sons, Ltd. Published 2015 by John Wiley & Sons, Ltd.

Appendix D: Data of the Motors Used in Simulations and Experiments

Table D.2 Parameters of induction motor type 2Sg 90L–4

Parameter	Value
Power, P_n	1.5 kW
Voltage, U_n	300 V, star connected
Frequency, f_n	50 Hz
Current, I_n	4.7 A
Power factor, φ_n	0.8
Efficiency, η_n	77%
Mechanical speed, n_n	1420 rpm
Stator resistance, R_s	2.1 Ω
Rotor resistance, R_r	2.1 Ω
Mutual inductance, L_m	254.4 mH
Stator inductance, L_s	263.2 mH
Rotor inductance, L_r	263.2 mH

Table D.3 Parameters of induction motor type 2Sh 90L–4

Parameter	Value
Power, P_n	1.5 kW
Voltage, U_n	400 V, star connected
Frequency, f_n	50 Hz
Current, I_n	3.5 A
Power factor, φ_n	0.78
Efficiency, η_n	79%
Mechanical speed, n_n	1410 rpm
Stator resistance, R_s	4.75 Ω
Rotor resistance, R_r	4.76 Ω
Mutual inductance, L_m	303.2 mH
Stator inductance, L_s	320.1 mH
Rotor inductance, L_r	320.1 mH

Table D.4 Parameters of induction motor type FSg 112M–8

Parameter	Value
Power, P_n	1.5 kW
Voltage, U_n	400 V, star connected
Frequency, f_n	50 Hz
Current, I_n	4 A
Power factor, φ_n	0.71
Efficiency, η_n	76.8%
Mechanical speed, n_n	720 rpm
Stator resistance, R_s	5.17 Ω
Rotor resistance, R_r	3.82 Ω
Mutual inductance, L_m	253.2 mH
Stator inductance, L_s	273.1 mH
Rotor inductance, L_r	273.1 mH

Table D.5 Parameters of induction motor type FSg 112M-8 with rotor fault (two broken bars)

Parameter	Value
Power, P_n	1.5 kW
Voltage, U_n	400 V, star connected
Frequency, f_n	50 Hz
Current, I_n	4 A
Power factor, φ_n	0.71
Efficiency, η_n	76.8%
Mechanical speed, n_n	720 rpm
Stator resistance, R_s	5.15 Ω
Rotor resistance, R_r	5.92 Ω
Mutual inductance, L_m	255.8 mH
Stator inductance, L_s	283.2 mH
Rotor inductance, L_r	283.2 mH

Table D.6 Parameters of induction motor type FSLg 132S-4

Parameter	Value
Power, P_n	5.5 kW
Voltage, U_n	400 V, star connected
Frequency, f_n	50 Hz
Current, I_n	11 A
Power factor, φ_n	0.84
Efficiency, η_n	85.5%
Mechanical speed, n_n	1450 rpm
Stator resistance, R_s	0.99 Ω
Rotor resistance, R_r	0.97 Ω
Mutual inductance, L_m	142.7 mH
Stator inductance, L_s	148.7 mH
Rotor inductance, L_r	148.7 mH

Appendix E: Adaptive Backstepping Observer

Marcin Morawiec

E.1 Introduction

In this appendix, the observer for flux and speed estimation for induction motor and LC filter control is presented (as an alternative to Chapter 5). The solution is the adaptive backstepping observer that can be named *F type* because of the filter structure.

The backstepping method was proposed in Krstić, Kanellakopoulos, and Kokotović.[1] According to the definition presented for the system $\dot{x} = f(x) + g(x)u$, the feedback control law exists $u = \alpha(x)$ and positive radially unbounded function $V(x)$. Accordingly, the system $\dot{x} = f(x) + g(x)u$ can be augmented by the integrator structure such that:

$$\dot{x} = f(x) + g(x)\xi, \tag{E.1}$$

$$\dot{\xi} = u, \tag{E.2}$$

where ξ is the control in the system and u can be chosen as the virtual control.

The first step in the backstepping procedure is to define the new tracking error between the virtual control $u = \alpha(x)$ and the desired ξ. The tracking error is defined as:

$$z = \xi - \alpha(x). \tag{E.3}$$

Calculation of the derivatives of Equation E3 gives:

$$\dot{z} = \dot{\xi} - \dot{\alpha}(x). \tag{E.4}$$

Variable Speed AC Drives with Inverter Output Filters, First Edition. Jaroslaw Guzinski, Haitham Abu-Rub and Patryk Strankowski.
© 2015 John Wiley & Sons, Ltd. Published 2015 by John Wiley & Sons, Ltd.

Using Equation E4, the system (Equations E1 and E2) can be transformed to the (x,ξ,z) coordinates. The system in (x,ξ,z) will be stable if the Lyapunov condition is satisfied:

$$\dot{V}(\xi, x, z) \leq 0. \quad (E.5)$$

From Equation E5, the control variable u can be obtained. The control u guarantees the asymptotical stability of the system (Equations E1 and E2).

The same procedure can be implemented with the exponential speed observer for induction machine. Assuming that the only one measured value is the stator current vector components and that the machine control variables $(u_{s\alpha,\beta})$ are known, the integrators take the form (2, 3):

$$\dot{\tilde{\xi}}_{\alpha,\beta} = \tilde{i}_{s\alpha,\beta}, \\ \tilde{i}_{s\alpha,\beta} = \hat{i}_{s\alpha,\beta} - i_{s\alpha,\beta}, \quad (E.6)$$

where $\hat{i}_{s\alpha,\beta}$ are the estimated stator current vector components.

In the next section, the primary observer structure of induction machine model with the filter will be augmented by the integrators. The next step in the backstepping procedure is to determine which of the feedback coupling controls v (correction terms) can stabilize the observer structure such that the Lyapunov condition will be fulfilled (Equation E5).

E.2 LC Filter and Extended Induction Machine Mathematical Models

The mathematical model of LC filter was presented in Abu-Rub, Iqbal, and Guzinski.(4) The LC filter scheme is presented in Figure E.1. Based on this scheme, the mathematical model is determined by four differential equations in the stationary coordinate system $(\alpha\beta)$:

$$\frac{di_{1\alpha}}{d\tau} = \frac{1}{L_f}(u_{1\alpha} - u_{c\alpha}) - \frac{R_c}{L_f}(i_{1\alpha} - i_{s\alpha}), \quad (E.7)$$

$$\frac{di_{1\beta}}{d\tau} = \frac{1}{L_f}(u_{1\beta} - u_{c\beta}) - \frac{R_c}{L_f}(i_{1\beta} - i_{s\beta}), \quad (E.8)$$

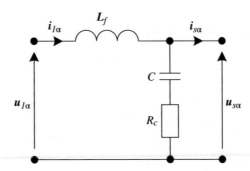

Figure E.1 The LC filter scheme for α component.

Appendix E: Adaptive Backstepping Observer

$$\frac{du_{c\alpha}}{d\tau} = \frac{1}{C}(i_{1\alpha} - i_{s\alpha}), \qquad (E.9)$$

$$\frac{du_{c\beta}}{d\tau} = \frac{1}{C}(i_{1\beta} - i_{s\beta}), \qquad (E.10)$$

where:

$u_{1\alpha}, u_{1\beta}$ is the input filter voltage vector components;
$i_{1\alpha}, i_{1\beta}$ is the inductor current vector components,
$u_{c\alpha}, u_{c\beta}$ is the capacitor voltage vector components,
$i_{s\alpha}, i_{s\beta}$ is the stator current vector components,
R_c is the damping resistance,
C is the capacitance, and
L_f is the inductance.

The induction machine model is presented in Chapter 3. This model can be extended to additional variables that will be treated as the state variables. In this section the same variables as in Krzemiński (5) are named Z and determined as follows:

$$Z_\alpha = \omega_r \psi_{r\alpha}, \qquad (E.11)$$

$$Z_\beta = \omega_r \psi_{r\beta}. \qquad (E.12)$$

Taking into account Equations E11 and E12 in the structure (3.22) to (3.26), differentiating Equations E11 and E12, the extended induction machine model is obtained:

$$\frac{di_{s\alpha}}{d\tau} = -\frac{R_s L_r^2 + R_r L_m^2}{L_r w_\sigma} i_{s\alpha} + \frac{R_r L_m}{L_r w_\sigma} \psi_{r\alpha} + \frac{L_m}{w_\sigma} Z_\beta + \frac{L_r}{w_\sigma} u_{s\alpha}, \qquad (E.13)$$

$$\frac{di_{s\beta}}{d\tau} = -\frac{R_s L_r^2 + R_r L_m^2}{L_r w_\sigma} i_{s\beta} + \frac{R_r L_m}{L_r w_\sigma} \psi_{r\beta} - \frac{L_m}{w_\sigma} Z_\alpha + \frac{L_r}{w_\sigma} u_{s\beta}, \qquad (E.14)$$

$$\frac{d\psi_{r\alpha}}{d\tau} = -\frac{R_r}{L_r}\psi_{r\alpha} - Z_\beta + \frac{R_r L_m}{L_r} i_{s\alpha}, \qquad (E.15)$$

$$\frac{d\psi_{r\beta}}{d\tau} = -\frac{R_r}{L_r}\psi_{r\beta} + Z_\alpha + \frac{R_r L_m}{L_r} i_{s\beta}, \qquad (E.16)$$

$$\frac{dZ_\alpha}{d\tau} = \frac{d\omega_r}{d\tau}\psi_{r\alpha} - \omega_r\left(Z_\beta - \frac{R_r L_m}{L_r} i_{s\alpha}\right) + \frac{R_r}{L_r} Z_\alpha, \qquad (E.17)$$

$$\frac{dZ_\beta}{d\tau} = \frac{d\omega_r}{d\tau}\psi_{r\beta} + \omega_r\left(Z_\alpha + \frac{R_r L_m}{L_r} i_{s\beta}\right) + \frac{R_r}{L_r} Z_\beta, \qquad (E.18)$$

$$\frac{d\omega_r}{d\tau} = \frac{L_m}{JL_r}\left(\psi_{r\alpha} i_{s\beta} - \psi_{r\beta} i_{su}\right) - \frac{1}{J} T_L, \qquad (E.19)$$

where $w_\sigma = L_r L_s - L_m^2$.

In the next section, the speed observer will be proposed for the LC filter (Equations E7 to E10) and induction machine model (Equations E13 to E18).

E.3 Backstepping Speed Observer

The mathematical model (E13) to (E19) and (E7) to (E10) contains 11 differential equations. In the backstepping procedure, Equation E19 can be omitted because the rotor angular speed is treated as an estimated parameter. Proceeding in accordance with the adaptive estimator with the integrator backstepping concept, one can derive the formula for the observer, where only the state variables are estimated as well as the rotor angular speed as an additional estimation parameter. The estimated values will be indexed by ^ and their deviations (prediction errors [1]) by ~.

Treating the inductor current vector components $i_{1\alpha,\beta}$ in Equations E13 to E18 and E7 to E10 as the measured values and the input filter voltage vector components $u_{1\alpha,\beta}$ as the known values, the standard exponential observer structure is obtained:

$$\frac{d\hat{i}_{s\alpha}}{d\tau} = -\frac{R_s L_r^2 + R_r L_m^2}{L_r w_\sigma}\hat{i}_{s\alpha} + \frac{R_r L_m}{L_r w_\sigma}\hat{\psi}_{r\alpha} + \frac{L_m}{w_\sigma}\hat{Z}_\beta + \frac{L_r}{w_\sigma}\hat{u}_{c\alpha} + \frac{R_c L_r}{w_\sigma}i_{1\alpha} - \frac{R_c L_r}{w_\sigma}\hat{i}_{s\alpha} + v_\alpha, \quad (E.20)$$

$$\frac{d\hat{i}_{s\beta}}{d\tau} = -\frac{R_s L_r^2 + R_r L_m^2}{L_r w_\sigma}\hat{i}_{s\beta} + \frac{R_r L_m}{L_r w_\sigma}\hat{\psi}_{r\beta} - \frac{L_m}{w_\sigma}\hat{Z}_\alpha + \frac{L_r}{w_\sigma}\hat{u}_{c\beta} + \frac{R_c L_r}{w_\sigma}i_{1\beta} - \frac{R_c L_r}{w_\sigma}\hat{i}_{s\beta} + v_\beta, \quad (E.21)$$

$$\frac{d\hat{\psi}_{r\alpha}}{d\tau} = -\frac{R_r}{L_r}\hat{\psi}_{r\alpha} - \hat{Z}_\beta + \frac{R_r L_m}{L_r}\hat{i}_{s\alpha} + v_{\psi\alpha}, \quad (E.22)$$

$$\frac{d\hat{\psi}_{r\beta}}{d\tau} = -\frac{R_r}{L_r}\hat{\psi}_{r\beta} + \hat{Z}_\alpha + \frac{R_r L_m}{L_r}\hat{i}_{s\beta} + v_{\psi\beta}, \quad (E.23)$$

$$\frac{d\hat{Z}_\alpha}{d\tau} = \frac{d\hat{\omega}_r}{d\tau}\hat{\psi}_{r\alpha} - \hat{\omega}_r\left(\hat{Z}_\beta - \frac{R_r L_m}{L_r}\hat{i}_{s\alpha}\right) + \frac{R_r}{L_r}\hat{Z}_\alpha + v_{Z\alpha}, \quad (E.24)$$

$$\frac{d\hat{Z}_\beta}{d\tau} = \frac{d\hat{\omega}_r}{d\tau}\hat{\psi}_{r\beta} + \hat{\omega}_r\left(\hat{Z}_\alpha + \frac{R_r L_m}{L_r}\hat{i}_{s\beta}\right) + \frac{R_r}{L_r}\hat{Z}_\beta + v_{Z\beta}, \quad (E.25)$$

$$\frac{d\hat{i}_{1\alpha}}{d\tau} = \frac{1}{L_f}\left(u_{1\alpha} - \hat{u}_{c\alpha}\right) - \frac{R_c}{L_f}\left(i_{1\alpha} - \hat{i}_{s\alpha}\right) + v_{1\alpha}, \quad (E.26)$$

$$\frac{d\hat{i}_{1\beta}}{d\tau} = \frac{1}{L_f}\left(u_{1\beta} - \hat{u}_{c\beta}\right) - \frac{R_c}{L_f}\left(i_{1\beta} - \hat{i}_{s\beta}\right) + v_{1\beta}, \quad (E.27)$$

$$\frac{d\hat{u}_{c\alpha}}{d\tau} = \frac{1}{C}\left(i_{1\alpha} - \hat{i}_{s\alpha}\right) + v_{c\alpha}, \quad (E.28)$$

$$\frac{d\hat{u}_{c\beta}}{d\tau} = \frac{1}{C}\left(i_{1\beta} - \hat{i}_{s\beta}\right) + v_{c\beta}, \quad (E.29)$$

Appendix E: Adaptive Backstepping Observer

whereby in Equations E20 to E29, the correction terms $v_{\alpha\beta}$, $v_{\psi\alpha\beta}$, $v_{Z\alpha\beta}$, $v_{1\alpha}$, $v_{1\beta}$, $v_{c\alpha}$, $v_{c\beta}$ and

$$\hat{Z}_\alpha = \hat{\omega}_r \hat{\psi}_{r\alpha}, \tag{E.30}$$

$$\hat{Z}_\beta = \hat{\omega}_r \hat{\psi}_{r\beta}. \tag{E.31}$$

Proceeding in accordance to backstepping approach (Section E.1), the observer model should be extended to the integrator structure in following manner:

$$\dot{\xi}_\alpha = \hat{i}_{s\alpha} + \hat{i}_{1\alpha}, \tag{E.32}$$

$$\dot{\xi}_\beta = \hat{i}_{s\beta} + \hat{i}_{1\beta}. \tag{E.33}$$

The observer correction terms will be determined by the backstepping procedure in such a way as to satisfy the Lyapunov condition (Equation E5). The variables Equations E30 to E31 are the new induction machine state variables, which will be used to reconstruct the induction machine angular speed.

Assuming the strict output-feedback form in which the inductor current vector components $i_{1\alpha,\beta}$ are only measured using Equations E20 to E29 and the extended model (E13) to (E18) and (E7) to (E10), the deviations model has the form:

$$\frac{d\tilde{i}_{s\alpha}}{d\tau} = -\frac{R_s L_r^2 + R_r L_m^2 + R_c L_r^2}{L_r w_\sigma}\tilde{i}_{s\alpha} + \frac{R_r L_m}{L_r w_\sigma}\tilde{\psi}_{r\alpha} + \frac{L_m}{w_\sigma}\tilde{Z}_\beta + \frac{L_r}{w_\sigma}\tilde{u}_{c\alpha} + v_\alpha, \tag{E.34}$$

$$\frac{d\tilde{i}_{s\beta}}{d\tau} = -\frac{R_s L_r^2 + R_r L_m^2 + R_c L_r^2}{L_r w_\sigma}\tilde{i}_{s\beta} + \frac{R_r L_m}{L_r w_\sigma}\tilde{\psi}_{r\beta} - \frac{L_m}{w_\sigma}\tilde{Z}_\alpha + \frac{L_r}{w_\sigma}\tilde{u}_{c\beta} + v_\beta, \tag{E.35}$$

$$\frac{d\tilde{\psi}_{r\alpha}}{d\tau} = -\frac{R_r}{L_r}\tilde{\psi}_{r\alpha} - \tilde{Z}_\beta + \frac{R_r L_m}{L_r}\tilde{i}_{s\alpha} + v_{\psi\alpha}, \tag{E.36}$$

$$\frac{d\tilde{\psi}_{r\beta}}{d\tau} = -\frac{R_r}{L_r}\tilde{\psi}_{r\beta} + \tilde{Z}_\alpha + \frac{R_r L_m}{L_r}\tilde{i}_{s\beta} + v_{\psi\beta}, \tag{E.37}$$

$$\frac{d\tilde{Z}_\alpha}{d\tau} = \frac{d\tilde{\omega}_r}{d\tau}\hat{\psi}_{r\alpha} - \hat{\omega}_r \tilde{Z}_\beta - \tilde{\omega}_r \hat{Z}_\beta + \tilde{\omega}_r \tilde{Z}_\beta + (\hat{\omega}_r \tilde{i}_{s\alpha} + \tilde{\omega}_r \hat{i}_{s\alpha} - \tilde{\omega}_r \tilde{i}_{s\alpha})\frac{R_r L_m}{L_r} + \frac{R_r}{L_r}\tilde{Z}_\alpha + v_{Z\alpha}, \tag{E.38}$$

$$\frac{d\tilde{Z}_\beta}{d\tau} = \frac{d\tilde{\omega}_r}{d\tau}\hat{\psi}_{r\beta} + \hat{\omega}_r \tilde{Z}_\alpha + \tilde{\omega}_r \hat{Z}_\alpha - \tilde{\omega}_r \tilde{Z}_\alpha + (\hat{\omega}_r \tilde{i}_{s\beta} + \tilde{\omega}_r \hat{i}_{s\beta} - \tilde{\omega}_r \tilde{i}_{s\beta})\frac{R_r L_m}{L_r} + \frac{R_r}{L_r}\tilde{Z}_\beta + v_{Z\beta}, \tag{E.39}$$

$$\frac{d\tilde{i}_{1\alpha}}{d\tau} = -\frac{1}{L_f}\tilde{u}_{c\alpha} + \frac{R_c}{L_f}\tilde{i}_{s\alpha} + v_{1\alpha}, \tag{E.40}$$

$$\frac{d\tilde{i}_{1\beta}}{d\tau} = -\frac{1}{L_f}\tilde{u}_{c\beta} + \frac{R_c}{L_f}\tilde{i}_{s\beta} + v_{1\beta}, \tag{E.41}$$

$$\frac{d\tilde{u}_{c\alpha}}{d\tau} = -\frac{1}{C}\tilde{i}_{s\alpha} + v_{c\alpha}, \tag{E.42}$$

$$\frac{d\tilde{u}_{c\beta}}{d\tau} = -\frac{1}{C}\tilde{i}_{s\beta} + v_{c\beta}. \tag{E.43}$$

and the integrator equations:

$$\dot{\xi}_\alpha = \tilde{i}_{s\alpha} + \tilde{i}_{1\alpha}, \tag{E.44}$$

$$\dot{\xi}_\beta = \tilde{i}_{s\beta} + \tilde{i}_{1\beta}, \tag{E.45}$$

where

$$\tilde{i}_{1\alpha} = \hat{i}_{1\alpha} - i_{1\alpha}, \tag{E.46}$$

$$\tilde{i}_{1\beta} = \hat{i}_{1\beta} - i_{1\beta}, \tag{E.47}$$

$$\tilde{u}_{c\alpha} = \hat{u}_{c\alpha} - u_{c\alpha}, \tag{E.48}$$

$$\tilde{u}_{c\beta} = \hat{u}_{c\beta} - u_{c\beta}, \tag{E.49}$$

$$\tilde{i}_{s\alpha} = \hat{i}_{s\alpha} - i_{s\alpha}, \tag{E.50}$$

$$\tilde{i}_{s\beta} = \hat{i}_{s\beta} - i_{s\beta}, \tag{E.51}$$

$$\tilde{\psi}_{r\alpha} = \hat{\psi}_{r\alpha} - \psi_{r\alpha}, \tag{E.52}$$

$$\tilde{\psi}_{r\beta} = \hat{\psi}_{r\beta} - \psi_{r\beta}, \tag{E.53}$$

$$\tilde{Z}_\alpha = \hat{Z}_\alpha - Z_\alpha, \tag{E.54}$$

$$\tilde{Z}_\beta = \hat{Z}_\beta - Z_\beta, \tag{E.55}$$

$$\tilde{\omega}_r = \hat{\omega}_r - \omega_r. \tag{E.56}$$

The first step in the backstepping procedure is to stabilize the integrators. The stabilizing function should be chosen so as to satisfy the Lyapunov condition. The Lyapunov function is determined:

$$V_1(\xi_{\alpha,\beta}) = \frac{1}{2}(\xi_\alpha^2 + \xi_\beta^2). \tag{E.57}$$

Appendix E: Adaptive Backstepping Observer

The derivative of Equation E57 takes the form:

$$\dot{V}_1\left(\tilde{\xi}_{\alpha,\beta}\right) = -c_\alpha \tilde{\xi}_\alpha^2 - c_\alpha \tilde{\xi}_\beta^2 + \tilde{\xi}_\alpha\left(\tilde{i}_{s\alpha} + \tilde{i}_{1\alpha} + c_\alpha \tilde{\xi}_\alpha\right) + \tilde{\xi}_\beta^2\left(\tilde{i}_{s\beta} + \tilde{i}_{1\beta} + c_\alpha \tilde{\xi}_\beta\right) \leq 0. \quad (E.58)$$

The condition will be fulfilled if the stabilizing function $\sigma_{\alpha,\beta}$:

$$\sigma_\alpha = \left(\tilde{i}_{s\alpha} + \tilde{i}_{1\alpha}\right) = -c_\alpha \tilde{\xi}_\alpha, \quad (E.59)$$

$$\sigma_\beta = \left(\tilde{i}_{s\beta} + \tilde{i}_{1\beta}\right) = -c_\alpha \tilde{\xi}_\beta, \quad (E.60)$$

where $c_\alpha > 0$.

The second step in the backstepping procedure is introducing the deviation variable z. The desired values are the stabilizing functions σ_α, σ_β and the virtual control is $\left(\tilde{i}_{s\alpha} + \tilde{i}_{1\alpha}\right)$ or $\left(\tilde{i}_{s\beta} + \tilde{i}_{1\beta}\right)$. The deviations are defined:

$$z_\alpha = \left(\tilde{i}_{s\alpha} + \tilde{i}_{1\alpha}\right) + c_\alpha \tilde{\xi}_\alpha, \quad (E.61)$$

$$z_\beta = \left(\tilde{i}_{s\beta} + \tilde{i}_{1\beta}\right) + c_\alpha \tilde{\xi}_\beta. \quad (E.62)$$

Using Equations E61 and E62, the integrator dependencies, Equations E44 and E45, take the form:

$$\frac{d\tilde{\xi}_\alpha}{d\tau} = z_\alpha - c_\alpha \tilde{\xi}_\alpha, \quad (E.63)$$

$$\frac{d\tilde{\xi}_\beta}{d\tau} = z_\beta - c_\alpha \tilde{\xi}_\beta. \quad (E.64)$$

Defining the deviations z and taking into account Equations E63 and E64, the back-step through the integrator was achieved.(1–3) Calculating the Equations E61 and E62 deviation derivatives results in:

$$\dot{z}_\alpha = -\frac{R_s L_r^2 + R_r L_m^2 + R_c L_r^2}{L_r w_\sigma}\tilde{i}_{s\alpha} + \frac{R_r L_m}{L_r w_\sigma}\tilde{\psi}_{r\alpha} + \frac{L_m}{w_\sigma}\tilde{Z}_\beta + \frac{L_r}{w_\sigma}\tilde{u}_{c\alpha} + v_\alpha - \frac{1}{L_f}\tilde{u}_{c\alpha}$$
$$+ \frac{R_c}{L_f}\tilde{i}_{s\alpha} + v_{1\alpha} + c_\alpha\left(\tilde{i}_{s\alpha} + \tilde{i}_{1\alpha}\right), \quad (E.65)$$

$$\dot{z}_\beta = -\frac{R_s L_r^2 + R_r L_m^2 + R_c L_r^2}{L_r w_\sigma}\tilde{i}_{s\beta} + \frac{R_r L_m}{L_r w_\sigma}\tilde{\psi}_{r\beta} - \frac{L_m}{w_\sigma}\tilde{Z}_\alpha + \frac{L_r}{w_\sigma}\tilde{u}_{c\beta} + v_\beta - \frac{1}{L_f}\tilde{u}_{c\beta}$$
$$+ \frac{R_c}{L_f}\tilde{i}_{s\beta} + v_{1\beta} + c_\alpha\left(\tilde{i}_{s\beta} + \tilde{i}_{1\beta}\right). \quad (E.66)$$

To determine the stabilizing terms, which guarantee the stability of the speed observer structure (Equations E20 to E29), the following Lyapunov function is proposed:

$$V\left(\tilde{\xi}_\alpha, \tilde{\xi}_\beta, z_\alpha, z_\beta, \tilde{\psi}_\alpha, \tilde{\psi}_\beta, \tilde{Z}_\alpha, \tilde{Z}_\beta, \tilde{i}_{1\alpha}, \tilde{i}_{1\beta}, \tilde{u}_{c\alpha}, \tilde{u}_{c\beta}, \tilde{\omega}_r\right) = \frac{1}{2}\left(\tilde{\xi}_\alpha^2 + \tilde{\xi}_\beta^2 + z_\alpha^2 + z_\beta^2 + \tilde{\psi}_\alpha^2 + \tilde{\psi}_\beta^2 + \tilde{Z}_\alpha^2 + \tilde{Z}_\beta^2 + \right.$$

$$\left. + \tilde{i}_{1\alpha}^2 + \tilde{i}_{1\beta}^2 + \tilde{u}_{c\alpha}^2 + \tilde{u}_{c\beta}^2 \right) + \frac{1}{2\gamma}\tilde{\omega}_r^2$$

(E.67)

The derivative of Equation E67 takes the form:

$$\dot{V} = -c_\alpha \tilde{\xi}_\alpha^2 - c_\alpha \tilde{\xi}_\beta^2 - c_\beta z_\alpha^2 - c_\beta z_\beta^2 - \frac{R_r}{L_r}(\tilde{\psi}_{r\alpha}^2 + \tilde{\psi}_{r\beta}^2)$$

$$+ z_\alpha \left(-\frac{R_s L_r^2 + R_r L_m^2 + R_c L_r^2}{L_r w_\sigma} \tilde{i}_{s\alpha} + \frac{R_r L_m}{L_r w_\sigma} \tilde{\psi}_{r\alpha} + \frac{L_m}{w_\sigma} \tilde{Z}_\beta + \frac{L_r}{w_\sigma} \tilde{u}_{c\alpha} + v_\alpha - \frac{1}{L_f}\tilde{u}_{c\alpha} \right.$$

$$\left. + \frac{R_c}{L_f}\tilde{i}_{s\alpha} + v_{1\alpha} + c_\beta z_\alpha + c_\alpha(\tilde{i}_{s\alpha} + \tilde{i}_{1\alpha}) \right)$$

$$+ z_\beta \left(-\frac{R_s L_r^2 + R_r L_m^2 + R_c L_r^2}{L_r w_\sigma} \tilde{i}_{s\beta} + \frac{R_r L_m}{L_r w_\sigma} \tilde{\psi}_{r\beta} - \frac{L_m}{w_\sigma} \tilde{Z}_\alpha + \frac{L_r}{w_\sigma} \tilde{u}_{c\beta} + v_\beta - \frac{1}{L_f}\tilde{u}_{c\beta} \right.$$

$$\left. + \frac{R_c}{L_f}\tilde{i}_{s\beta} + v_{1\beta} + c_\beta z_\beta + c_\alpha(\tilde{i}_{s\beta} + \tilde{i}_{1\beta}) \right).$$

$$+ \tilde{\psi}_{r\alpha}\left(-\frac{R_r}{L_r}\tilde{\psi}_{r\alpha} - \tilde{Z}_\beta + \frac{R_r L_m}{L_r}\tilde{i}_{s\alpha} + v_{\psi\alpha} \right) + \tilde{\psi}_{r\beta}\left(-\frac{R_r}{L_r}\tilde{\psi}_{r\beta} + \tilde{Z}_\alpha + \frac{R_r L_m}{L_r}\tilde{i}_{s\beta} + v_{\psi\beta} \right)$$

$$+ \tilde{u}_{c\alpha}\left(-\frac{1}{C}\tilde{i}_{s\alpha} + v_{c\alpha} \right) + \tilde{u}_{c\beta}\left(-\frac{1}{C}\tilde{i}_{s\beta} + v_{c\beta} \right)$$

(E.68)

To ensure asymptotic stability, the Lyapunov condition (Equation E5) must be satisfied. This condition implies the following speed observer correction terms:

$$v_\alpha = -\left(-\frac{R_s L_r^2 + R_r L_m^2}{L_r w_\sigma} - \frac{R_c L_r}{w_\sigma} + \frac{R_c}{L_f} \right)\left(-\tilde{i}_{1\alpha} - c_\alpha \tilde{\xi}_\alpha \right) - \frac{R_r L_m}{L_r w_\sigma}\tilde{\psi}_{r\alpha},$$

(E.69)

$$v_\beta = -\left(-\frac{R_s L_r^2 + R_r L_m^2}{L_r w_\sigma} - \frac{R_c L_r}{w_\sigma} + \frac{R_c}{L_f} \right)\left(-\tilde{i}_{1\beta} - c_\alpha \tilde{\xi}_\beta \right) - \frac{R_r L_m}{L_r w_\sigma}\tilde{\psi}_{r\beta},$$

(E.70)

$$v_{c\alpha} = k_c\left(-\left(\frac{L_r L_f - w_\sigma}{w_\sigma L_f} \right)(\tilde{i}_{s\alpha} + \tilde{i}_{1\alpha} + c_\alpha \tilde{\xi}_\alpha) + \frac{1}{C}\tilde{i}_{s\alpha} \right),$$

(E.71)

Appendix E: Adaptive Backstepping Observer

$$v_{c\beta} = k_c \left(-\left(\frac{L_r L_f - w_\sigma}{w_\sigma L_f} \right) \left(\tilde{i}_{s\beta} + \tilde{i}_{1\beta} + c_\alpha \tilde{\xi}_\beta \right) + \frac{1}{C} \tilde{i}_{s\beta} \right), \quad (E.72)$$

$$v_{1\alpha} = -k_o \left(\tilde{i}_{s\alpha} (c_\alpha + c_\beta) + \tilde{i}_{1\alpha} (c_\alpha + c_\beta) + \tilde{\xi}_\alpha (1 + c_\alpha c_\beta) \right), \quad (E.73)$$

$$v_{1\beta} = -k_o \left(\tilde{i}_{s\beta} (c_\alpha + c_\beta) + \tilde{i}_{1\beta} (c_\alpha + c_\beta) + \tilde{\xi}_\beta (1 + c_\alpha c_\beta) \right), \quad (E.74)$$

$$v_{\psi\alpha} = k_\psi \left(\tilde{Z}_\beta - \frac{R_r L_m}{L_r} \tilde{i}_{s\alpha} \right), \quad (E.75)$$

$$v_{\psi\beta} = -k_\psi \left(\tilde{Z}_\alpha + \frac{R_r L_m}{L_r} \tilde{i}_{s\beta} \right), \quad (E.76)$$

$$v_{Z\alpha} = k_Z \left(-\frac{d\tilde{\omega}_r}{d\tau} \hat{\psi}_{r\alpha} - \hat{\omega}_r \tilde{i}_{s\alpha} - \frac{R_r L_m}{L_r} - \frac{R_r}{L_r} \tilde{Z}_\alpha + \frac{L_m}{w_\sigma} \left(\tilde{i}_{s\beta} + \tilde{i}_{1\beta} + c_\alpha \tilde{\xi}_\beta \right) \right), \quad (E.77)$$

$$v_{Z\beta} = k_Z \left(-\frac{d\tilde{\omega}_r}{d\tau} \hat{\psi}_{r\beta} - \hat{\omega}_r \tilde{i}_{s\beta} - \frac{R_r L_m}{L_r} - \frac{R_r}{L_r} \tilde{Z}_\beta - \frac{L_m}{w_\sigma} \left(\tilde{i}_{s\alpha} + \tilde{i}_{1\alpha} + c_\alpha \tilde{\xi}_\alpha \right) \right), \quad (E.78)$$

where c_α, $c_\beta > 0$ and $(k_c, k_o, k_z, k_\psi) > 0$ are introduced gains.

The speed observer correction terms cause the Lyapunov condition to take the following form:

$$\dot{V} = \begin{pmatrix} -c_\alpha \tilde{\xi}_\alpha^2 - c_\alpha \tilde{\xi}_\beta^2 - c_\beta \tilde{z}_\alpha^2 - c_\beta \tilde{z}_\beta^2 - \frac{R_r}{L_r} \left(\tilde{\psi}_{r\alpha}^2 + \tilde{\psi}_{r\beta}^2 \right) + \\ +\tilde{\omega}_r \left(\frac{1}{\gamma} \dot{\hat{\omega}}_r + \tilde{Z}_\alpha \left(-\hat{Z}_\beta + (\hat{i}_{s\alpha} - \tilde{i}_{s\alpha}) \frac{R_r L_m}{L_r} \right) + \tilde{Z}_\beta \left(\hat{Z}_\alpha + (\hat{i}_{s\beta} - \tilde{i}_{s\beta}) \frac{R_r L_m}{L_r} \right) \right) \end{pmatrix} \leq 0. \quad (E.79)$$

If the value of rotor angular speed is estimated from the dependence

$$\dot{\hat{\omega}}_r = -\gamma \left(\tilde{Z}_\alpha \left(-\hat{Z}_\beta + (\hat{i}_{s\alpha} - \tilde{i}_{s\alpha}) \frac{R_r L_m}{L_r} \right) + \tilde{Z}_\beta \left(\hat{Z}_\alpha + (\hat{i}_{s\beta} - \tilde{i}_{s\beta}) \frac{R_r L_m}{L_r} \right) \right), \quad (E.80)$$

then the Lyapunov function takes the form

$$\dot{V} = \left(-c_\alpha \tilde{\xi}_\alpha^2 - c_\alpha \tilde{\xi}_\beta^2 - c_\beta \tilde{z}_\alpha^2 - c_\beta \tilde{z}_\beta^2 - \frac{R_r}{L_r} \left(\tilde{\psi}_{r\alpha}^2 + \tilde{\psi}_{r\beta}^2 \right) \right) \leq 0. \quad (E.81)$$

In Equation E80, it was assumed that $\dot{\tilde{\omega}}_r = \dot{\hat{\omega}}_r$ because $\dot{\omega}_r = 0$ and $\gamma > 0$.

The value of the rotor speed can be estimated from nonadaptive equation (2, 3):

$$\hat{\omega}_r = \frac{\hat{Z}_\alpha \hat{\psi}_{r\alpha} + \hat{Z}_\beta \hat{\psi}_{r\beta}}{\hat{\psi}_{r\alpha}^2 + \hat{\psi}_{r\beta}^2}. \tag{E.82}$$

The correction terms, Equations E77 and E78, can be transformed to the following form (because $\tilde{\omega}_r = \hat{\omega}_r$):

$$v_{Z\alpha} = k_Z \left(\frac{R_r L_m}{L_r} \left[\left(\tilde{Z}_\alpha \left(-\hat{Z}_\beta + \left(\hat{i}_{s\alpha} - \tilde{i}_{s\alpha} \right) \right) + \tilde{Z}_\beta \left(\hat{Z}_\alpha + \left(\hat{i}_{s\beta} - \tilde{i}_{s\beta} \right) \right) \right) \hat{\psi}_{r\alpha} - \hat{\omega}_r \tilde{i}_{s\alpha} \right] \right. \\ \left. - \frac{R_r}{L_r} \tilde{Z}_\alpha + \frac{L_m}{w_\sigma} \left(\tilde{i}_{s\beta} + \tilde{i}_{1\beta} + c_\alpha \tilde{\xi}_\beta \right) \right), \tag{E.83}$$

$$v_{Z\beta} = k_Z \left(\frac{R_r L_m}{L_r} \left[\left(\tilde{Z}_\alpha \left(-\hat{Z}_\beta + \left(\hat{i}_{s\alpha} - \tilde{i}_{s\alpha} \right) \right) + \tilde{Z}_\beta \left(\hat{Z}_\alpha + \left(\hat{i}_{s\beta} - \tilde{i}_{s\beta} \right) \right) \right) \hat{\psi}_{r\beta} - \hat{\omega}_r \tilde{i}_{s\beta} \right] \right. \\ \left. - \frac{R_r}{L_r} \tilde{Z}_\beta - \frac{L_m}{w_\sigma} \left(\tilde{i}_{s\alpha} + \tilde{i}_{1\alpha} + c_\alpha \tilde{\xi}_\alpha \right) \right), \tag{E.84}$$

for $\gamma = 1$.

Finally, the speed backstepping observer structure is determined by Equations E20 to E29, correction terms, Equations E69 to E76, E83, E84, and E80 or E82.

In the next section, the stability analysis of the speed backstepping observer system will be carried out.

E.4 Stability Analysis of the Backstepping Speed Observer

The presented backstepping approach is based on the Lyapunov function, control of Lyapunov function (CLF). The correction terms Equations E69 to E76, E83, and E84 are chosen in such a manner to satisfy the Lyapunov condition (Equation E5). This condition guarantees the asymptotic stability of the observer system if the constant gains are $c_{\alpha,\beta}$, $\gamma_{1,2} > 0$. If these gains are $c_{\alpha,\beta}$, $\gamma_{1,2} \gg 0$, then the observer is stiff (1, 3, 6–17) but the estimated rotor speed has more oscillations. The oscillations can lead to the observer working unstably. The gains move the system trajectory (the real poles) near to zero on the real axis (Re) or keep it away from the zero point (each real pole must be <0 for the system to be stable). These observer gains influence the speed observer convergences. In the nonlinear system literature, the backstepping observer convergence is proved by the Lassale-Yoshizawa theorem or Barbalat's Lemma (1, 3, 5–17). These convergence theories are based on sets theory. If the set of solutions of the backstepping observer is bounded and the Lyapunov condition (Equation E5) is satisfied, then each solution is converged to the positive bounded solution set.

Appendix E: Adaptive Backstepping Observer

The other situation is with the $k_{\psi,z}$ gains. The $k_{\psi,z}$ should be $k_{\psi,z} > 0$ but only $k_\psi < 1$ pu. The k_ψ must be smaller than 1.0 because for each $k_\psi \geq 1.0$ the flux subsystem is unstable.

The estimator system is oriented with the rotor flux vector $\bar{\psi}_r$ (dq coordinate system), so $\psi_{rd} = |\bar{\psi}_r|$ and $\psi_{rq} = 0$ and the stator current vector components and ω_ψ can be treated as follows:

$$i_{sd} = \frac{\psi_{rd}}{L_m}, \quad i_{sq} = \frac{L_r T_L}{L_m \psi_{rd}}, \quad \omega_\psi = \frac{R_r L_m}{L_r}\left(\frac{i_{sq}}{\psi_{rd}} + \omega_r\right). \tag{E.85}$$

The differential equations of the speed observer backstepping in (dq) system have the form:

$$\frac{d\tilde{i}_{sd}}{d\tau} = -a_1 \tilde{i}_{sd} + a_5 \tilde{\psi}_{rd} + a_6 \tilde{Z}_q + a_7 \tilde{u}_{cd} - (a_1 - a_2)(\tilde{i}_{1d} + c_\alpha \tilde{\xi}_d + \omega_\psi \tilde{\xi}_q), \tag{E.86}$$

$$\frac{d\tilde{i}_{sq}}{d\tau} = -a_1 \tilde{i}_{sq} + a_5 \tilde{\psi}_{rq} - a_6 \tilde{Z}_d + a_7 \tilde{u}_{cq} - (a_1 - a_2)(\tilde{i}_{1q} - \omega_\psi \tilde{\xi}_d + c_\alpha \tilde{\xi}_q), \tag{E.87}$$

$$\frac{d\tilde{\psi}_{rd}}{d\tau} = a_3(1-k_\psi)\tilde{i}_{sd} - a_9 \tilde{\psi}_{rd} - \tilde{Z}_q(1-k_\psi) + \omega_\psi \tilde{\psi}_{rq}, \tag{E.88}$$

$$\frac{d\tilde{\psi}_{rq}}{d\tau} = a_3(1-k_\psi)\tilde{i}_{sq} - a_9 \tilde{\psi}_{rq} + \tilde{Z}_d(1-k_\psi) - \omega_\psi \tilde{\psi}_{rd}, \tag{E.89}$$

$$\frac{d\tilde{Z}_d}{d\tau} = (1-k_Z)(\omega_r \tilde{i}_{sd} a_3 + a_9 \tilde{Z}_d) - (\omega_r - \omega_\psi)\tilde{Z}_q + k_Z a_6(\tilde{i}_{sq} + \tilde{i}_{1q} - \omega_\psi \tilde{\xi}_d + c_\alpha \tilde{\xi}_q) + \tilde{\omega}_r(\hat{i}_{sd} a_3 - \hat{Z}_q), \tag{E.90}$$

$$\frac{d\tilde{Z}_q}{d\tau} = (1-k_Z)(\omega_r \tilde{i}_{sq} a_3 + a_9 \tilde{Z}_q) + (\omega_r - \omega_\psi)\tilde{Z}_d - k_Z a_6(\tilde{i}_{sd} + \tilde{i}_{1d} + c_\alpha \tilde{\xi}_d + \omega_\psi \tilde{\xi}_q) + \tilde{\omega}_r(\hat{i}_{sq} a_3 + \hat{Z}_d), \tag{E.91}$$

$$\frac{d\tilde{i}_{od}}{d\tau} = -\frac{1}{L_f}\tilde{u}_{cd} + \tilde{i}_{sd} a_8 - \tilde{i}_{1d} k_o(c_\alpha + c_\beta) - \tilde{\xi}_q k_o \omega_\psi(c_\alpha + c_\beta) - \tilde{\xi}_d k_o(c_\beta c_\alpha + 1), \tag{E.92}$$

$$\frac{d\tilde{i}_{oq}}{d\tau} = -\frac{1}{L_f}\tilde{u}_{cq} + \tilde{i}_{sq} a_8 - \tilde{i}_{1q} k_o(c_\alpha + c_\beta) + \tilde{\xi}_d k_o \omega_\psi(c_\alpha + c_\beta) - \tilde{\xi}_q k_o(c_\beta c_\alpha + 1), \tag{E.93}$$

$$\frac{d\tilde{u}_{cd}}{d\tau} = -\tilde{i}_{sd} a_{10} + \omega_\psi \tilde{u}_{cq} - a_4(\tilde{i}_{1d} + c_\alpha \tilde{\xi}_d + \omega_\psi \tilde{\xi}_q), \tag{E.94}$$

$$\frac{d\tilde{u}_{cq}}{d\tau} = -\tilde{i}_{sq} a_{10} - \omega_\psi \tilde{u}_{cd} - a_4(\tilde{i}_{1q} - \omega_\psi \tilde{\xi}_d + c_\alpha \tilde{\xi}_q), \tag{E.95}$$

$$\frac{d\tilde{\xi}_d}{d\tau} = \tilde{i}_{sd} + \tilde{i}_{1d} + \omega_\psi \tilde{\xi}_q - c_\alpha \tilde{\xi}_d, \tag{E.96}$$

$$\frac{d\tilde{\xi}_q}{d\tau} = \tilde{i}_{sq} + \tilde{i}_{1q} - \omega_\psi \tilde{\xi}_d - c_\alpha \tilde{\xi}_\beta, \tag{E.97}$$

$$\frac{d\tilde{\omega}_r}{d\tau} = \gamma \hat{Z}_q \tilde{Z}_d - \gamma \hat{Z}_d \tilde{Z}_q - \gamma \frac{R_r L_m}{L_r} \hat{i}_{sd} \tilde{Z}_d - \gamma \frac{R_r L_m}{L_r} \hat{i}_{sq} \tilde{Z}_q, \tag{E.98}$$

where: $a_1 = \dfrac{R_s L_r^2 + R_r L_m^2 + R_c L_r^2}{L_r w_\sigma}$, $a_2 = \dfrac{R_c}{L_f}$, $a_3 = \dfrac{R_r L_m}{L_r}$, $a_4 = k_c \left(\dfrac{L_r L_f - w_\sigma}{w_\sigma L_f} \right)$, $a_5 = \dfrac{R_r L_m}{L_r w_\sigma}$,

$a_6 = \dfrac{L_m}{w_\sigma}$, $a_7 = \dfrac{L_r}{w_\sigma}$, $a_8 = R_c / L_f - k_o (c_\alpha + c_\beta)$, $a_9 = R_r / L_r$, $a_{10} = a_4 + \dfrac{1}{C}(1 - k_c)$.

To examine the impact of the observer gains, the nonlinear system is linearized near the equilibrium point. The linearized system has the general form (6):

$$\frac{d}{d\tau}\tilde{x} = A\tilde{x}, \quad \tilde{x} = \hat{x} - x \tag{E.99}$$

where **A, B** are the Jacobian matrices.

In the estimator structure, Equations E86 to E98 can be linearized near the equilibrium point and the matrix **A** is defined as:

$$A = \begin{bmatrix}
-a_1 & 0 & a_5 & 0 & 0 & a_6 & -(a_1-a_2) & 0 & 0 & 0 & -c_\alpha(a_1-a_2) & -\omega_\psi(a_1-a_2) & 0 \\
0 & -a_1 & 0 & a_5 & -a_6 & 0 & 0 & -(a_1-a_2) & 0 & 0 & \omega_\psi(a_1-a_2) & -c_\alpha(a_1-a_2) & 0 \\
a_3(1-k_\psi) & 0 & -a_9 & \omega_r & 0 & -(1-k_\psi) & 0 & 0 & 0 & 0 & 0 & 0 & 0 \\
0 & a_3(1-k_\psi) & -\omega_r & -a_9 & 1-k_\psi & 0 & 0 & 0 & 0 & 0 & 0 & 0 & 0 \\
\omega_r a_3(1-k_z) & k_z a_6 & 0 & 0 & a_8(1-k_z) & -\omega_r + \omega_\psi & 0 & k_z a_6 & 0 & 0 & -k_z a_6 \omega_\psi & k_z c_\alpha a_6 & i_{sd} a_3 \\
-k_z a_6 & \omega_r a_3(1-k_z) & 0 & 0 & \omega_r - \omega_\psi & a_9(1-k_z) & -k_z a_6 & 0 & 0 & 0 & -k_z a_6 c_\alpha & -k_z a_6 \omega_\psi & i_{sq} a_3 + Z_d \\
a_8 & 0 & 0 & 0 & 0 & 0 & -k_o(c_\alpha+c_\beta) & 0 & -\dfrac{1}{L_f} & 0 & -k_o(c_\alpha c_\beta+1) & -k_o\omega_\psi(c_\alpha+c_\beta) & 0 \\
0 & a_8 & 0 & 0 & 0 & 0 & 0 & -k_o(c_\alpha+c_\beta) & 0 & -\dfrac{1}{L_f} & k_o\omega_\psi(c_\alpha+c_\beta) & -k_o(c_\alpha c_\beta+1) & 0 \\
-a_{10} & 0 & 0 & 0 & 0 & 0 & -a_4 & 0 & 0 & 0 & -a_4 c_\alpha & -a_4\omega_\psi & 0 \\
0 & -a_{10} & 0 & 0 & 0 & 0 & 0 & -a_4 & -\omega_\psi & 0 & a_4\omega_\psi & -a_4 c_\alpha & 0 \\
1 & 0 & 0 & 0 & 0 & 0 & 1 & 0 & 0 & 0 & -c_\alpha & \omega_\psi & 0 \\
0 & 1 & 0 & 0 & 0 & 0 & 0 & 1 & 0 & 0 & -\omega_\psi & -c_\alpha & 0 \\
0 & 0 & 0 & 0 & -\gamma a_3 i_{sd} + \gamma Z_q & -\gamma a_3 i_{sq} - \gamma Z_d & 0 & 0 & 0 & 0 & 0 & 0 & 0
\end{bmatrix}, \tag{E.100}$$

where i_{sd}, i_{sq} and ω_ψ, ω_r, Z_d, $Z_q = 0$ are determined for the equilibrium point.

After taking into account Equation E99, the properties of the speed observer, in stationary state, are dependent on the LC filter, induction machine parameters, c_α, c_β, k_ψ, k_M, k_o, k_c gains, module of rotor flux vector, and electromagnetic torque.

Appendix E: Adaptive Backstepping Observer

The values of the rotor angular speed are estimated from the adaptive Equation E80 and nonadaptive Equation E82.

The analysis of the speed observer is performed as a result the following parameters and the equilibrium point changes:

Case 1. $\omega_r = -1.0$ to 1.0 pu, for $c_\alpha = 1$, $c_\beta = 1$, $\gamma = 1$, $k_\psi = 0.9$, $k_Z = 1$, $k_o = 1$, $k_c = 1$, $T_L = 0.7$ pu (Figure E.2);

Case 2. $T_L = 0$–0.7 pu, for $c_\alpha = 1$, $c_\beta = 0.1$, $\gamma = 1$, $k_\psi = 0.8$, $k_Z = 1$, $k_o = 1$, $k_c = 1$, $\omega_r = 1$ pu (Figure E.3);

Case 3. $c_\beta = 0.01 - 10$ pu, for $c_\alpha = 1$, $T_L = 0.7$, $\gamma = 1$, $k_\psi = 0.9$, $k_Z = 1$, $k_o = 1$, $k_c = 1$, $\omega_r = 1$ pu (Figure E.4);

Case 4. $k_\psi = 0.1$–1 pu, for $c_\alpha = 1$, $c_\beta = 1$, $T_L = 0.7$, $\gamma = 1$, $k_Z = 1$, $k_o = 1$, $k_c = 1$, $\omega_r = 1$ pu (Figure E.5);

Case 5. $k_o = 0.1$–1 pu, for $k_Z = 1$, $c_\alpha = 1$, $c_\beta = 1$, $T_L = 0.7$, $\gamma = 1$, $k_\psi = 0.9$, $k_c = 1$, $\omega_r = 1$ pu (Figure E.6);

Case 6. $k_c = 0.1$–1 pu, for $k_Z = 1$, $c_\alpha = 1$, $c_\beta = 1$, $T_L = 0.7$, $\gamma = 1$, $k_\psi = 0.9$, $k_c = 1$, $k_o = 1$, $\omega_r = 1$ pu (Figure E.7).

Figure E.2 presents case 1 while the rotor speed value is changing from $\omega_r = -1.0$ to 1.0 pu.

In Figure E.3 case 2 is presented. The load torque value is changed from 0 to 0.7 pu. The speed observer is stable.

In Figure E4 case 3 is shown. The gain cβ is changing from 0.01 to 10 pu.

In Figure E.5 case 4 is presented. The gain k_ψ is changing from 0.1 to 1 pu

If k_o is smaller than 0.3 for adaptive speed estimation or 0.15 for nonadaptive estimation, then speed observer is unstable. This case is shown in Figure E.6 (case 5).

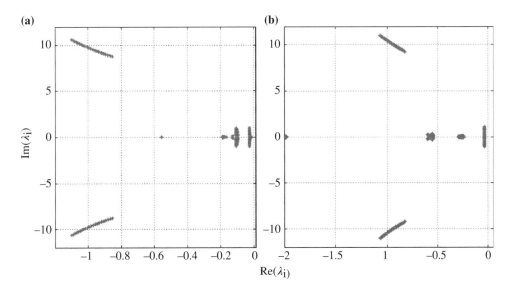

Figure E.2 The spectrum of matrix A of the linearized observer system for the rotor speed changes from -1.0 to 1.0 for $c_\alpha = 1$, $c_\beta = 1$, $\gamma = 1$, $k_\psi = 0.9$, $k_o = 1$, $k_c = 1$, $T_L = 0.7$ pu: **a,** adaptive speed estimation and **b,** nonadaptive speed estimation (case 1).

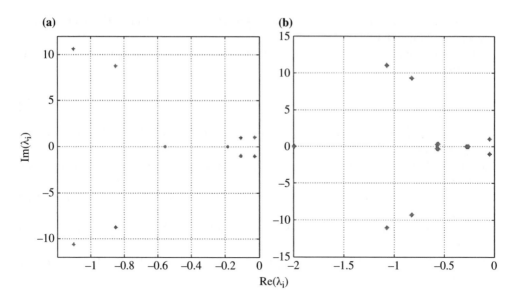

Figure E.3 The spectrum of matrix A of the linearized observer system for load torque changes from $T_L = 0$–0.7 and $c_\alpha = 1$, $c_\beta = 1$, $\gamma = 1$, $k_\psi = 0.9$, $k_o = 1$, $k_c = 1$, $\omega_r = 1.0$ pu, **a,** adaptive speed estimation and **b,** nonadaptive speed estimation (case 2).

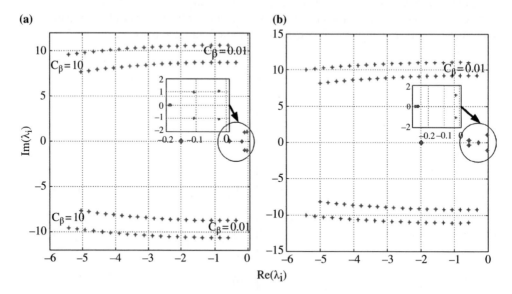

Figure E.4 The spectrum of matrix A of the linearized observer system for $c_\beta = 0.01$–10 and $c_\alpha = 1$, $\gamma = 1$, $k_\psi = 0.9$, $k_z = 1$, $k_o = 1$, $k_c = 1$, $\omega_r = 1.0$, $T_L = 0.7$ pu, **a,** adaptive speed estimation and **b,** nonadaptive speed estimation (case 3).

Appendix E: Adaptive Backstepping Observer

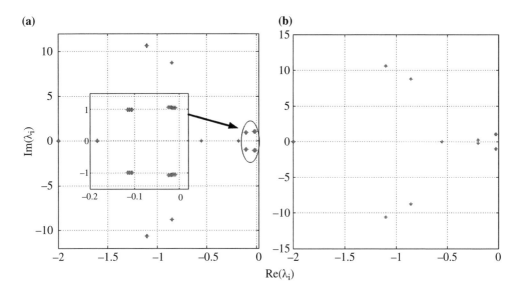

Figure E.5 The spectrum of matrix A of the linearized observer system for $k_\psi = 0.1–1$, oraz $c_\beta = 1$, $c_\alpha = 1$, $\gamma = 1$, $k_z = 1$, $k_o = 1$, $k_c = 1$, $\omega_r = 1.0$, $T_L = 0.7$ pu, **a**, adaptive speed estimation and **b**, nonadaptive speed estimation (case 4).

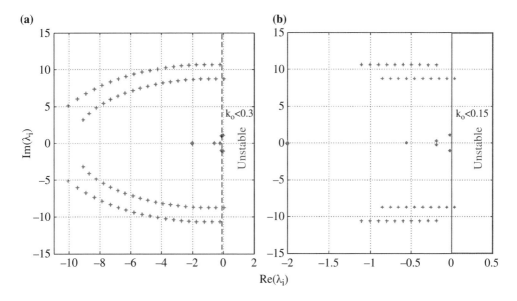

Figure E.6 The spectrum of matrix A of the linearized observer system for $k_o = 0.1–1$, oraz $c_\beta = 1$, $c_\alpha = 1$, $\gamma = 1$, $k_\psi = 0.9$, $k_z = 1$, $k_c = 1$, $\omega_r = 1.0$, $T_L = 0.7$ pu, **a**, adaptive speed estimation and **b**, nonadaptive speed estimation (case 5).

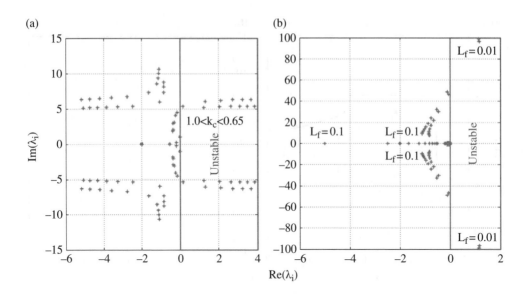

Figure E.7 The spectrum of matrix A of the linearized observer system for (a) $k_c = 0.1$–1 and (b) $L_f =$ 0.01–0.1 oraz $c_\beta = 1$, $c_\alpha = 1$, $\gamma = 1$, $k_\psi = 0.9$, $k_z = 1$, $k_o = 1$, $k_c = 1$, $\omega_r = 1.0$, $T_L = 0.7$ j.w., **a,** adaptive speed estimation and **b,** nonadaptive speed estimation (case 6).

For $1.0 < k_c < 0.65$ (case 6) the speed observer is unstable (Figure E.7a).
If L_f is 0.01 or smaller than 0.01 pu (case 6) then the speed observer is unstable (Figure E.7b).
The speed estimation method influences the stability range of the speed observer structure. The proposed F structure of the speed observer has more stability range for nonadaptive speed estimation from Equation E82. The stability analysis proved that the special rules for selection of observer gains should be applied.

E.5 Investigations

The tests were carried out for a 5.5-kW drive system with numerical simulation and experimental set up. The motor drive system parameters are given in Appendix D. The control system was implemented with a *DSP Sharc* ADSP21363 floating point signal processor and Altera Cyclon 2 FPGA. The signal processor had 3 Mb SRAM, 333 MHz, 666 MIPS, and 2GFLOPS. The transistor switching frequency was 10 kHz. The *Runge Kutta IV* integration method was implemented for the speed observer. The control system calculations were about 18 μs without code machine optimization. In this control system, the speed observer backstepping was implemented.

In Figure E.8, the following transients from simulation are shown: estimated rotor speed error ω_r; capacitor voltage vector component error $\tilde{u}_{c\alpha}$; inductor current vector component error $\tilde{i}_{1\alpha}$; and rotor flux vector component error $\tilde{\psi}_{r\alpha}$ while the machine is reversing from 1 to −1 pu. The estimated rotor speed error is smaller than 0.05 pu and the rotor flux vector error $\tilde{\psi}_{r\alpha}$ is about 0.05 pu. The speed value is estimated from adaptive dependence.

In Figure E.9 the experimental results (transients) are shown. The rotor speed is changing from 0.005 to −0.005 pu. The rotor speed error is about 0.003 pu in dynamic state. The rotor speed value is estimated from the nonadaptive dependence (Equation E82).

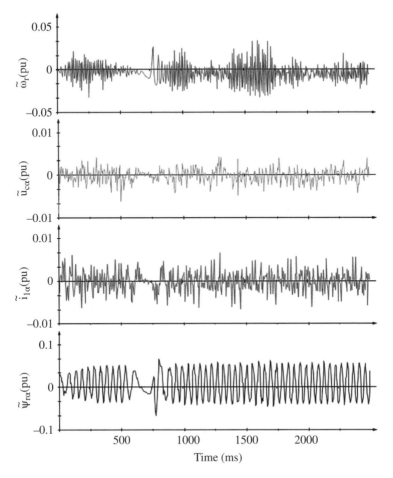

Figure E.8 Simulation results while machine is reversing from 1 to −1 pu, estimation error variables are presented, and rotor speed value is estimated from the adaptive dependence (Equation E80).

In Figure E.9, the estimated rotor speed, rotor speed error, rotor flux vector module, and rotor flux vector component are shown. Rotor speed value is estimated from the nonadaptive dependence (Equation E82)

E.6 Conclusions

In this appendix, the novel structure of rotor angular speed observer, named F, is presented. The observer is for an induction machine supplied by voltage source inverter with the LC output filter. This structure was obtained by the use of backstepping synthesis and an adaptive mechanism. The proposed estimators are characterized by small rotor angular speed estimation error <0.01 pu in dynamic states and for nonadaptive speed estimation. This error value depends on machine operating point. The back step was proposed through the integrator structure. The integrator structure is used to obtain the F type adaptive observer backstepping.

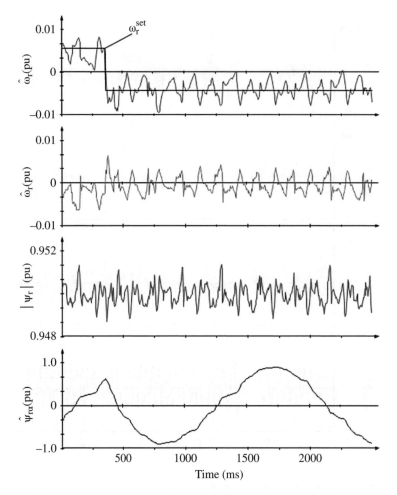

Figure E.9 Experimental results while machine is reversing from 0.005 to −0.005 pu.

The F type observer structure has additional stabilizing elements in the differential equations, which makes it less oscillatory. In the observer structure, the rotor speed can be estimated in two ways: adaptive and non-adaptive. The adaptive with the F type observer structure is characterized by higher state of variables oscillation for small speed (smaller than 0.1 pu).

The main advantages of the proposed F-type observer structure are:

- The rotor speed is more accurately determined than in standard observer backstepping, especially at a low rotor speed;
- The stability is guaranteed by the Lyapunov criteria;
- The F type observers are relatively easy to tune using methods;
- The sensorless control system with the F type observer backstepping works stably for the zero speed command;
- The proposed F type observer structure is more complicated than the standard observer backstepping, but stability of the control system is provided over the whole speed range.

Appendix E: Adaptive Backstepping Observer

The proposed methodology, based on the extended observer structure and adaptive backstepping, gives new possibilities for the development of observer theory in electric drive systems. The F type observer can be successfully used in industrial applications.

References

[1] Krstić M, Kanellakopoulos I, Kokotović P. Nonlinear and adaptive control design. Hoboken, NJ: Wiley-Interscience Publication; 1995.

[2] Morawiec M. Z type observer backstepping for induction machines. IEEE Transactions on Industrial Electronics. 2014; 62 (4):2090–2102. doi: 10.1109/TIE.2014.2355417..

[3] Morawiec M, Guziński J. Sensorless control system of an induction machine with the Z-type backstepping observer. IEEE 23rd International Symposium on Industrial Electronics (ISIE), 2014, Istanbul, Turkey. June 1–4, 2014.

[4] Abu-Rub H, Iqbal A, Guzinski J. High performance control of AC drives with MATLAB/Simulink models. Chichester, UK: John Wiley & Sons, Ltd; 2012.

[5] Krzemiński Z. Speed observers for sensorless control of AC machines. Polish Electrical Review. 2014; **5** (90).

[6] Sastry S, Bodson M. *Adaptive control: Stability. Convergence and robustness*. Englewood Cliffs, NJ: Prentice Hall; 1989.

[7] Ioannou PA, Sun J. *Robust adaptive control*. Englewood Cliffs, NJ: Prentice Hall; 1995.

[8] Goodwin GC. Robust adaptive control. Proceedings of the IFAC Workshop Newcastle, Australia. August 22–24, 1988.

[9] Khalil HK. *Nonlinear systems*. 2nd ed. Upper Saddle River, NJ: Prentice Hal; 1996.

[10] Marino R, Peresada S, Tomei P. Global adaptive output feedback control of induction motors with uncertain rotor resistance. *IEEE Transactions on Automatic Control*. 1998; **44** (5): 967–983.

[11] Marino R, Peresada S, Tomei P. Output feedback control of current-fed induction motors with unknown rotor resistance. *IEEE Transactions on Control Systems Technology*. 1996; **4** (4):336–347.

[12] Jeon SH, Oh KK, Choi, J-Y. Flux observer with online tuning of stator and rotor resistances for induction motors. *IEEE Transactions on Industrial Electronics*. 2002; **49** (3):653–664.

[13] Lin J-S, Kanellakopoulos I. Nonlinearities enhance parameter convergence: the strict feedback case. Proceeding of 35th IEEE Conference on Decision and Control, Kobe, Japan. December 11–13, 1996.

[14] Lin J-S, Kanellakopoulos I. Nonlinearities enhance parameter convergence: the strict feedback case. *IEEE Transactions on Automatic Control*. 1999; **44** (1): 89–94.

[15] Popov VM. *Hyperstability of control systems*. New York: Springer-Verlag; 1973.

[16] Slotine J-J, Weiping L. *Applied nonlinear control*. Englewood Cliffs, NJ: Prentice Hall; 1991.

[17] Marino R, Peresada S, Tomei P. *Nonlinear control design—Geometric. Adaptive and robust*. Englewood Cliffs, NJ: Prentice Hall; 1995.

Appendix F: Significant Variables and Functions in Simulation Files

To ensure a good readability of the C programs, the used variables were written with the following indexes:

Index	Example	Description
d	isd	d-component of fixed reference frame
E	Ex11	Error variable
f/F	isxF	Filtered value or filter model variable
i	ki21	Integral component of PI controller
o	Ls_o	Observer variable
p	kp11	Proportional component of PI controller
q	Usd	q-component of fixed reference frame
r	fr	Rotor variable
ref	usyref	Reference value
reg	x12reg	Controlled variable (appears in the controller feedback)
s	is	Stator variable
so	isx_so	Speed observer variable
sof	omega_soF	Filtered value of speed observer
x	usx	x-component of rotating reference frame
y	fry	y-component of rotating reference frame
z	x11z	Reference value

Variable Speed AC Drives with Inverter Output Filters, First Edition. Jaroslaw Guzinski, Haitham Abu-Rub and Patryk Strankowski.
© 2015 John Wiley & Sons, Ltd. Published 2015 by John Wiley & Sons, Ltd.

Appendix F: Significant Variables and Functions in Simulation Files

Program variables:

Variable	Description
`DD[8][4]`	Right side of differential equations
`E,ie,JLpp,JCpp,Lpp,Cpp,Rdiod`	DC link variables
`EMFx,EMFy`	Electromotive force (EMF) alpha and beta components
`Ex[1], Ey[1]`	EMF vector, last sample
`Ex12,Ex22,Ex11,Ex21`	Control error
`frx, fry`	Components of rotor linked flux
`frx_so, fry_so`	Rotor flux components of speed observer
`h, ht`	Integrating time p.u. and real (ht) for motor model
`i1dF, i1qF`	Measurement current filter i1
`i1dz, i1qz`	Reference values of current filter i1
`i1x,i1y`	Filter variables
`ia, ib, ic`	Stator currents in three phase
`id`	DC link current
`impulse`	Actual value of PWM counter (from 0 to Timp with h step)
`isx, isy`	Stator current component in $x\text{–}y$ system
`isx_so, isy_so`	Stator current vector components of speed observer
`isx0, isy0`	Measured currents of the observer
`J`	Loop counter
`k1_so,k2_so,k3_so,k4_so`	Gain variables of observer
`ki11,kp11,ki21,kp21`	PI controller settings
`ki12,kp12,ki22,kp22`	PI controller settings
`Lm`	Mutual inductance
`Lr`	Rotor inductance
`Ls`	Stator inductance
`m0`	Static load torque
`me`	Electromagnetic torque
`mMech`	Initial load torque for fan type load
`omega_so`	Calculated rotor angular speed
`omega_sof`	Filtered calculated rotor angular speed
`omegaR`	Motor speed
`omegaU`	Stator electrical pulsation
`roU`	Desired voltage output vector position angle
`roU`	Angle position
`Rr`	Rotor resistance
`Rs`	Stator resistance
`Rs_so,Rr_so,Lm_so,Ls_so,Lr_so`	Motor parameters for the observer
`Sb,Sbf`	Internal variables of the speed observer
`sector`	Sector number of inverter output vector position
`START. START_REG`	Control variables
`tau`	Total time in per unit
`Td`	PWM dead time
`timee`	Real time
`Timp`	PWM switching period
`trapH`	Calculation step of control system

Variable	Description
ud	DC link voltage
ukx, uky	Filter variables
US, USX, USY	Inverter reference voltage
UsdF, UsqF	Filtered value of the stator voltage Us
Usdz, Usqz	Reference values of stator voltage vectors
Usx, Usy	LC filter output voltage
usx, usy	Stator components in x–y system
usx0, usy0	Measured stator voltages of the observer
vector	Inverter output vector number
w, a1, a2, a3, a4, a5, a6, a7	Motor coefficients
w_so, a1_so, a2_so, a3_so, a4_so	Motor parameter coefficients in observer equations
X[32]	Internal variables for controllers integration function
x11, x12, x21, x22	Variables of multiscalar model
x11z, x12z, x21z, x22z	Reference values of multiscalar model
Y[32]	Internal variables for integration
YY[32]	Left side of differential equations
zetaX, zetaY	Disturbances (corresponding to motor EMF)

Functions:

Function	Description
void F4DERY(...)	The right side of differential equations of motor model
void SIMULATION(...)	Main simulation function
void F4(...)	Differential equation solver with Runge–Kutta 4
double limit(...)	Limits the value to lower and upper bound
double nzero(...)	Zero division protection
void PWMU_2XY(...)	Space vector PWM
void VPWM1(...)	Voltage symmetrical PWM, rotating voltage vector method
double angle(...)	Angle calculation (0- to 2-π range)
void PWM_0W(...)	PWM control with simulation of common mode
void TRAPDERY(...)	Differential equations for control system-PI controller
void TRAPEZ(...)	Integration method
double angle_pll(...)	Angle calculation ($\pm\pi$ range)

Index

abcde coordinates, 243, 246
abcde reference frame, 243
Active Systems for Reducing CM current, 27
active zero voltage control (AZVC), 32
actual filter output voltage, 142
adaptive algorithms, 176
adaptive flux observer, 126–127
adaptive neuro-fuzzy inference system (ANFIS), 231, 232
adaptive speed observer, 125
additional feedback, 176
adjustable speed drive, 1–3, 9
angular misalignment, 222, 224
angular rotor speed, 221
angular speed, 170
artificial imbalance, 227, 230
automatic control system, 143
AZVC-1, 32
    modulation, 35
AZVC-2, 32–34
    algorithm, 45
    control, 38
    modulation, 34

base quantities, 54–56
bearing
    capacitive current, 15
    circulating current, 15, 16
    currents, 9, 14
    dominant current, 17
    failure, 9
    protection rings, 26

cable parameters, 12
calculation errors, 119
capacitive bearing current, 15
capacitor voltage $u_{c\alpha}$, $u_{c\beta}$, 146
cascaded structure, 143
C_b, motor bearings equivalent capacitance, 13
C_c, cable capacitance, 12
characteristic filter impedance, 77
characteristic impedance, 281
choke
    current $i_{1\alpha}$, $i_{1\beta}$, 146
    inductance, 76
circulating bearing current, 15, 16
Clarke transformation, 94
classical SVM, 31
close loop
    system, 97
    sensorless system, 185
    control, 233
    torque control, 182
    sensorless mode, 271
common mode (CM)
    analysis, 10
    capacitance, 20

Variable Speed AC Drives with Inverter Output Filters, First Edition. Jaroslaw Guzinski, Haitham Abu-Rub and Patryk Strankowski.
© 2015 John Wiley & Sons, Ltd. Published 2015 by John Wiley & Sons, Ltd.

common mode (CM) (cont'd)
  choke, 21, 22, 25, 69
    core, 22, 23
  current, 20, 36, 65–67
    reduction, 21, 46
    relation, 25
  filter, 65–66, 80
    function, 81
    output voltage, 142
  parameters, 18, 19
  series circuit, 10
  transformer, 13, 25, 26
  voltage, 9, 18
common mode voltage (CMV), 9, 12
  frequency, 32
  peak-to-peak value, 12
  reduction, 28
  waveform, 37–39, 42
concentrated windings, 257
control
  function, 156
  signals, 151
  variables, 141
converter short-circuit protection, 71
C_{rf}, motor rotor to frame capacitance, 13
current
  angular speed, 168
  samples, 275
current-flux dependencies, 51, 123
C_{wf}, motor winding to frame capacitance, 13
C_{wr}, motor winding to rotor capacitance, 13

d component, 145
$d1$–$q1$ reference plane, 263
$d3$–$q3$ reference plane, 265
damping
  coefficient ξ_d, 79
  resistance, 90, 160
DC midpoint, 11
dead time compensation algorithm, 27
decoupled subsystems, 158
decoupling, 157
design process, 280
detection of the drive mechanical part faults, 109
diagnostic
  indicator, 232
  rotor faults of, 233
  system, 232
differential mode (DM) choke, 23, 24
direct torque control (DTC), 156, 178
disturbance compensation, 143, 152
disturbance
  signal, 121
  state observer, 98
disturbed drive system, 226

dominant bearing current, 17
dq components, 246
dq observer, 109, 110
drive diagnosis, 218
drive system imbalance, 222
DTC-SVM, 178, 179, 180
dV/dt filter, 65, 83

eddy current, 65
electrical discharge machining, 15
electromagnetic
  subsystem, 151, 159
  torque analysis, 219
estimated rotor flux, 144
estimated torque, 230
estimation error, 119
extended disturbance model, 106

fast Fourier transformation, 226
fast voltage rises, 83
fault
  diagnosis, 231, 271
  drive, 224, 225
  motor, 235
  types, 231
field oriented control (FOC), 141, 143, 144, 156, 271
  structure, 145, 147, 157
  system, 121, 122, 123
field weakening, 165
  region, 161, 167
filter
  characteristic impedance, 279
  damping resistor, 77
  elements, 66, 74
  equivalent electrical circuits in $\alpha\beta xy$, 247
  parameter transformations, 284
  resonant frequency, 77, 79
  system, 67
filtered motor speed, 129
finite elements method, 257, 258
five-phase
  drive, 257
  frame, 243
  induction motor, 242, 243, 259, 260
    control, 253
    model, 245
  LC filter, 246
    $\alpha\beta xy$ coordinates, 248
  PWM inverter, 266
  sensorless drive, 257
  sine-wave filter, 247
  speed sensorless induction motor drive, 254
  two-level voltage source inverter, 247, 248, 249
  voltage source inverter, 247

Index

flux
  control, 183
  estimation error, 173
  linkage vector, 124
  observer, 255
    structure, 123, 124
four active vectors, 248, 251, 252, 253
four-terminal network sinusoidal filter, 78, 79
full disturbance models, 221

Gopinath observer, 239

H bridge, 274
harmonic analysis, 236, 237
harmonic distribution, 223, 224, 225
healthy drive, 223, 225

induction motor
  model, 49–54
    gamma, 54—5
    T, 50–53
  multiscalar control, 186
inverse
  Clarke transformation, 94
  Park transformation, 145
inverter, 75
  input current, 94
  output current, 141, 273
  output filter, 71, 95, 235, 238
    design, 74
  output vectors, 29
  output voltage, 11, 87, 141, 275
  switching frequency, 20
  switching states, 29

KL
  coordinates, 112
  reference system, 114
Krzeminski observer, 101

LC filter, 99, 100, 103, 178, 180, 186, 241, 242, 253, 257, 260, 272, 277
  design, 277
  model, 129
  simulator, 99, 100
L_c, cable inductance, 12
linearization, 157
load angle control, 272
load torque
  analysis, 219
  estimation, 235, 238
  observer, 221
load angle
  control, 166, 167, 174, 176, 177
  controller, 169, 170

Luenberger
  flux observer, 129
  gains, 137
  observer, 129, 138
  variables, 137

machine discharging current, 17
MATLAB/Simulink simulation, 5–6
measured vibrations, 227
mechanical
  faults, 218
  loses, 57
  subsystem, 152, 158
motor
  2Sh 90L–4 type, 287
  capacitances, 20
  choke, 13, 85, 167, 171, 174, 176, 177
  common mode CM circuit, 18
  current and voltages sensors, 97
  electromagnetic time constant, 158
  electromotive force, 103
  filters, 65
  flux estimation error, 180
  FSg 112M–8 type, 287
  FSg 112M–8 type, 288
  FSLg 132S–4 type, 288
  load torque estimation system, 220
  model in the dq axis, 151
  parameters, 13, 51, 56-59
  reversing, 173
  Sg 90L–4 type, 286, 287
  shaft, 223
  speed observer, 126, 127
  supply voltage, 141
  terminal voltage, 88, 89
  torque, 159, 219
    analysis, 218
    transmission, 222
  voltage
    observer, 143
    oscillations, 86
motors data, 286–288
multiloop
  control, 186
  filter control, 262
  principle, 187
multiphase drives, 241
multiscalar control, 141, 156, 159, 160
  structure, 157, 161
  variables, 157

natural variables, 157
neuro-fuzzy inference system, 231
nonlinear field oriented control (NFOC), 148, 152–155

nominal load the motor current ripple, 277
nonlinear
　components, 158
　control
　　method, 156
　　structure, 162, 271
　　system, 172
　decoupling, 173
　field oriented control, 148
　induction motor drive system, 174–177
　system, 157
　vector control, 147
　　structure, 152
nonlinearities, 170

observer
　error, 184
　feedbacks, 99
　frequency characteristic, 112
　robustness, 182–185
　structure, 125
　system, 161
orthogonal frame, 246
output filter model, 71
overvoltage spikes, 86

$p\_SPEED\_OBSERVER()$, 132, 134
parasitic capacitances, 10, 14
Park transformation, 256
per unit system, 41, 54, 56
position angle, 117
power dissipation, 280
protection against high dV/dt, 84
protective earth PE, 9, 81
pulse width modulation (PWM), 273
　algorithms, 44
　　modifications, 28, 29
　inverter and CMV, 39
PWM_SFUN, 39

q component, 145
quality factor, 77–79, 279

R_b, motor bearings equivalent resistance, 13
R_c, cable resistance, 12
reference
　coordinate systems αβ, 160
　estimated angular speed, 128
　estimation errors, 120, 125
　filter output voltage, 195
　inverter output voltage, 142
resonant frequency, 278–280
Rogowski coil, 6, 15
rotor
　bar faults, 233, 234
　bars fault detection, 235

coupling factor, 114
faults, 233, 235
flux, 256
flux calculation, 129
flux Luenberger observer, 136
flux pulsation, 115, 117, 118
flux vector, 144, 166
frequencies, 229
grounding current, 17
linked flux vector amplitude, 115
matrix, 243
pulsation, 127
speed, 115

saturation mode, 153
sensorless
　control, 175
　drive, 97, 166, 182
　fault detection, 228
　FOC system, 146, 149-150
　mode, 242
　motor control system, 273
　operation, 153, 163–165
　steady state operation, 181
shaft voltage measurement, 17
significant variables, 308
simplified disturbance model, 105
simulation files functions, 308
sine-wave filter, 261
single-phase inverter, 273, 274
sinusoidal filter, 65, 66, 69, 74, 75, 114, 117, 123,
　　126, 145, 149, 150, 153, 156, 160, 271
　control, 141
　damping resistances, 69
six-pulse rectifier, 76
slip
　angular speed, 168
　pulsation, 115
space vector
　modulation SVM, 11, 273
　　one zero vector method, 36
　pulse width modulation, 130, 178, 248
speed
　controller, 159
　estimation error, 175, 184
　observer, 135, 171, 250, 255
　　based on voltage model of induction
　　　motor, 114
　　operating for rotating coordinates, 109
　　complete model of disturbances, 107
　　dual model of stator circuit, 122
　　extended model of disturbances, 106
　　simplified model of disturbances, 103, 133
sensorless FOC, 264
sensorless induction motor drive, 98
$SPEED\_OBSERVER()$, 132

Index

START_REG, 134
state observer, 98–99, 103, 150
    LC filter simulator, 130
    sensorless control, 155–156
state variables, 146
    simulation, 97
stationary frame $\alpha\beta xy$, 244
stator
    current
        magnitude, 168
        vector, 166
    frequencies, 229
    matrix, 243
    voltage
        model, 116
        vector position, 160
steady state
    condition, 265
    operation, 259
subordinated control system, 160, 161, 168, 175
SVM 28, 34
    algorithm, 29, 45
    control, 38
    SVM1Z algorithm, 37–38, 44, 45
    SVM1Z PWM, 37
S_w, switch modeling breakdown of the bearing, 13
switching
    frequency, 277, 280
    time calculations, 33
    times, 34, 35, 251
synchronous sampling method, 273–275
system dq, 144

third harmonic currents, 262, 263, 265
three active vector modulation, 32
three nonparity active vectors, 30
three-phase
    ABC coordinate, 282

choke, 284
frequency converter, 9, 10
time constant
    magnitude of the rotor circuit, 116
    inertial filter of, 127
    stator circuit of, 114
torque, 151
    observer, 231
    transmission system, 219
total harmonic distortion (THD), 74
transformation matrix, 282, 283
transient speed control error, 175
transistor switching frequency, 77
two active vectors, 250, 251, 252
two-phase $\alpha\beta 0$, 282

uninterruptible power supply (UPS), 180

V/f control, 90
variable estimation methods, 97
vector
    3NPAV, 31
    components of the stator current, 115
void F4DERY, 63, 193
void p_SPEED_OBSERVER, 239
void PWM, 43
void PWM_0W, 92
void PWM5f, 267
void TRAPDERY, 192
void TRAPEZ, 191
voltage drop, 278
voltage frequency, 278

wave reflections, 83

xy components, 246

zero vector switching time, 251